Digital Video El

With 12 Complete Projects

Andrei Cernasov

Boston, Massachusetts   Burr Ridge, Illinois
Dubuque, Iowa   Madison, Wisconsin   New York, New York
San Francisco, California   St. Louis, Missouri

To Domnica, Andre, and Nathalie for their love,
understanding and patience

The **McGraw·Hill** *Companies*

Cataloging-in-Publication Data is on file with the Library of Congress.

2 3 4 5 6 7 8 9 0 BKM BKM 0 9 8 7 6

ISBN-13: 978-0-07-143715-8

ISBN-10: 0-07-143715-0

The sponsoring editor for this book was Stephen S. Chapman and the production
supervisor was Sherri Souffrance. It was set in Century Schoolbook by PV&M
Publishing Solutions. The art director for the cover was Anthony Landi.

McGraw-Hill books are available at special quantity discounts to use as premiums
and sales promotions, or for use in corporate training programs. For more infor-
mation, please write to the Director of Special Sales, McGraw-Hill Professional,
Two Penn Plaza, New York, NY 10121-2298. Or contact your local bookstore.

Contents

ABOUT THE AUTHOR

Andrei Cernasov is cofounder of Video Architects, Inc., developers of miniature video equipment for videoconferencing, security, and telemedical applications. He holds a B.S. in electrical enginering from City College of New York and a Ph.D. in physics from City University of New York. He currently teaches in the Business Engineering Masters Program at Bridgeport University. He lives in Ringwood, New Jersey.

Introduction

We live in an information-rich society, where a large part of our daily life is influenced, to a significant degree, by video images. Entertainment, news, and even wars are delivered to us, every minute of every day, via screens. Doctors look at CAT scans on high-resolution LCD monitors; financial analysts deliver presentations on micro-mirror DMD projectors; commuters pass time playing videogames on their OLED cell phone screens. And many more applications are being developed all the time.

Behind the screens there is a host of new and exciting display technologies. Flat panel LCDs are displacing CRTs from their dominant position in the PC monitor market. Plasma panels, from the R&D perspective, are by now "old news." Most of the current plasma work is focused on the manufacturing processes for wall size architectural displays. DMD (Digital Micro-mirror Device) technologists (TI) are looking over their shoulders as the LCoS (Liquid Crystal on Silicon) developers are closing in. And the first products using OLED (Organic LED) displays are just entering the consumer marketplace.

Semiconductor houses are also evolving. Once the realm of giant corporations with bottomless pockets, ICs can now be developed by visionary entrepreneurs on shoestring budgets. Even big companies are going "fab-less" as globalization of semiconductor foundries and an endless appetite for new consumer and industrial ICs overwhelm their local production capacities. To entice manufacturers into using their chips, producers are doing all they can to reduce product development cycles. There is an abundance of development boards, reference designs, and software development kits for simple ICs (amplifiers, filters...) as well as for the most complex chips (Codecs, CMOS cameras...).

The engineering profession itself is changing. If forty years ago electronics engineering students had to master vacuum tubes and slide rules, in the seventies they had to learn about semiconductors and the use of scientific calculators, in the eighties their attention turned to software design and computers, while the nineties were dominated by wireless devices and everything-Internet. In the new century what electronics professionals have to adjust to is global competition and continuous change.

Generic skills which can be easily sub-contracted are no longer in demand. However, people who can add high value to products designed for local markets such as medical, security, or defense, are more sought after than ever. All these markets have a high level of video technology content. Engineers in developing countries are well aware that video is an integral part of most consumer products, and as such it must be mastered. This book provides a fast track method of acquiring the needed knowledge and skills.

Book Contents

After an introduction to human vision (Chapter 1) the reader is presented the structure of current digital video signals and standards (Chapter 2). The book then details the operation of a number of video systems in terms of a relatively small number of function blocks (Chapter 3). Each such block generally consists of a major function IC and a small number of external components.

Chapter 3 provides for general methods of connecting various blocks together, including voltage translations, bus width and speed matching, and multi-level voltage supplies. In Chapters 5 through 10 we cover the most commonly used function blocks, starting with their theory of operation (where applicable) and ending with circuit schematics.

Three special purpose blocks are the subjects of Chapters 11, 12, and 13; Multi-Image Processors, Scalers, and JPEG/MPEG Image Compression Processors. Although here, as in some of the preceding chapters, theoretical constructs are presented first, readers may choose to skip the theory and go directly to the practical sections of the book.

In Chapter 14 we address the issue of laying out video printed circuit boards, one of the "black arts" associated with this field. Finally an extensive projects section (Chapter 15) brings many of the concepts into the experimental realm. The reader is presented with the theory of operation and the design files for five function modules which can be interconnected together to prototype a number of standard resolution digital video systems. In addition, the chapter includes the files for a turnkey scan converter and a general purpose IR remote and data link.

Book Web Site

The Gerber files for all the circuits presented in Chapter 15, together with the ViewMate Gerber viewer from PentaLogix (http://pentalogix.com), can be downloaded from the book web site:

http://www.digital-video-electronics.com

Another package you can download is the 8x51 Development Tool Set from the Keil Software, Inc (http://www.keil.com). It allows the reader to develop software routines for the 8051 board, although this demo version limits the size of the code to 2K.

The book web site provides datasheets for the parts used in this book, as well as links to part suppliers and service providers (PCB design houses, PCB manufacturers, and board assemblers).

Also check for new projects and experiments, new video ICs, new firmware, assembled board suppliers, links to periodicals, and so on.

CPLD Design Software

The CPLD firmware was developed using the Max+plus II Baseline/Quartus II Web edition design software available free from Altera (www.altera.com).

PAL Designers

All the circuits in this book will work both with PAL and NTSC systems; although in order to maintain focus we have emphasized the NTSC standard. For PAL operation some of the circuits may need minor changes in register settings. Details are available at http://www.digital-video-electronics.com.

College and Vocational School Instructors

The layout of this book makes it ideal for teaching both college and vocational school level digital video courses. Many of the chapters start with a thorough theoretical analysis, which is then followed by the operational description of actual circuits. Senior college courses would cover the theoretical background while vocational schools would not. Chapter 15 can be used by both as the basis for a laboratory course with minimal equipment requirements.

1

Human Vision

1.1 Fundamentals of Vision

The visual system consists of three major interconnected sections: the *eyes*, the *visual pathways*, and the *visual processing centers* of the brain. The function of the eyes is to capture external video images and convert them into two-dimensional information messages.

These messages are transported by the *optic nerve bundle* to the brain. There, in ways still not fully understood, the image messages are used to recreate a coherent visual model of the world.

The model is three-dimensional, a property that is conferred to it by the stereo nature of our eye arrangement. It serves as further proof of our closer relation to predator and scavenger species rather than to prey.

The binocular positioning of the eyes gives us depth of field, but it restricts how much of the environment we are able to see at any given time. This is partially compensated by the unusual amount of mobility that our eyes have.

Enhanced eye mobility in humans throughout evolution also reduced the need for neck movement, which in turn allowed for the development of a larger cranium. This was a definite plus for early primates—they experienced a rapid growth in brain size over a relatively short period of time (evolutionarily speaking).

Our eyes enjoy the benefits of both color vision and high resolution, especially in the center of the field of view. We also have maintained our peripheral vision, a vestige of our reptilian distant past. Once a peripheral visual stimulus alerts us of possible danger, such as a rapidly approaching novice skier, the *extraocular muscles* can steer the image to the high-resolution center in less than 200 milliseconds.

The human eye, just like any other complex device, needs protection and maintenance. Unlike any other device, our eyes are capable of self-repair. Our eyes are nestled in a protective bone cavity called the *orbit*. The eyelids, those movable tissues covering the optical "face" of our eyes, provide both physical protection and "window cleaning" services. This mechanical protection spares the eyes the inevitable cuts and bruises of everyday life. As optical protectors, the eyelids supplement the *pupil* in controlling the amount of light that passes through. They also stop most light during sleep.

About once every four seconds, involuntary blinks clean the front surfaces of the eyes and spread tiny amounts of tear liquid over the cornea to moisturize it. A little known fact is that tears have antimicrobial properties and therefore actively prevent eye infections.

1.2 Eye Anatomy

The human eye is roughly spherical with a diameter of a bit less than an inch. The eyeball is protected on the outside by the *fibrous tunic*, or what we call the "white" of the eye, or *sclera*. This tough layer consists of tightly packed and interwoven fibers, strong enough to maintain an internal pressure of twice the atmospheric pressure.

The sclera bulges outward at the front of the eye to form the *cornea*. In the same area, the organization of sclera fibers becomes so orderly that they turn from white to optically clear.

Right below the sclera we find the *vascular tunic* or *choroid*, which surrounds the eye except for its optical front end. The choroid consists of a network of blood vessels and capillaries that feed the light sensitive inner layer, the *retina*, with oxygen and nutrients. Due to its dark red color the choroid

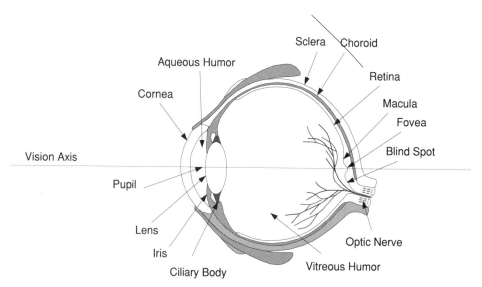

Figure 1.1 Eye Anatomy.

also functions as a light absorber. It minimizes the impact of multiple internal reflections on the contrast of the retinal image.

Toward the front of the eye, in the proximity of the lens, the choroids spins off a spongy tissue called the *ciliary body*. Its function is to produce the all-important *aqueous humor*, the liquid that fills the corneal cavity, or *anterior chamber*. This remarkable liquid has largely the same characteristics as blood, except it is clear! It feeds oxygen and nutrients to the tissues in the area and expels their metabolic waste products. Embedded in the ciliary body is the *ciliary muscle*, the motive force behind the focusing mechanism of the eye lens.

The main body of the eye consists of the *vitreous chamber* which is filled with a gelatinous, transparent fluid called the *vitreous humor*. Unlike the aqueous humor, the vitreous humor is not continuously replenished, although it is protected against evaporation by its own protective membrane.

A tissue called the *iris* brings blood supplies up and around the ciliary body to a set of circular muscles located in front of the lens. The surface layer of the iris is strongly pigmented giving humans a wonderful variety of eye colors. Iris colors are created by the balance between *eumelanin* (brown/black) and *pheomelanin* (red/yellow) produced by iris *melanocyte* cells.

The circular muscle within the iris controls the size of the optical opening of the eye. This aperture, also known as the *pupil,* opens up in the dark to allow more light in and constricts in a bright environment to reduce the amount of light that reaches the retina.

The main element of the optical front end of the eye is the lens. A structure softer than common contact lenses, the eye lens is suspended by ligaments called *zonule fibers* to the front section of the ciliary body. The contraction or relaxation of the ciliary muscles, as relayed by the ligaments, changes the shape of the lens, a process called *accommodation*. As a result of accommodation we can change, at will, the focal length of the eye lens and therefore maintain image focus.

The innermost layer of the eye is the *retina*. Images are formed on the surface of the retina, with the cells and tissues of the retina converting light to nerve impulses.

As a sensory tissue, the retina has a lot in common with the brain, with most of it consisting of nerve or nerve-like cells. And just like brain tissue, damage to the retinal tissue is irreparable.

The retina is mostly transparent, except in the area of the *optic disk,* or "blind spot." This is the where the optic nerve bundle leaves the orbit on its way to the brain, and where arteries and veins enter the eye. Since the optic disk contains no photodetectors, any light impinging on it will not be seen by the eye, creating a "blind spot."

The area of the retina where images are perceived in greatest detail is called the *macula*. Not only does the macula exhibit the highest density of photoreceptors, but at its center (the *foveal depression)*, it has the top layers of the retina peeled off in order to reduce the amount of light scattering that such layers may introduce.

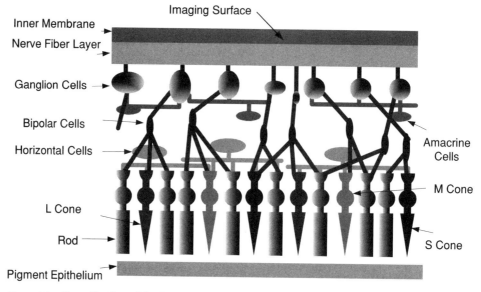

Figure 1.2   Cross Section of the Retina.

A cross section of the retina reveals a three-layer structure. The outer layer contains photoreceptor cells that are functionally divided into color cones and broad spectrum rods.

The cones are sensitive to relatively narrow bands of wavelengths and therefore colors. The L cones are responsive to red, the M cones to green, and the S cones to blue. The sensitivity of the rods, however, is much higher, but rods do not discriminate wavelengths.

The inner layer of the retina consists of *ganglion nerve cells* with their axons merged into a sublayer of optical nerve fibers. Although there are over 130 million photoreceptors in each eye, there are only about 1 million ganglion cells ferrying the information out, a process known as *convergence.*

At this time it is unclear how such information is compressed. However, it is unlikely that, in view of evolutionary pressures, any of it is useless or discarded. What it is known is that for the macula the ratio of photoreceptors to ganglion cells is almost one, while at the periphery it drops to one ganglion to several hundred photoreceptors. The retina area (including photoreceptors) in the vicinity of a ganglion cell—which can influence its response—is called the *receptive field* of the cell. It follows that the size of a ganglion's receptive field is much larger at the periphery than in the macula region.

The connections between the photoreceptors and ganglion cells are made by a number of parallel networks of cells forming the middle retinal layer. It is clear that the middle layer is not just "wiring," since this network could be achieved more economically by directly connecting the inner and outer layers.

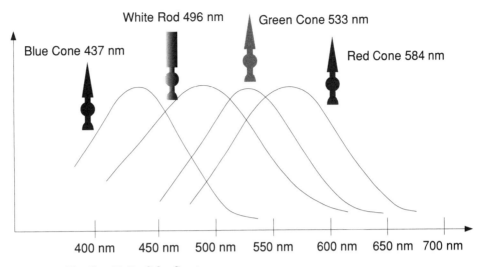

Figure 1.3   Eye Sensitivity Color Spectrum.

Its main function is the selection and processing of information. The middle layer is composed of the cell bodies of the *bipolar*, *horizontal*, and *amacrine* cells, although few of their operational functions are understood.

As we have seen, the distribution of photoreceptors on the surface of the retina is not uniform but peaks at the fovea and drops to zero in the optic disk region. Furthermore, the composition of the photoreceptor mix also changes from the fovea, which is cone dominated, to the periphery, which has more rods than cones.

1.3 Spatial and Temporal Effects

The understanding of human vision is indispensable for any designer of video equipment, as the eye is the ultimate "client" for their efforts. Our short introduction to the anatomy of the eye provides a start. However, key questions regarding the quantified response of our vision system to varied and changing stimuli still need to be answered.

1.3.1 Ricco's Law of Spatial Summation

It is common experience that, as it gets darker, we have more and more difficulty reading small print and distinguishing small objects. What is the relationship between the size of an object and the threshold illumination level needed to perceive it? If we standardize our objects to a set of small white disks of different areas, and we view each disk for a fixed period of time, we find that the threshold intensity, I, is inversely proportional to the area, A, of the disk, a relationship known as Ricco's Law of Spatial Summations:

$$I \times A = \text{constant}$$

It appears that the retinal response is an integral or summation of the photon flux over small retinal areas. In relative darkness, the threshold response disk area spans a 0.5-degree angle, while under good lighting conditions the angle is found to be 0.06 degrees.

Past a certain disk size (known as the *area of complete summation*) the threshold light intensity remains constant, which may indicate a dynamic pixilation of the retina. Once a certain detail size is reached, that size becomes a "pixel equivalent" for the rest of the visual processing chain. There is considerable evidence that each area of complete summation is processed by a single ganglion cell.

We should note that the area of complete summation quantifies the two-dimensional resolution of the eye. If we measure its one-dimensional (line to line) performance, we find it to be three times better or approximately 0.02 degrees in good lighting. This may be due to signal processing inferences rather than optics.

1.3.2 Sensitivity versus Resolution

Revisiting the concept of convergence, we may note that high convergence results in large receptive fields, which, according to Ricco's law, corresponds to high sensitivity: the larger the receptive field, the lower the intensity threshold. In low convergence areas the size of individual receptive fields is significantly smaller. Because these fields are closely packed, the resolution of the eye is much higher, but the intensity threshold increases accordingly.

To accommodate both sensitivity and resolution, the eye developed two overlapping (duplex) sensory networks—one for high sensitivity, the other for increased resolution. The high-sensitivity network consists of a mesh of rods and ganglion cells with large receptive fields. The vision provided by this network is known as *scotopic* vision and dominates the way we see at night.

The second network has a much higher percentage of cones than rods and is associated with ganglion cells with very small receptive fields. Vision using this network is called *photopic* vision and provides most of our visual input under bright light conditions.

As mentioned before, there are over 130 million photoreceptors in each eye. Among them we find roughly 3 million L (red) cones, 3 million M (green) cones, and only 1 million S (blue) cones, with the balance consisting of achromatic rods. At the center of the macula we find mostly cones—about 150,000 of them packed into an area of one square millimeter.

We conclude that photopic visual acuity is associated mostly with the cones, for they occupy most of the macula. The low density of S cones also indicates that our eyes have a much lower resolution for blue color detail than for red or green. Still, the brain must somehow integrate the S cone signals into a continuous blue image since we do not see any gaps when we look at a blue object.

1.3.3 Bloch's Law of Temporal Summation

Similar behavior is found when disk sizes and intensity levels are kept constant, but the duration of each observation is varied. The time, T, it takes to perceive an object of a given (small) size is in inverse proportion to the luminosity, I, on that object. The relationship between I and T is given by Bloch's law:

$$I \times T = \text{constant}$$

When the duration is increased past a critical value (called the *duration of complete summation*), the threshold intensity becomes constant. For rods, Bloch's law levels off at about 100 milliseconds; for cones, however, it stops being accurate after 50 milliseconds.

The formal similarity between Ricco's law and Bloch's law suggests that what determines the threshold intensity level is the *total energy of the photons* contributed to a ganglion cell, regardless if it is obtained by summation over time or integration over local retinal area.

1.3.4 Weber's Law

The receptive fields associated with ganglion cells are not all the same. The most common type is the *center-surround receptive field*. Its distinguishing feature is the presence of two distinct concentric subfields with opposite stimuli responses: a center circle and a doughnut.

Depending on their type, there are *center-on/doughnut-off* and *center off/doughnut-on* fields. The first responds stronger when their center is illuminated more than the periphery, while the second responds stronger when the periphery is illuminated more. This arrangement makes the associated ganglion cell indifferent to uniform illumination but sensitive to high illumination gradients, such as those we find in high-contrast images or around the edges of objects. What happens if we start with a uniform illumination I_0 on top of which we superimpose an illumination gradient ΔI? The answer is given by *Weber's law;* it states that the minimally noticeable change in light intensity is a constant proportion of the background intensity.

$$\frac{\Delta I}{I_0} = \text{constant}$$

In other words, an object illuminated by a local source shows more detail when seen against a dark backdrop than when placed against a well-lit background, a phenomenon known as *lightness constancy.*

However, when both the object and the background are illuminated by the same source, the degree of detail evidenced by the object is the same regardless of scene illumination (for the same background). In this case, both ΔI and I_0 are the result of surface reflectance and therefore their ratio remains the same for all illumination levels. This is known as *lightness contrast*.

A related parameter is maximum linear eye resolution which, under photopic conditions, is found to be about 0.02 degrees.

1.3.5 The Broca-Sulzer Effect

An interesting side effect of Bloch's law is the *Broca-Sulzer effect*. It occurs when the eye is exposed to bright flashes of light comparable in length to the Bloch critical duration (50 to 100 milliseconds). Here, there is a transient apparent amplification of the intensity of the light followed by a normalization decay. It appears that there is a secondary gain mechanism in place during the temporal summation period that subsides when the summation is complete. This is similar in many ways to the transient response of an underdamped feedback system.

1.3.6 Critical Flicker Frequency

The critical flicker frequency (CFF) is that frequency value of an alternating illumination source above which the light appears to be continuous. There are many factors influencing our perception of flicker. One of them is the size of the source. A large source such as a TV monitor may not exhibit flicker if looked at directly, but the flicker may be quite pronounced if viewed at a slant. This points to a higher CFF for the peripheral retina (rod-dominated) than for the macula (cone-dominated).

CFF is also higher for larger objects than for smaller ones. There also seems to be a trained tolerance to flicker which is most apparent when looking at old 50 Hz PAL monitors, the kind found in Europe. Although most Europeans do not see the flicker, many visitors who grew up watching 60 Hz NTSC monitors are quite aware of it.

1.3.7 The Ferry-Porter Law

According to the *Ferry-Porter law*, the CFF is a linear function of the logarithm of the source luminance, I.

$$CFF = A \log (I) + B$$

where A and B are constants.

The Ferry-Porter law indicates that the higher the light intensity is, the higher the CFF. Stated differently: for a given stimulus frequency, the lower the intensity, the lower the CFF. The flickering of a PAL monitor, for example, becomes less evident if the luminosity is lowered or if we move the monitor into a well-lit area (unless the lights are also powered by a 50 Hz grid!).

Above CFF, there is no distinction between the perceived luminosity of a continuous source and that of an intermittent source of equivalent emissive power (the *Talbot Plateau law*).

1.3.8 The Brücke-Bartley Effect

The perceived brightness of a source operating below CFF is significantly higher than the brightness of an equivalent continuous source. Peak perceived brightness is reached when the "on" time of the flicker approaches the Bloch critical duration. This explains the high effectiveness of colored flashing warning signs operating in the 15 to 20 Hz range. The Brücke-Bartley effect is clearly a steady-state variation of the Broca-Sulzer effect discussed earlier.

However, this maximum sensitivity peak holds only for photopic (cone-dominated) vision. Then the CFF is close to our familiar 60 Hz and the flicker contrast threshold is around 1 percent. For scotopic vision the CFF drops to 15 Hz and the contrast threshold increases to 20 percent.

Video Signals and Standards

2.1 Designing Television

The way our eyes process video information is highly parallel in nature. Video images come into focus on the surface of the retina all at once. Then, after a short integration process, they are converted to nerve impulses and passed on to the ganglion cells on their way to the central nervous system.

The same basic principles also apply to the operation of a video camera but only up to a point. Images are formed on the surface of a photosensitive array, where they are integrated over each pixel area for a user-defined period of time (exposure time). The output of each elementary photodetector is a function of the light intensity reaching its surface.

Here the similarities end. Capturing video information from hundreds of thousands of photodetectors simultaneously is technologically impractical. The continuous transmission of such amounts of data from the camera to remote displays or video processors is nearly impossible.

Because we cannot simply duplicate the intricate details of natural vision, a television device must use a different approach in the collection and transmission of video images.

We do not need to match the information bandwidth of our apparatus to the vast resources of the vision system. After all, a monitor screen occupies a very small section of our field of view, is two-dimensional, and operates under a fixed set of operating parameters. The eye, however, handles a field of view 30 times larger, provides three-dimensional image fusion, is adaptive, and likely executes numerous forms of image processing algorithms.

Furthermore, closer analysis of the eye's sensor mechanisms reveals that although the space bandwidth of the eye is very wide, the temporal bandwidth is narrow. The integration time is a relatively long 50 to 100 msec (Bloch duration), and there is no perceptible difference between the effects of a pulsed source operating beyond CFF and a continuous source of equal average power (*Talbot Plateau law*).

Therefore our eyes will perceive as continuous a sequence of bright still images, if separated by an interval of time shorter than the CFF period. The still images themselves do not need to be rendered all at once, but they can be drawn line by line, picture element by picture element (pixel by pixel), as long as the rate at which individual elements are updated (refreshed) exceeds the CFF. These straightforward observations formed the basis for scanned-line television as we know it today.

As for the screen format, in the early days of television the 4×3 window represented a compromise between the natural "panoramic" view of our eyes and what display tube technology could produce at that time. The decision regarding the number of lines was also straightforward. At the low end we need a certain minimal number of lines to reproduce a decent quality, recognizable image. At the high end we are limited by the bandwidth we want to invest in the communication channel between the image source and display. Finally, accounting for the size of our living quarters, we should be able to share a screen with others in relative comfort. A two-foot-high display viewed from ten feet away will look spatially continuous (no discernable scan lines) if it operates close to the one-dimensional resolution limit of the eye, or about 0.02 degrees (Ricco's law). This requires about 570 lines of resolution, as derived from the formula

$$N = \frac{H}{D \tan(\alpha)}$$

where H is the height of the display, D the distance to the display, α the angle measure of the one-dimensional resolution of the eye, and N the resulting number of lines.

2.2 Analog Television Standards

In 1941, the National Television Standards Committee (NTSC) formalized all these observations in a unified standard for the broadcast transmission and reception of television signals in the United States. This standard, known as NTSC, was further refined in 1953. The Europeans experimented with a number of different television protocols that were maintained until their consolidation in the early 1980s into the PAL/SECAM standard.

Both NTSC and PAL/SECAM use an *interlaced scanning* technique as a low-bandwidth method to collect and render images. The interlacing process consists in the separation of a source image *frame* into two separate image fields—one comprised of all the odd rank lines (the *odd field*), the other one comprised of all the even rank lines (the *even field*). Each field is read pixel by pixel and line by line, from the top left to the bottom right of the image.

On the transmitter side, individual fields are sent out in sequential order: an odd field followed by an even field followed by an odd field and so on. The

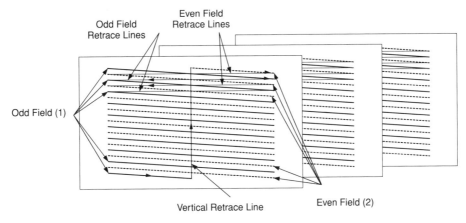

Figure 2.1   Interlaced Scanning.

same order is kept on the receiver side. Since the two fields are interleaved and separated vertically by only one scan line they appear visually continuous (Figure 2.1). The odd and even field display rates are 30 Hz for NTSC and 25 Hz for PAL/SECAM, both far below the CFF. Therefore if we suppress one field and just view the other we experience a very strong flicker effect. What integrates the two fields into a stable image is our vision system. The integration mechanism is related to the slight overlap in the receptive fields of neighboring ganglion cells.

Besides having different field rates, NTSC and PAL/SECAM also have different numbers of scan lines; NTSC mandates 525 lines (262½ per field) while PAL/SECAM requires 625 lines (312½ per field). Not all the scan lines and not all the pixels on a given scan line are visible. For NTSC, only 480 of the lines carry video information, the others being used to relay synchronization, closed caption, and other data.

NTSC lines are approximately 63.5 microseconds (μsec) long which corresponds to a line frequency of 15.718 Hz. Each line starts with a negative horizontal synchronization pulse (HSYNC), followed by a sinusoidal segment called the *color burst* (Figure 2.2). HSYNC is 4.7 μsec long and marks the beginning of each new scan line.

The color burst consists of nine full cycles of a 3.5795458 MHz *color subcarrier* signal; this serves as a phase reference for the color encoding scheme used by the NTSC standard. A negative vertical synchronization pulse (VSYNC) is used to indicate the beginning of each new field. Its length must be at least three scan lines long. New frames always start with an odd field, and all odd fields start with a full scan line and end with a half line. The exact field frequency is specified at 59.94 Hz. At the beginning of each new frame the phase of the color burst segment is inverted (a change of 180 degrees).

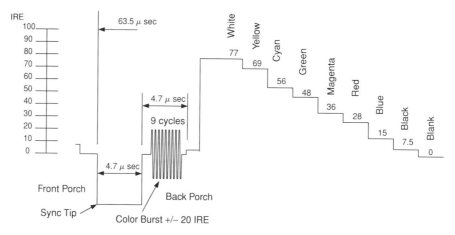

Figure 2.2 NTSC Horizontal Line Detail (not to scale).

The brightness of the image is given by the overall amplitude of the video signal. Color, however, is conveyed by the amplitude and phase of the modulated 3.579545 MHz subcarrier. Saturation is amplitude modulated while hue is phase modulated with respect to the color burst reference (see Chapter 8).

Looking at the timing diagrams of NTSC signals (Figure 2.3), we can distinguish a few other features such as serration and equalization pulses. The purpose of serration pulses is to assist in the maintenance of a constant horizontal scanning rate (mostly in older cathode ray tubes or CRT), while equalization pulses serve as reference for the half-line retrace in the middle of a frame and to zero any DC component buildup on the video line.

The total video bandwidth of NTSC signals is 4.2 MHz and includes a section where the intensity and color spectrums overlap (Figure 2.4). Upon closer inspection we find that the spectral lines for luma and chroma in the common area are carefully interleaved, so they can be easily separated at the receiver site by an appropriately named *comb filter* (see Chapter 8).

The PAL/SECAM standard specifies a 625-line frame with 576 active lines. For most PAL/SECAM variants the color subcarrier frequency is 4.43361875 MHz. The HSYNC and VSYNC widths are 4.7 μsec and three horizontal lines, respectively. The line length is 64 μsec, with a ten cycle color burst. This translates into a line frequency of 15.625 Hz and a field frequency of 50 Hz.

Video signals that encode intensity, color, and synchronization information on a single line are known as *composite* or *baseband* signals. Video amplitudes are measured in IRE units (named for the Institute of Radio Engineers), with each unit $\frac{1}{140}$ of full scale (1.020 V) or about 7.3 mV.

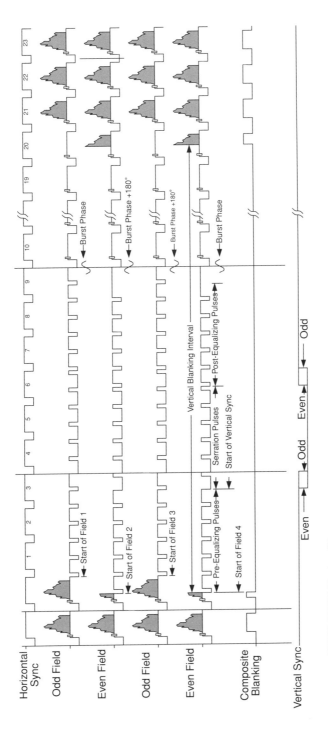

Figure 2.3 NTSC NTSC Timing Diagram.

15

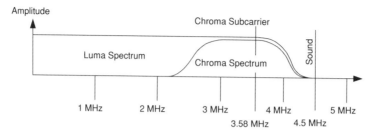

Figure 2.4   NTSC Frequency Spectrum (macro view).

Further information regarding the analog NTSC and PAL/SECAM standards can be obtained from a number of good sources, some of which are listed in the Bibliography section.

Besides interlaced scanning, another popular technique for serializing an image in the analog domain is *progressive* scanning (Figure 2.5). This approach simply scans the whole image, left to right and top to bottom, in one pass. Since it must operate above the CFF, the frame rate of a progressive scan system has to be at least twice that of an interlaced system (60 frames per second for NTSC, 50 for PAL/SECAM). Despite its higher bandwidth require-

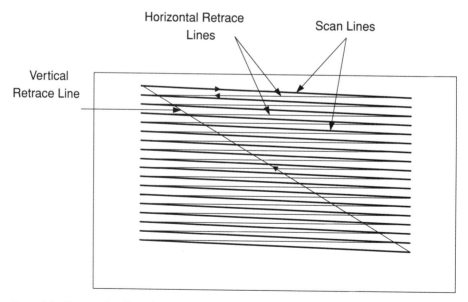

Figure 2.5   Progressive Scanning.

ments, progressive scanning offers a better image quality with fewer motion artifacts, and it is easier to process. It is extensively used in computer monitors and in high-end television sets.

2.3 Color Spaces

2.3.1 RGB

We start our forays into digital video by analyzing the structure of a relatively simple device such as the image capture appliance in Figure 2.6. Its function is to convert the analog video output of a camera into a standard serial digital stream. The camera selected for this example is RGB type. It may contain a single color sensor array with individual detectors covered with red, green, and blue filters, or may have three separate sensor arrays, one for each primary color. Either way this camera is equivalent to three separate cameras, each of them capturing a different color image. To recreate the original image we only need to superimpose the images encoded by the three analog outputs of the camera. This is done by bombarding the red, green, and blue phosphorous dots of a CRT pixel or by overlapping the three separate color images of a projection display.

The horizontal and vertical sync information for the camera may be embedded in the green color signal or may be provided as separate digital outputs.

If we map each of the three color signals into an 8-bit digital space, with the total absence of that color marked as 0 and the highest value marked as 255 we are, by definition, digitizing that signal to an 8-bit resolution. The device that performs the actual mapping is called an analog-to-digital converter (also known as an A/D converter or ADC). The synthesis of any color by mixing the right amounts of primary colors is a matter of elementary physics. The colors of the familiar *color bars* pattern, for example, are obtained by mixing red, green, and blue in the proportions shown in Table 2.1.

Because all three primary colors are used at their maximum digital values of 255 (which correspond to 1.02 V analog values), this table illustrates the composition of maximum brightness or 100 percent color bars. To obtain less intense colors we only need to reduce the values of all three primary colors in equal proportion.

TABLE 2.1   Color Bars 100% Saturation

	White	Yellow	Cyan	Green	Magenta	Red	Blue	Black
R	255	255	0	0	255	255	0	0
G	255	255	255	255	0	0	0	0
B	255	0	255	0	255	0	255	0

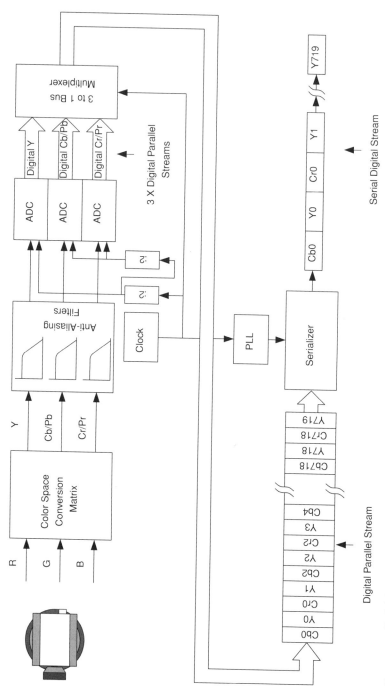

Figure 2.6 Serial Digital Image Capture Appliance.

2.3.2 YCbCr

Although the red, green, and blue primary colors form a convenient color system for a camera sensor and even a graphics workstation, it is less than ideal for video transport or video storage applications. The main reason is the fact that all three colors have equally wide bandwidths. Since the human eye has a much lower resolution for the color blue is it necessary to capture and process as much blue information as we do red or green? Certainly not. In fact, our eyes have a higher resolution for white detail than for any particular color component, despite the fact that visual acuity is due almost entirely to color cones. To take advantage of these physiological traits the YCbCr color space uses luma (Y) and two color vectors—Cb (variant of blue) and Cr (variant of red)—as intensity and color descriptors.

For standard definition TV applications (SDTV), the equations that relate YCbCr to the RGB color system are given by

$$Y = 0.257R + 0.504G + 0.098B + 16$$
$$Cb = -0.148R - 0.291G + 0.439B + 128$$
$$Cr = 0.439R - 0.368G - 0.071B + 128$$

When the RGB values cover their full 0-to-255 span, Y varies between 16 and 235 and the range of Cb and Cr extends from 16 to 240. The reverse transformation matrix is found to be

$$R = 1.164(Y - 16) + 1.596(Cr - 128)$$
$$G = 1.164(Y - 16) - 0.813(Cr - 128) - 0.391(Cb - 128)$$
$$B = 1.164(Y - 16) + 2.018(Cb - 128)$$

A common restriction imposed on RGB systems is to limit their component values to a 16-to-235 range. Under these conditions, the equations become

$$Y = 0.299R + 0.587G + 0.114B$$
$$Cb = -0.172R - 0.339G + 0.511B + 128$$
$$Cr = 0.511R - 0.428G - 0.083B + 128$$

$$R = Y + 1.371(Cr - 128)$$
$$G = Y - 0.698(Cr - 128) - 0.336(Cb - 128)$$
$$B = Y + 1.732(Cb - 128)$$

Table 2.2 lists the values for the YCbCr components required to produce 75 percent intensity, fully saturated color bars.

TABLE 2.2   SDTV YCbCr Color Bars

SDTV	White	Yellow	Cyan	Green	Magenta	Red	Blue	Black
Y	180	162	131	112	84	65	35	16
Cb	128	44	156	72	184	100	212	128
Cr	128	142	44	58	198	212	114	128

High-definition television (HDTV) uses slightly different expressions for the YCbCr space. For RGB with a 0-to-255 value range, they are

$$Y = 0.183R + 0.614G + 0.062B + 16$$
$$Cb = -0.101R - 0.338G + 0.439B + 128$$
$$Cr = 0.439R - 0.399G - 0.040B + 128$$

$$R = 1.164(Y - 16) + 1.793(Cr - 128)$$
$$G = 1.164(Y - 16) - 0.534(Cr - 128) - 0.213(Cb - 128)$$
$$B = 1.164(Y - 16) + 2.115(Cb - 128)$$

For RGB with a 16-to-235 value range, the transformations become

$$Y = 0.213R + 0.715G + 0.072B$$
$$Cb = -0.117R - 0.394G + 0.511B + 128$$
$$Cr = 0.511R - 0.464G - 0.047B + 128$$

$$R = Y + 1.540(Cr - 128)$$
$$G = Y - 0.459(Cr - 128) - 0.183(Cb - 128)$$
$$B = Y + 1.816(Cb - 128)$$

The new color bars YCbCr values are shown in Table 2.3.

TABLE 2.3   HDTV YCbCr Color Bars

HDTV	White	Yellow	Cyan	Green	Magenta	Red	Blue	Black
Y	180	168	145	133	63	51	28	16
Cb	128	44	147	63	193	109	212	128
Cr	128	136	44	52	204	212	120	128

2.3.3 YPbPr

Besides YCbCr, a newer, 10-bit color space is finding use both in SDTV and HDTV applications. This space also consists of a luma (Y) and two chroma vectors (Pr and Pb), which are scaled and shifted versions of the YCbCr vectors.

$$Y = 0.625(Y - 64)$$
$$Pb = 0.612(Cb - 512)$$
$$Pr = 0.612(Cr - 512)$$

YPbPr can also be used in 8-bit applications if we process only the eight most significant bits of each component.

All color space conversions we presented here can be performed either in the analog domain using summing amplifiers and constant gain blocks, or in the digital domain using color conversion chips, look-up tables, or Field Programmable Gate Arrays (FPGAs) or Complex Programmable Logic Devices (CPLDs).

2.3.4 Gamma Correction

The light intensity generated by a CRT pixel dot is not a linear function of the input signal, but it is proportional to the input signal raised to a power approximately equal to 2.2 (known as *gamma*). To compensate for this CRT nonlinearity, a corrective function is applied to the signal at the camera or transmission site. This function is the inverse of the display transfer function and the transformation is referred to as *gamma correction*.

Most current standard and high-definition systems have replaced the uniform gamma relationship with a more accurate piecewise function. When describing CRT nonlinearity, this transfer function is given by

$$\begin{cases} (Ro, Go, Bo) = (Ri, Gi, Bi)/4.5 & \text{if } (Ri, Gi, Bi) < 0.0812 \\ (Ro, Go, Bo) = \{[(Ri, Gi, Bi) + 0.099]/1.099\}^{1/0.45} & \text{if } (Ri, Gi, Bi) \geq 0.0812 \end{cases}$$

where (Ri, Gi, Bi) are the values of the color component signals applied to the inputs of the CRT and (Ro, Go, Bo) are the resulting light intensities. At the transmission site the corresponding gamma correction function is then

$$\begin{cases} (Ro, Go, Bo) = 4.5 \times (Ri, Gi, Bi) & \text{if } (Ri, Gi, Bi) \leq 0.018 \\ (Ro, Go, Bo) = 1.099 \times (Ri, Gi, Bi)^{0.45} - 0.099 & \text{if } (Ri, Gi, Bi) > 0.018 \end{cases}$$

Since broadcast signals are all gamma corrected, displays other than CRTs have to first cancel out the transmitter gamma correction and then compensate for the nonlinearity of their own transfer function locally.

2.4 Sampling Formats

2.4.1 4:4:4

Once we separate the video information into intensity (luma) and color (chroma) components we proceed with the sampling process. From the sample distribution standpoint there are four major sampling formats in use today. The highest resolution of all is the 4:4:4 format (Figure 2.7). In this case all three components (Y, Cb, Cr) are sampled during each pixel period. Although the 4:4:4 format offers the highest possible performance, it does not provide any space or bandwidth savings over RGB.

2.4.2 4:2:2

Using the fact that our eyes have a lower resolution for color detail than intensity detail, we can sample only the chroma components of every other pixel. This format is known as 4:2:2 and is illustrated in Figure 2.8. Because our brain integrates each color individually, the perceived image degradation over the 4:4:4 format is rather minimal except when the image is viewed up close.

2.4.3 4:1:1

When compression efficiency is more important than image fidelity we can reduce the number of color samples even further. In the 4:1:1 format we sample one chroma pair for every fourth pixel (Figure 2.9).

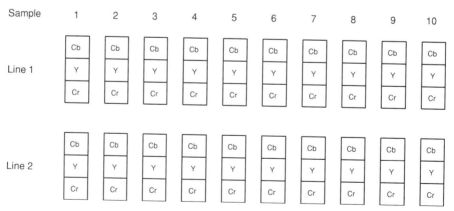

Figure 2.7   4:4:4 Sampling Format.

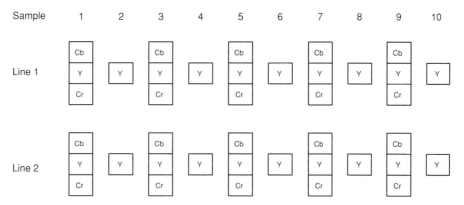

Figure 2.8   4:2:2 Sampling Format.

2.4.4 4:2:0

The same overall density of chroma samples but a more uniform two-dimensional arrangement is achieved by the 4:2:0 format (Figure 2.10). The color samples are now located in between the scan lines, each color sample being associated with a 2×2 array of luma samples.

There is a price to pay; the color samples are interpolated values that require line storage and digital processing to calculate. The 4:2:0 format is used by the MPEG (DVD, HDTV...) compression standards (see Chapter 14).

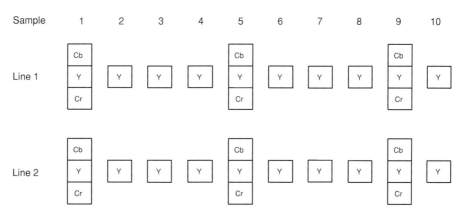

Figure 2.9   4:1:1 Sampling Format.

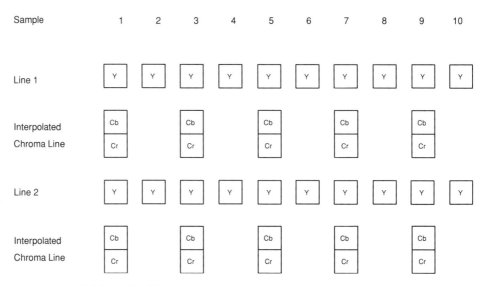

Figure 2.10   4:2:0 Sampling Format.

2.5 Introduction to Digital Video

2.5.1 Aspect Ratios

The quality of an image is directly related to its number of pixels. The more pixels we use the closer the tracking is between the image in front of the camera and the one displayed by the monitor. A more "panoramic" format also adds to the attractiveness of an image. But the new 16:9 screen formats (*aspect ratios*) have 33 percent more area and require a corresponding increase in the number of pixels per line (Figure 2.11). The inescapable conclusion is that superior image quality requires more pixels, more signal bandwidth, and more processing power.

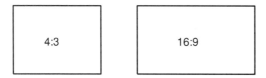

Figure 2.11   Common Screen Aspect Ratios.

2.5.2 Why Digitize Video?

Besides increasing the number of pixels, another approach to image enhancement is to reduce the effects of noise. In an analog system noise is difficult to control; it can be environmental, thermal, and even mechanical in nature. The use of digital processing eliminates noise entirely (although it may introduce occasional digital *errors*). To give analog noise little opportunity for interference, digitization must take place as close to the image capture site as possible. Conversion back to analog should be close to the display site but that is less critical.

Digitizing video is not without cost. Each sample generates between 8 and 30 bits of information, depending on sampling format and ADC resolution, while encoding the synchronization information adds three more bits to the stream (horizontal blanking, vertical blanking, and field). The digital infrastructure required to handle it all can be quite complex and expensive.

Since the number of paths we can take to improve image quality is practically unlimited, a new set of standards have emerged to organize and regulate this process. Without standards, no universally compatible scheme would ever materialize. To give this matter urgency the American government has mandated the complete migration to digital television by 2006.

Figure 2.12 and Table 2.4 summarize the most dominant of the proposed digital television formats. They fall into one of two categories: standard-definition television (SDTV) and high-definition television (HDTV).

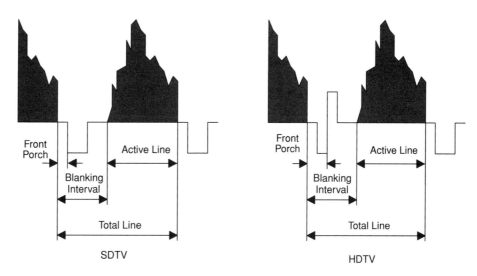

Figure 2.12   SDTV and HDTV Formats.

TABLE 2.4   SDTV and HDTV Formats

	Porch	Blank	Active Pixels	Total Pixels/ Line	Sample Rate	Aspect	Format Ratio	Refresh Rate	Clock
525 Line Interlaced Analog	16	138	720	858	13.5 MHz	4:3	720 × 480	29.97 Hz	13.5 MHz
525 Line Progressive Analog	16	138	720	858	27 MHz	4:3	720 × 480	59.94 Hz	27 MHx
525 Line Interlaced Analog	21.5	184	960	1144	18 MHz	16:9	960 × 480	29.97 Hz	18 MHz
525 Line Progressive Analog	21.5	184	960	1144	36 MHz	16:9	960 × 480	59.94 Hz	36 MHz
750 Line Progressive Analog	114	370	1280	1650	74.176 MHz	16:9	1280 × 720	59.94 Hz	74.176 MHz
1125 Line Interlaced Analog	88	280	1920	2200	74.176 MHz	16:9	1920 × 1080	29.97 Hz	74.176 MHz
1125 Line Progressive Analog	88	280	1920	2200	148.35 MHz	16:9	1920 × 1080	59.94 Hz	148.35 MHz

2.5.3 Digital Video Standards

SDTV is currently envisioned as a transitional set of standards between NTSC and full HDTV. Although its vertical resolution is equal to that of NTSC, its horizontal resolution is higher—720 pixels per line compared to 452 for NTSC. Its digital nature also makes it noise-free, contributing to a significant increase in image quality. There are two separate SDTV formats with 4:3 aspect ratios; one is interlaced (480i—Figure 2.13) and the other progressive (480p—Figure 2.17).

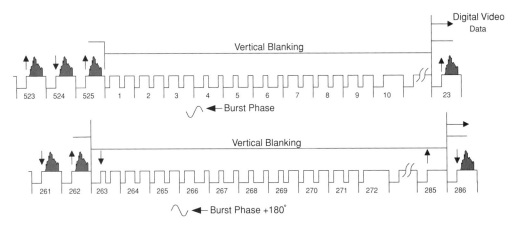

Figure 2.13   SDTV 480i Vertical Raster.

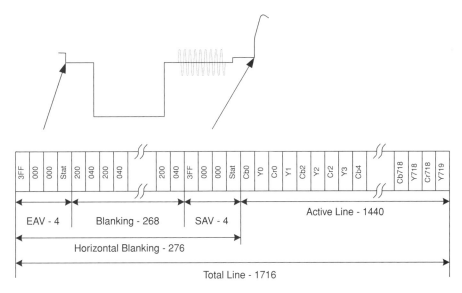

Figure 2.14   SDTV 480i Horizontal Digital Header.

2.5.3.1 SDTV 480i

The conversion of an analog image to digital comes pretty close to a direct mapping process. At the beginning of each new digital line we find a header framed by an EOV *(end of video)* four-word sequence and an SOV *(start of video)* sequence, also four words long. The EOV is issued at the end of the previous active line and coincides with the start of the analog front porch, while the SOV marks the start of the active portion of the current line (Figure 2.14). The first three words of the EOV and SOV sequences are invariably FF:00:00 for 8-bit systems and 3FF:000:000 for 10-bit systems. The fourth is a status word and its contents are shown in Figure 2.15.

In brief, H and V are horizontal and vertical blanking bits, and F is the field flag. The values of these bits are different for different regions of the digitized image, as seen in Figure 2.16.

Bits d0 through d5 of the status word are error correction bits and implement a simple exclusive OR algorithm for HVF bit recovery.

	d9	d8	d7	d6	d5	d4	d3	d2	d1	d0
Status	1	F	V	H	V⊕H	F⊕H	F⊕V	F⊕V⊕H	0	0

Figure 2.15   EAV—SAV Status Word.

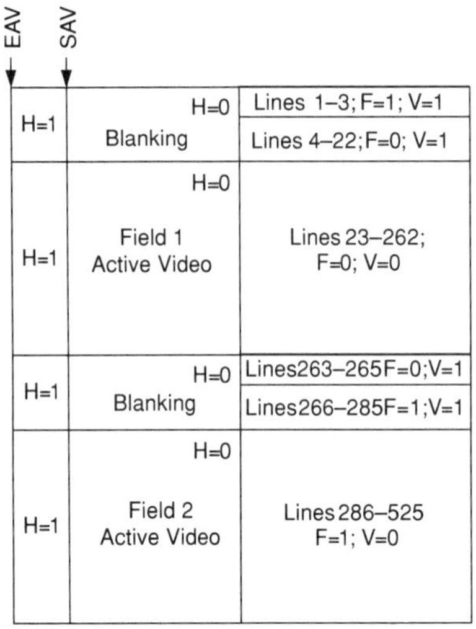

EAV	SAV		
H=1		H=0	Lines 1–3; F=1; V=1
		Blanking	Lines 4–22; F=0; V=1
H=1		H=0	
		Field 1 Active Video	Lines 23–262; F=0; V=0
H=1		H=0	Lines 263–265 F=0; V=1
		Blanking	Lines 266–285 F=1; V=1
H=1		H=0	
		Field 2 Active Video	Lines 286–525 F=1; V=0

Figure 2.16   SDTV 480i Reference Flags.

2.5.3.2 SDTV 480p

The image quality of progressive scan systems is inherently superior to that of an equivalent interlaced system. However, it also operates at a clock speed twice that of 480i, which makes the hardware implementation of 480p more difficult. The associated SDTV 480p reference flags are defined in Figure 2.18.

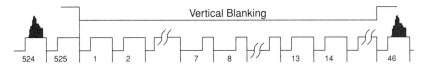

524 | 525 | 1 | 2 | 7 | 8 | 13 | 14 | 46

Vertical Blanking

Figure 2.17   SDTV 480p Vertical Raster.

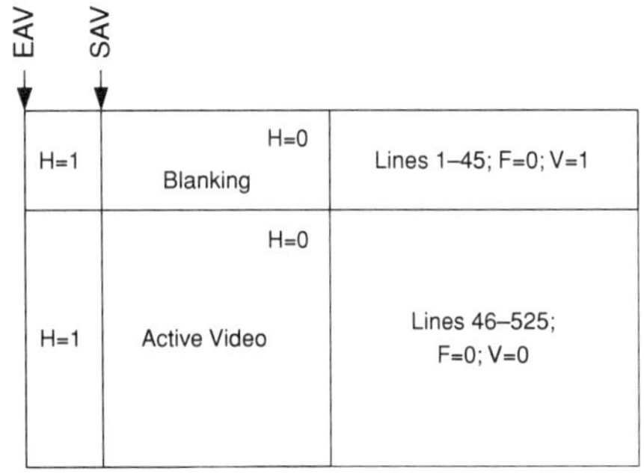

EAV	SAV		
H=1		H=0	Lines 1–45; F=0; V=1
		Blanking	
H=1		H=0	
		Active Video	Lines 46–525; F=0; V=0

Figure 2.18   SDTV 480p Reference Flags.

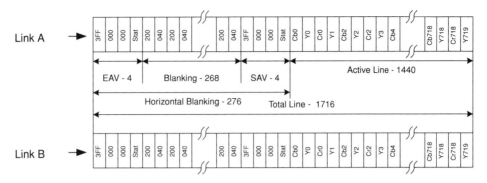

Figure 2.19 SDTV 480p Digital Stream.

To accommodate the higher information rate, 480p uses two separate digital streams as seen in Figure 2.19.

2.5.3.3 SDTV 480i 16:9
A variation on the 480i stretches each line from 720 pixels to 960 pixels (Figure 2.20). This effectively changes the screen aspect ration from 4:3 to 16:9. The vertical timing remains the same.

2.5.3.4 HDTV 720p
HDTV 720p is part of a midlevel group of standards designed to be less expensive than full-blown HDTV.

The 720p has 720 lines with 1280 pixels per line (Figures 2.21 and 2.22). Two separate data channels handle the luma and chroma information in

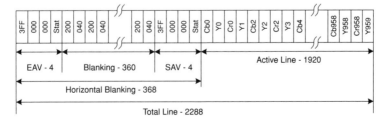

Figure 2.20 SDTV 480i with 16:9 Aspect Ratio.

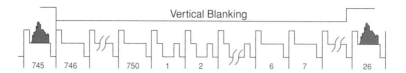

Figure 2.21   HDTV 720p Vertical Raster.

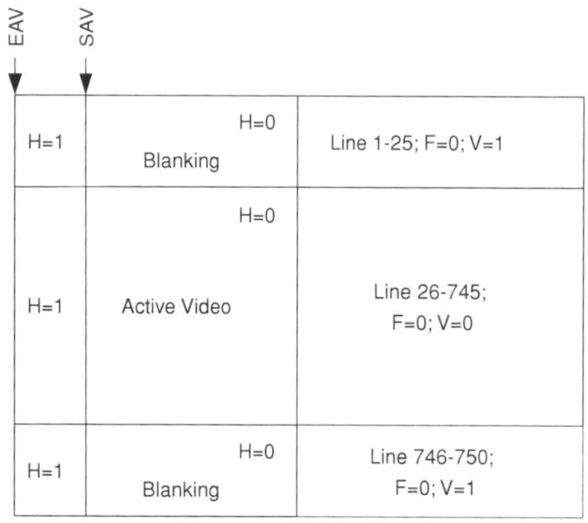

EAV	SAV		
H=1	Blanking	H=0	Line 1-25; F=0; V=1
H=1	Active Video	H=0	Line 26-745; F=0; V=0
H=1	Blanking	H=0	Line 746-750; F=0; V=1

Figure 2.22   HDTV 720p Reference Flags.

parallel, thus keeping the clock frequency at a manageable 74.176 MHz (Figure 2.23).

2.5.3.5 HDTV 1080i

The 1080i is truly a high-resolution format. Images consist of 1080 lines, each line being composed of 1920 pixels (Figures 2.24 and 2.25). The screen has a 16:9 aspect ratio but most displays will easily down-convert to 4:3 either by filling in black edges or by cropping the image. Although the 30 Hz frame rate

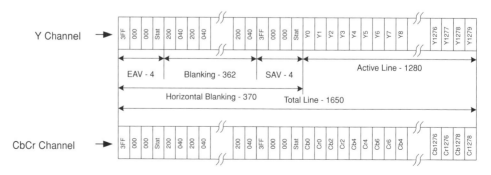

Figure 2.23   HDTV 720p Digital Stream.

Vertical Blanking
1123
1124
1125
1
2
5
6
7
21
560
561
562
563
564
567
568
569
584

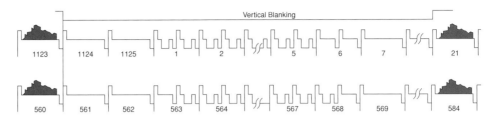

Figure 2.24 HDTV 1080i Vertical Raster.

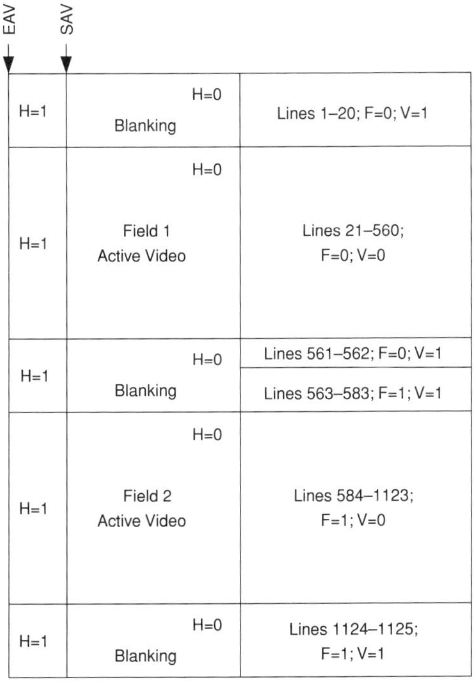

EAV
SAV
H=1
H=0
Blanking
Lines 1–20; F=0; V=1
H=1
H=0
Field 1
Active Video
Lines 21–560;
F=0; V=0
H=1
H=0
Blanking
Lines 561–562; F=0; V=1
Lines 563–583; F=1; V=1
H=1
H=0
Field 2
Active Video
Lines 584–1123;
F=1; V=0
H=1
H=0
Blanking
Lines 1124–1125;
F=1; V=1

Figure 2.25 HDTV 1080i Reference Flags.

Y Channel
3FF 000 000 Stat 040 040 040 040 040 040 3FF 000 000 Stat Y0 Y1 Y2 Y3 Y4 Y5 Y6 Y7 Y8 Y1916 Y1917 Y1918 Y1919
EAV - 4
Blanking - 272
SAV - 4
Active Line - 1920
Horizontal Blanking - 280
Total Line - 2200
CbCr Channel
3FF 000 000 Stat 200 200 200 200 200 200 3FF 000 000 Stat Cb0 Cr0 Cb2 Cr2 Cb4 Cr4 Cb6 Cr6 Cb4 Cb1916 Cr1916 Cb1918 Cr1918

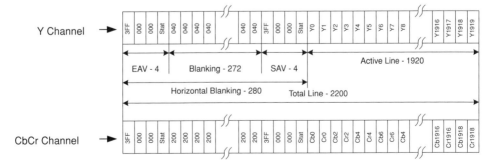

Figure 2.26 HDTV 1080i Digital Stream.

Vertical Blanking
1121
1122
1125
1
2
5
6
7
42

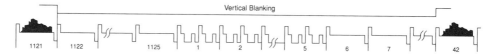

Figure 2.27 HDTV 1080p Vertical Raster.

keeps the clock at 74.176 MHz, we still have to use a dual-channel processing path (Figure 2.26).

2.5.3.6 HDTV 1080p

1080p represents the high end of HDTV. It is both a maximum resolution format (1920 × 1280) and progressive. To achieve such a feat the 1080i processing channels shown in Figure 2.26 are pushed to 148.35 MHz. The vertical raster arrangement and the reference flags of the HDTV 1080p standard are shown in Figures 2.27 and 2.28, respectively.

2.5.3.7 HDTV RGB

All the SDTV and HDTV formats covered so far operate in the YCbCr color space. However, in some applications the signals are neither stored nor transmitted to a far side. In machine vision, for example, images are processed immediately after they are captured. In such cases, the HDTV RGB progressive format shown in Figure 2.29 can be most convenient. Its resolution is 1280 × 720 and operates at a 74.176 MHz clock rate.

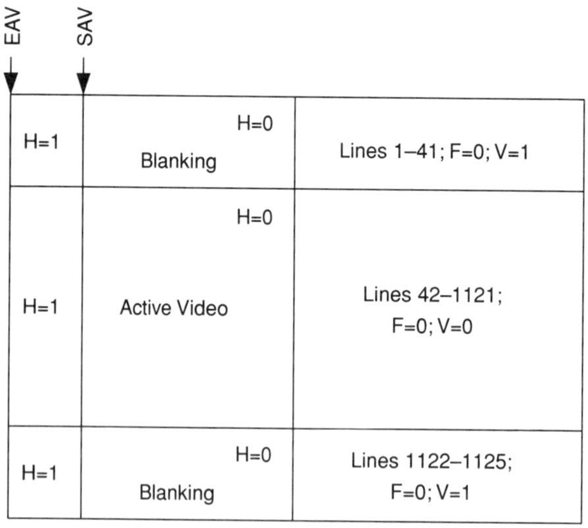

EAV
SAV
H=1
H=0
Blanking
Lines 1–41; F=0; V=1
H=1
H=0
Active Video
Lines 42–1121;
F=0; V=0
H=1
H=0
Blanking
Lines 1122–1125;
F=0; V=1

Figure 2.28 HDTV 1080p Reference Flags.

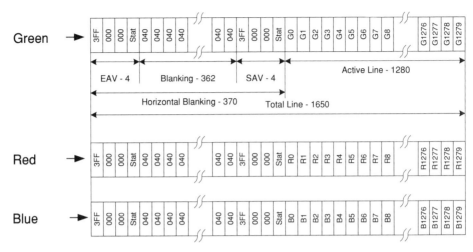

Figure 2.29   HDTV RGB Digital Stream.

2.5.4 Serial Digital Video—SDI and HD-SDI

Going back to the image capture appliance in Figure 2.6 we see how video information is transformed from the initial RGB color space to the more economical YCbCr (or YPbPr) space. We also explored several options when selecting a target digital domain for our analog inputs. We converted the three analog input signals into one, two, or three parallel digital streams.

The final step at the transmitter end is to reduce these eight to thirty lines of digital output to a much smaller number of very high-speed digital links. This conversion is done by a device called a *serializer* (Figure 2.30).

The serializer is a smart parallel-in/serial-out shift register. In order to convert a 10-bit parallel bus the serializer must be driven by a serial clock with a frequency ten times higher than the frequency of the parallel clock.

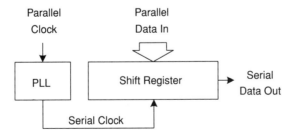

Figure 2.30   Serializer Structure.

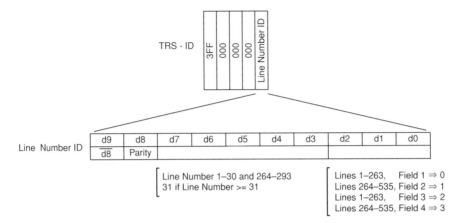

Figure 2.31   TRS-ID SDI Header.

The two clocks must also be in phase. To serialize 27-MHz and 36-MHz SDTV we need 270-MHz and 360-MHz serial clocks, respectively. The resulting stream is referred to as *serial digital interface* (SDI).

For HDTV formats, the serial clock rates jump up to 1.4845 GHz and produce a *high-definition* SDI stream, or HD-SDI. The circuit responsible for the generation of the serial clock is a phase locked loop, or PLL (see Chapter 4).

Before serializing a new line a specific synchronization and control sequence called the Timing Reference Signal and line number ID (TRS-ID) must be introduced into the data stream during the horizontal blanking period. Its structure is shown in Figure 2.31. Not only does the receiver use the TRS-ID to recognize the beginning of the new line but it also uses the line ID in the retrieval of transmitter-embedded data.

After TRS-ID insertion, the serializer organizes the parallel input information into data packets (Figure 2.32).

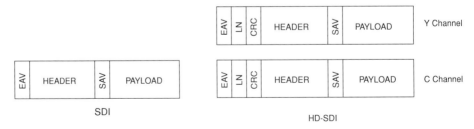

Figure 2.32   SDI Data Packet.

The header segment contains a number of link parameters including line number, line number cyclic redundancy code (CRC), length of payload (according to pixel format), destination and source addresses, block segmentation of the payload, payload CRC, and header CRC. For very high data rates, such as those necessary for HDTV 1080i and 1080p formats, the luma and chroma streams use separate serial digital channels.

2.5.5 SDTV and HDTV Voltage Level

While in the analog domain the voltage levels of the different SDTV and HDTV signals must be tightly controlled.

Looking at Figure 2.33 we may notice that the horizontal sync pulse of the HDTV signal is actually tri-level, with a new line starting at the zero-crossing point of the waveform. This presents a more precise reference point than the edges of a negative SDTV sync pulse.

The sync pulses are marked with dotted lines to indicate that sometimes the sync information is embedded into the color components (usually green) and sometimes is provided on separate lines. Figure 2.34 shows the voltage levels associated with a luma/chroma format (YPbPr in this case).

2.6 Special Video Connectors

Many of the signals discussed in this chapter enter and leave various devices using BNC and occasionally RCA connectors. There are a couple of exceptions. S-video connectors are such an exception. A standard consumer interface s-video currently keeps the luma and chroma signals separate. This eliminates

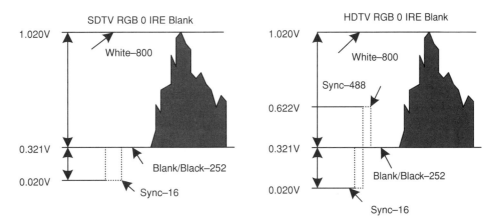

Figure 2.33   SDTV and HDTV Analog Voltage Levels.

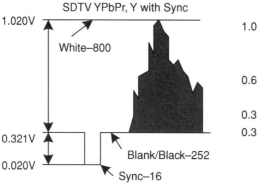
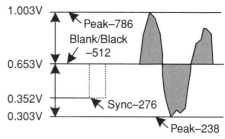

Figure 2.34   SDTV YPbPr Voltage Levels.

the need for Y/C separation (comb filtering) at the receiver end, a process that degrades to some extent the original signal. The "pin-out" representation of an s-video connector can be seen in Figure 2.35.

The DVI-HDTV connector found on many high-resolution monitors is a more complex device (Figure 2.36). It consists of a digital section and a separate RGB analog section. The digital section has five high-speed, individually shielded, differential lines, each capable of supporting serial rates of up to 165 MHz. Together with the differential clock line (CLK) they implement a com-

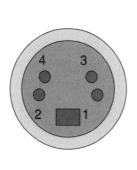

Pin	Signal
1	GND
2	GND
3	Y
4	C

Figure 2.35   S-Video Connector, "Pin-out".

Pin	Signal	Pin	Signal	Pin	Signal	Pin	Signal
1	D2–	9	D1–	17	D0–	C1	Red
2	D2+	10	D1+	18	D0+	C2	Green
3	Shield	11	Shield	19	Shield	C3	Blue
4	D4–	12	D3–	20	D5–	C4	HSYNC
5	D4+	13	D3+	21	D5+	C5	GND
6	SCL	14	+5V	22	Shield		
7	SDA	15	GND	23	CLK+		
8	VSYNC	16	Hot Plug	24	CLK–		

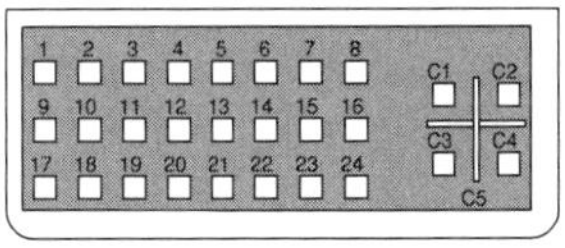

Figure 2.36   DVI – HDTV Connector Pin-out.

plete Transition-Minimized Differential Signaling (TMDS) data link. The analog section consists of three contacts corresponding to the three RGB color components. The associated HSYNC and VSYNC signals are accommodated by separate connector pins. A *hot plug detect* pin is used to sense live connection and disconnection and control the link's power supply accordingly. Finally, a local bus I^2C interface (SDA-data, SCL-clock) provides the access port into the device.

Digital Video Applications

If we analyze the internal architecture of most video devices we find a number of function blocks networked together by means of standard interfaces and ports. The total number of such blocks is small, but the variety of ways they can be interconnected is almost unlimited. Within each block we may find one or more integrated circuits (ICs), while at times a single IC may contain multiple blocks, a common occurrence in price sensitive applications such as consumer electronics. In this chapter we look at some current video systems, identify these blocks, and study their role in each design. In Chapter 4 we formulate a simple methodology for combining multiple blocks into new user defined arrangements.

In converting function blocks to circuitry we use the recent trend among semiconductor manufacturers towards "designer friendly" ICs; we select only well supported products with ample documentation and superior engineering tools, including development boards, reference designs, and software development kits (SDKs). In a sense, the emergence of highly integrated, easy-to-use ICs has democratized the industry by releasing many designers from the rigors of building video circuits at the level of discrete components. Indeed, very little of the "black art" tradition associated with video design is left.

3.1 Image Capture Cards

An integral part of any multimedia PC, the image capture function can be performed either by internal *image capture cards* or by external USB appliances. The first block along the video signal path of an elementary card (Figure 3.1) is the *video decoder-digitizer*. Its function is to first translate the composite input into intensity and color signals (YCbCr, YPbPr, or YUV) and then digitize them using on-board analog-to-digital converters (ADC). The processing speed of the digitizer must be sufficient to accommodate the rather brisk rate of the NTSC/PAL video formats (13.5 million pixels per second).

Using the timing information extracted by the decoder's *phase locked loop* (PLL), the digital video is stored in dual-ported registers where they become available to the host computer through a PC bus interface. The information passed on to the PC includes pixel color and intensity, horizontal and vertical syncs, odd/even field flag, blanking intervals, and so on.

Although all of the blocks in Figure 3.1 are now available in single IC packages, thus making this design very inexpensive, there is one major drawback. It requires significant PC overhead. The PC must deinterlace the image (PC monitors use progressive scanning), crop and scale the image from a standard NTSC/PAL format to a user-controlled size and shape, translate the color and intensity information into RGB (PC monitor standard), and finally position the image on the screen where the user wants it.

If a PC is fully dedicated to these tasks its operating system and software can easily handle the job. A case in point is the *security multiplexer*, a device that combines up to 16 NTSC/PAL inputs into a single VGA/XGA output, at a rate of about 2 images per second. In this case, a PC chassis sequentially captures and tiles together the inputs from a number of image capture cards, each card fed by one or more security cameras. Once in the memory, the images can be stored on the PC hard drive or archived on CD-ROM.

If "near live" video is not good enough or the PC has no resources available for video processing, a more complex image capture card is needed. In Figure 3.2 the down-scaling and de-interlacing tasks are performed locally by dedicated hardware. Complete video images are then stored in a frame-sized banked memory buffer. While the card stores a digitized image in one bank, the PC asynchronously retrieves the previous image from the second bank,

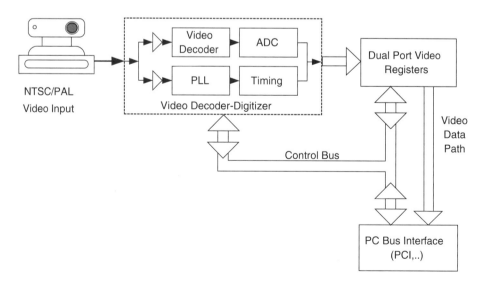

Figure 3.1   Minimal Image Capture Card.

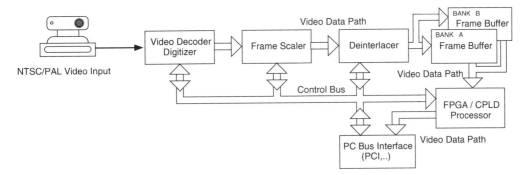

Figure 3.2   Expanded Image Capture Card.

generally in high-speed bursts. The card's added "muscle" resides in its *scaler, deinterlacer,* and *FPGA/CPLD processor* blocks. To make this design a stand-alone image capture appliance, one must add a local controller and a USB interface that connects it to the host PC.

3.2 Frame Synchronizer

Both the timing and pixel arrangement of standard video signals are rigidly organized. The visible part of an NTSC field, for example, always starts in the top left corner, and takes about 16.7 milliseconds to end up at the bottom right. The absolute time when the first scan line starts is random, and may depend on such factors as the time the camera turns on or the status of its battery. The pixel rates of different video sources are also slightly different, since they are driven by clocks with nonzero tolerances. This virtually guarantees that images generated by two different video devices are out of sync. At a given time the first may display the 103rd pixel of the twenty-sixth line of an even field, while the second displays the tenth pixel of the eighty-seventh line of an odd field.

There are consequences to this signal structure. If we want to combine, on a pixel-by-pixel basis, video from multiple sources we must first synchronize the frames so we mix the same rank pixels of each image. In analog switching applications, synchronizing the sources assures that there is no loss of image integrity as the monitors re-synchronize, after each switch, to the new source. Synchronization also minimizes the need for further image buffering in downstream video processors.

Figure 3.3 details the structure of a stand-alone *frame synchronizer*. Its key element, the *genlock* block, generates the all-important timing signals for the system. The front sync separator of the genlock block detects and reconstructs the horizontal sync (HSYNC) and vertical sync (VSYNC) pulses of the *reference video source*. It also recognizes its odd and even fields, as well as the burst and back porch segments. The PLL that follows uses the HSYNC pulses to cre-

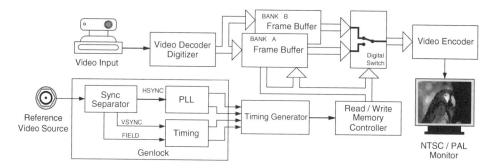

Figure 3.3   Frame Synchronizer.

ate an output pixel clock equal to that of the reference, and a second process-ing clock that is a multiple of the pixel clock.

On the main signal path, the analog video input is decoded, digitized, and stored in a dual-ported *frame buffer*. From there the digital video information is read by the *memory controller* and delivered to the *video encoder* block for conversion back to analog and insertion of the color burst.

The memory controller is driven by the clock and sync signals extracted from the reference, so it starts reading each field and each line exactly at the time the reference video source generates its own corresponding field and line. The resulting output of the frame synchronizer is an analog output identical to the video input signal except for sharing the video timing of the reference video source, including its color burst.

The *digital switch* block feeding the video encoder always reads the frame buffer bank that the decoder does not write to at that time. We thus avoid the "image tearing" that occurs when read and write pointers of slightly different speed cross each other while accessing the same memory buffer. In practice, the digital switch, the timing generator, and the memory controller are imple-mented in a single FPGA/CPLD chip driven by the processing clock.

3.3 Scan Converter/Video Scaler

Since we watch television and read e-mails on identical looking displays, we might assume that TVs and computer monitors are similar electronically or at least compatible. They are not. When we introduce a new TV standard, studios invest large sums of money to accommodate the new formats. These up-front expenditures plus requirements to maintain back-compatibility to legacy sys-tems make incremental technology improvements of TV monitors impractical. So TVs are predestined to fall behind in technology shortly after a new stan-dard is introduced. There are also bandwidth-related concerns. Because TV content is studio based, TV signals need to be transmitted to our homes over bandwidth-limited channels such as cable, satellite, or fiber; trade-offs must be made between image resolution and the number of content-carrying channels reaching the viewer.

PCs, by contrast, are local systems and their bandwidths are limited only by the speed of their internal buses (one notable exception is Internet access where the bandwidth of external communication channels comes into play). Another factor pushing PC monitor technology is the skyrocketing popularity of videogames. As a result, PC monitors have better colors, higher resolutions, and faster refresh rates than TV monitors.

It is still possible to display computer images on TV (composite or RGB) monitors but a *scan converter* is needed in the process (Figure 3.4). The reverse route, displaying full screen resolution NTSC/PAL video on a PC monitor, is handled by the *video scaler* processor (Figure 3.5).

Scan conversion starts with video signal digitizing. A color space converter (or matrix) translates the PC native RGB format to YCbCr or YPbPr, a color scheme more common to TV and video equipment. Another benefit of YCbCr and YPbPr is that, if 4:2:2 coding is adequate, any subsequent processing requires significantly reduced memory resources. The down-scaling sequence begins with the vertical filtering of the image by a dedicated *vertical scaler*. Using a multitap digital filter at its core, the scaler calculates output pixel values by calculating a weighted average of a number of pixels from adjacent lines of the source image.

Since the output resolution is fixed at NTSC/PAL values, the scaling ratios depend only on the original image resolution. Each ratio requires a different set of polynomial coefficients. Two memory buffers support the process: an upstream source image buffer for the input image and a downstream buffer for the vertically scaled image.

An optional line averaging "flicker filter" may also be used. The *horizontal scaler* that follows is also a multitap digital filter, but because it averages neighboring pixels on a given line, it needs only minimal buffering. The final image is then interlaced and encoded into NTSC, PAL, or related variants of composite video.

Video scalers are more complex than scan converters; they must contend with potentially fast-changing input images such as those of hockey games or car races. Problems can occur during deinterlacing when consecutive fields are merged into single noninterlaced frames. Then, fast vertical movements result in image blurring, and horizontal movements translate into fuzzy edges and "hair comb" contours.

High-end deinterlacers use a specialized detector for each potential problem; edge, vertical, horizontal, and interfield motions. In each case, a separate interpolator averages pixels from frame to frame in accordance with the data received from their respective detectors. This process requires significant memory and processing power, and is generally performed by dedicated hardware.

Once the video is deinterlaced it proceeds to the scaling sections which are similar to the ones used for scan converters. One exception: The video is up-scaled to PC monitor resolutions and scan rates, and not down-scaled as in

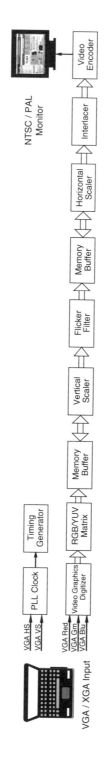

Figure 3.4 Scan Converter.

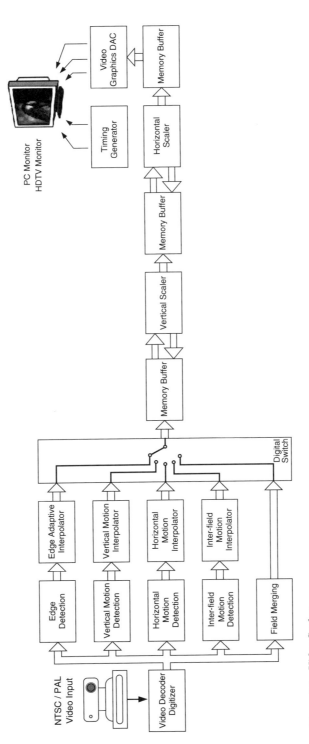

Figure 3.5 Video Scaler.

45

scan converters. Multiple buffering may also lead to enough pipeline delays to require audio resynchronization.

If 24 frames per second of film footage is processed, an additional rate conversion to 60 fields per second is necessary. It can be done by simply sourcing two consecutive video fields from a film frame followed by three consecutive video fields from the next film frame, and so on (known as a *3:2 pull-down process).*

3.4 Graphics, Logo, and Character Inserters

From channel IDs to crawling Energizer™ bunnies and crosshair calibration grids, more and more graphic information is being added to the videos we watch. The graphics can be generated "off-line" by advertising agencies and staff artists or in real time by clocks, news services, or instruments. Either way, the graphics need to be converted into a convenient electronic format and inserted into the live video stream at the proper time and in the proper position.

Assuming that standard bitmapped graphics files are used, they must be "prepped" by the host computer before they can be processed by the *graphics inserter.* After stripping the header and rounding off the data to eight to twelve bits (depending on system resolution) the RGB is converted to YCbCr or YPbPr. Then, irregularly shaped logos and symbols are superimposed on a "key" color background and enclosed in a geometric (most of the time rectangular) *graphics box.* For graphic animations a number of different graphics boxes are used, each with their own screen location, rank, and display time. All information is subsequently encapsulated in a file and transferred to the inserter on a CD-ROM or by means of standard LAN, USB, RS-232, or Firewire ports. There graphics, position, and script data are separated and processed. Under the control of script commands, images needed for immediate overlay are transferred to a fast *graphics memory buffer* and position data is sent to an *overlay template.* Permanent graphics and character fonts are stored in nonvolatile *graphics memory storage* (Figure 3.6).

As a new image scan starts, the *read processor* is reset and both its horizontal pixel and vertical line counters are cleared. When the count reaches the desired position of the graphics box (detected by comparison to the overlay template), the read processor starts reading pixels from the graphics memory buffer into the *color key detect* and the video encoder. If the "key" color is detected the *video switch* is connected to the *input video source;* if not, the pixel represents valid graphics and the switch is connected to the output of the video encoder. A *pipeline register* compensates for the delay incurred while the key detector tests for color key matching.

The video input is also processed by a genlock block that helps locate the graphics box with respect to the sync pulses. This sequence is repeated as long as the script data requires. A new sequence starts when the graphics memory buffer and the overlay template are updated by the *local controller.*

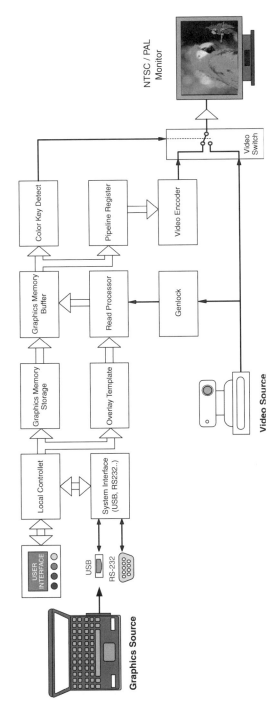

Figure 3.6 Graphics Overlay/Logo Inserter.

3.5 PiPs, SbSs, and PoPs with Genlock

Besides being used to generate "video preview" windows in newer TVs, Picture-in-Picture (PiP) processors can be found in many other applications including broadcasting, videoconferencing, distance learning, "tele-justice," and "tele-medicine." A typical design for a stand-alone PiP appliance is shown in Figure 3.7. Meant for studio applications, the unit overlays an *inset video* image on top of a *background video* image, with the output "genlocked" to a reference video signal.

Operationally, there are two input video paths. The background video is digitized and stored in a *background video buffer*. The inset video is digitized, then scaled by a user-programmable *down-scaler* and then stored in an *inset video buffer*. The *local controller* reads the desired position and scaling factor either from a host computer (via *system interface*) or from the local *user interface* panel. It then generates matching overlay template and down-scaler instructions. As the *overlay processor* (read/write memory controller) compares its pixel and column counts with the overlay template, it commands the digital switch to select the appropriate source for the output pixel—for example, background image, inset image, or border color (a fixed YCbCr combination inserted by the overlay processor itself).

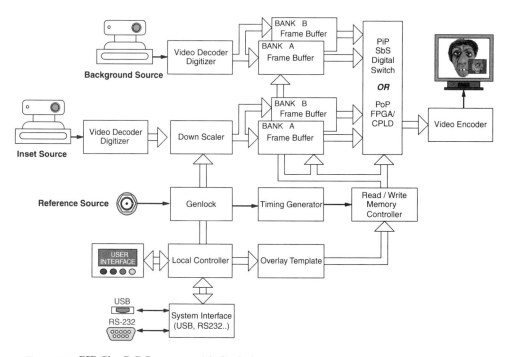

Figure 3.7   PIP, Sbs, PoP Processor with Genlock.

The synchronization of the overlay processor (that is, pixel clock and vertical and horizontal syncs) to the reference source is done by a genlock block. Since the background, inset, and reference signals are all random (different pixel clocks and syncs), we must take precautions so the read pointers to the background and inset buffers (synchronized to the reference) and the write pointers (synchronized to the background and inset sources) are not allowed to pass each other when accessing the same memory bank.

This would result in image tearing—the pixels that are read before the pointers cross each other belong to a different frame than the ones that are read afterwards. To avoid this, output pixels should always be read from the memory bank to which nothing is being written at that time. For example, as the background decoder writes to Bank B of its buffer, the inset processor directs the digital switch to read background pixels from Bank A. Bank selection is done at the end of each frame and occasionally results either in the dropping or repeating of a frame. Essentially the same architecture can be used to implement Side-by-Side (SbS) and Picture-on-Picture (PoP) formats. In the first case, neither of the input images is scaled but both are cropped, centered, and positioned next to each other by the overlay processor. In the PoP format, two images of reduced luminance are placed on top of each other, mostly to highlight minute differences. This can be useful in industrial quality control operations or in clinical diagnosis. Its execution requires a weighted addition of the two full-size video input streams.

3.6 QUAD Security Processor

Combining four scaled videos onto a single screen could be easily done just by expanding the PiP design to four inputs. However, the security industry is extremely price sensitive, and no QUAD solution is viable if it costs more than the combined cost of the three monitors it replaces. So how can we minimize cost while preserving performance? The key rests in the nature of the QUAD video output. The output consists of four images, each exactly a quarter screen in size. This allows for the use of much simpler scaling techniques, such as pixel, line, and field decimation.

Data organization for a pixel/field decimation QUAD is shown in Figure 3.8. In Figure 3.8 (a), the even field is totally discarded, while the odd field is "even pixel decimated," by keeping only the odd rank pixels for each line. Then the odd lines are grouped into a "pseudo odd-field" and the even lines are grouped into a "pseudo even-field," both a quarter size of the original fields. The result is a quadrant-size image based entirely on the information contained in the odd field of the input.

Similar decimation processes lead to the scaled video data configurations in Figure 3.8 (b) (odd field/even pixel decimation), (c) (even field/odd pixel decimation), and (d) (odd field/odd pixel decimation). Most QUAD devices use at least two of these decimation methods per channel. The reason is performance. When used alone all of the above methods result in relatively poor image

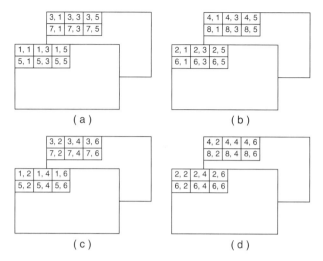

Figure 3.8 QUAD Processor Pixel Organization.

quality. Single field processing makes the refresh rate a "jumpy" 15 frames per second while throwing out every other pixel reduces detail. To dramatically improve the output image QUAD designers alternate pseudo-fields like the ones seen in (a) and (d), (b) and (c), or use all four. As seen in Figure 3.9, this design is channel symmetric, with all its buffers combined adding up in size to a single-frame, dual-bank memory.

The hardware required is very simple (Figure 3.10): four video decoder-digitizers, one FPGA/CPLD, one SDRAM video memory, one video encoder, a single chip CPU controller, and a system interface chip. One note: SDRAM timing can be satisfied only because the average pixel rate of each video quadrant is a quarter of the pixel rate for the video output. So in all, the SDRAM is accessed at twice NTSC pixel rates (calculated by averaging write cycles for four quadrants plus output read cycles), or 27 megapixels per second. For a 16-bit wide SDRAM, if we operate the QUAD in 4:2:2 mode (16 bits per pixel, 8 for Y and 8 for CrCb), we still have ample bandwidth to insert colorful borders, text, and simple graphic overlays.

3.7 Videophone

Videophones were introduced to the general public during the 1963 World's Fair and quickly became an integral part of most science fiction novels and movies. They are now a commercial reality. You can buy them from your local computer store or from most online telecom outfits. They come in many shapes and (bandwidth) sizes, from telephone-like sets with built-in LCD screens to PC-attached webcams.

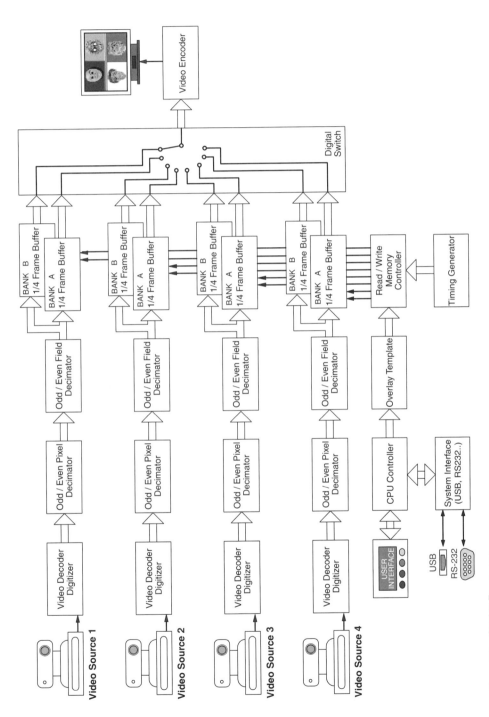

Figure 3.9 QUAD Processor.

51

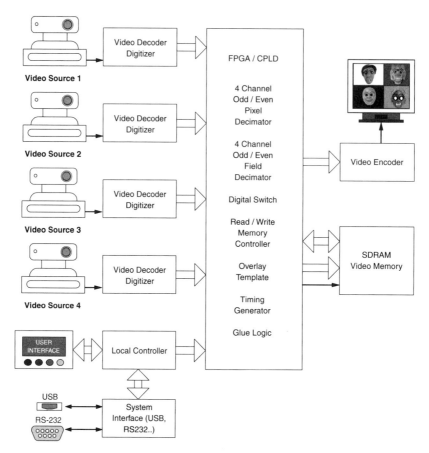

Figure 3.10   QUAD Processor Hardware Design.

At the core of each videophone is a pair of *CODEC* devices, one used for video and the other for audio processing (Figure 3.11). CODECs are digital compression and decompression engines that can be realized either in hardware or software. A common data compression standard used in videophones as well as in DVD and HDTV technology, is *MPEG-2* (Chapter 13). MPEG-2 replaces a stream of full resolution frames with a sequence of full resolution "key" frames (called *intraframes* or *I frames*) followed by lower resolution correction frames. This data architecture recognizes the fact that video content changes little between consecutive frames. A type of correction frames called *predictive* (P) frames approximate future frame elements by comparing the current I frame to the previous I frame.

Another type called *bilinear* (B) frames compare and record differences between the last I frame and subsequent P frames. *Discrete cosine transformation* (DCT) compression is applied to both reference and correction frames.

In a typical arrangement, one key frame is followed by 11 to 14 P and B correction frames. MPEG-2 was designed to be easy to decode, so compressed files can be played back (decompressed) by a myriad of inexpensive consumer devices. Live encoding is a different matter, requiring the use of powerful digital signal processors (DSPs) or PC hardware.

The lower bandwidth of sound makes the audio compression algorithms used by videophones less demanding. A number of audio standards are also covered by MPEG-2, including the popular *MP3*. Although MPEG-2 is the standard of choice for most wideband applications, it is not the only standard available for videophones. Older systems use simpler standards such as *H.320* for ISDN, *H.324* for POTS (plain old telephone service), and *H.323* for LAN (local area network) service.

A set of channel-specific interfaces connects the videophone to the appropriate transmission media (PCs, Ethernet POTS, or cable). The user controls in most cases consist of an infrared (IR) remote and a front panel for stand-alones, or a graphical user interface (GUI) for PC-based webcams. Essential to

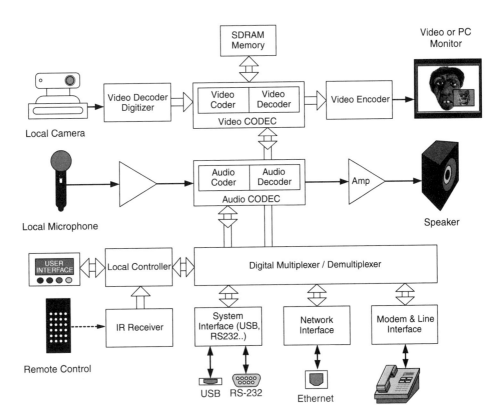

Figure 3.11 Videophone.

videophone functionality are also a good quality auto-focus camera, an echo-free mike, a clear sound speaker, and a high-contrast color monitor.

3.8 Videowall Controller

From a technology standpoint, big-audience displays are the domain of large light-emitting diode (LED) arrays while small individual systems are dominated by single-unit CRTs and flat panel displays. In the middle, for point of purchase (POP) video ads and meeting hall information centers, videowalls are the systems of choice.

A videowall consists of an array of horizontally and vertically stacked video monitors, driven by a dedicated *videowall controller*. The monitors are of a high brightness type and, by necessity, have a narrow frame. The drive to minimize edges has led to the development of enclosed projector and screen assemblies referred to as videowall "cubes." The obvious solution of single screens illuminated by *projection arrays* has proven impractical due to light leaks and image overlaps from adjacent projectors. Other display-related problems are nonuniform center-to-edge illumination in projection cubes, and different monitor appearance due to poor calibration and aging.

By now we are familiar with most of the blocks that make up the videowall controller (Figure 3.12). A front-end *analog matrix switch* selects which channels are to be displayed, and routes them through video decoder-digitizers into upstream *input frame buffers*.

Images are scaled up according to the requirements of the controller script, segmented consistent with the size and position of monitors, and routed through a *digital matrix switch* into downstream *output frame buffers*.

At this point each output buffer contains the subimage associated with a particular monitor. *Video graphics* digital to analog converters (DACs) or *encoders* read the output frame buffers and convert them to RGB or composite signals.

When working with videowalls a great deal of care must be taken in selecting the scaling, cropping, and positioning factors for each image, since the output must appear as seamless as possible.

3.9 Video to LED Array Processor

The development path to reliable large-area outdoor displays has been long and, well, interesting! With a huge ready market, sign producers tried everything from the ridiculous to the sublime: colored bulbs, arrays of red, green, and blue CRT tubes, rotating painted cubes, black pistons filled with white liquid, half-white/half-black rotating balls, and many others too strange to mention. As Times Square and stadiums everywhere can attest to, a solution now exists and it is the long awaited RGB LED color array. Its commercial viability has as much to do with technological advances (in the form of the elusive blue LED) as with the advent of low cost global labor.

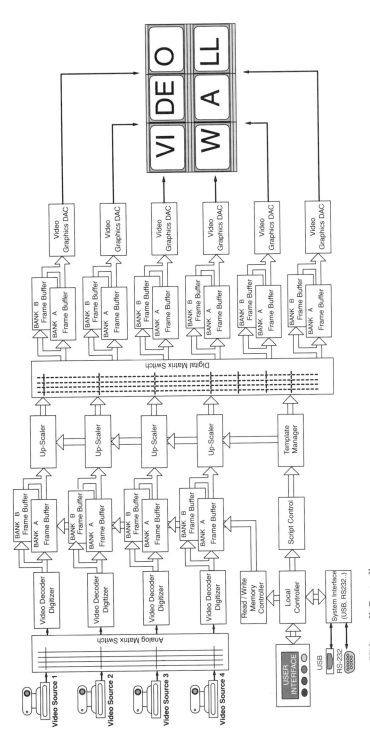

Figure 3.12 Videowall Controller.

To ensure utmost compatibility with current video sources, the best approach to LED array design is to convert from a standard video format (such as NTSC/PAL or VGA) to a system-specific, LED custom format. Driving LEDs is a very different process than what we have encountered so far. Since the brightness of LEDs is determined by their average forward current, good control demands the use of the more complex constant current drives instead of the simpler constant voltage type. With no pixel persistence and only a limited peak current capacity, LEDs can be multiplexed only at relatively low duty cycles. For example, a LED requiring 25 mA for peak output brightness and with a peak current rating of 500 mA needs a minimum duty cycle of 5 percent. If we drive one line of LEDs at a time, it follows that, with the given diodes, we can design panels with at most 20 rows (in this arrangement we simultaneously constant current sink all the columns).

If we want 32 intensity levels per LED at a 30 Hz refresh rate, our pulse width modulated (PWM) multiplexing circuit must be able to drive each row with pulse widths between about 1 and 32 milliseconds. For a 20 column tile design, each row will need to accommodate a peak demand of 10 A (20×500 mA peak PWM LED current). It is obvious that the current-carrying capacity and switching characteristics of the LED row drivers need to be quite robust. So does the heat sinking!

In Figure 3.13 we break the overall LED array into a number of individual *tiles*, each with its own logic control, drivers, and power supply. The only data interface between each tile and the *master controller* is a *local area network* (LAN), which supplies both the video and the tile control information to the *tile controller*. Once on the LED tile, the video data is decompressed (if needed), converted to RGB (if needed), rounded off, converted to PWM (see below), and stored in a *video buffer*.

Because the LEDs are pulse width modulated (PWM), a brighter color is achieved by keeping the LED on for a longer time (wider PWM pulse). To implement 32 brightness levels at 30 frames per second (each frame 33.3 milliseconds), we divide the frame time into 32 time slots each about 1.1 milliseconds in duration (Figure 3.14). In this example, the process of PWM conversion creates 32 tables, each bit within a table marking one of the LEDs. All tables are read sequentially within each frame time, and the data is serialized, formatted, and delivered to the row and column shift registers. An OFF lamp will have its corresponding bit in all PWM tables set to zero, a maximum brightness LED will have them all set to one, with in-between output levels having a correspondingly proportional number of ones logged. Much of the PWM conversion and subsequent processing is done by FPGA/CPLD firmware. The faster the FPGA/CPLD, the larger the size of the LED tile it can drive.

The *row and column shift registers* must have latching outputs, thus keeping the attached drivers on while new *time slot* data is loaded. Each LED is turned on when the matrix line on which it is located is driven high, and its corresponding matrix column is driven low. Separate *watchdog* circuits monitor power supply current levels and shift register outputs for signs of any fail-

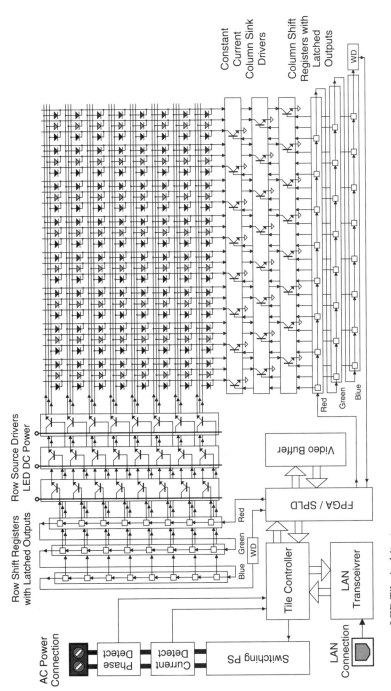

Figure 3.13 LED Tile Architecture.

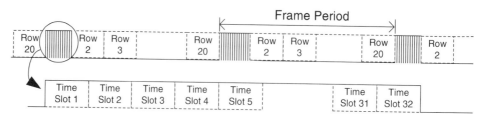

Figure 3.14 Time Division of LED Frames.

ure which may compromise the LEDs. If failure is detected, the tile power supply is clamped and the master controller is flagged.

Synchronization among different tiles is achieved by tracking the time elapsed since the last zero crossing of the input AC power line. This approach (and the low cost of memory) allows for frame "time tagging" and the local storage of whole video clips and scripts. A temperature-controlled cooling fan and a switching power supply complete the tile.

The master controller design is straightforward (Figure 3.15). A video decoder-digitizer and a graphics digitizer process incoming composite or XGA/VGA video and a CPLD/FPGA-based *memory controller* stores the images in a dual-ported *master video memory*. From there a CPLD/FPGA-based *pixel server* allocates different pixel blocks, through *local video networks*, to different LED tiles.

3.10 Alpha Channel Mixer and Cross-Fader

As the Information Age progresses, our screens carry more of the responsibility of delivering the information to us in a time- and space-efficient manner. Our commercial TV channels have changed into visual mosaics filled with translucent logos and crawling text. Movie scenes do not change suddenly anymore but blend into each other or gently fade away. The means to produce these effects are provided by the *alpha channel mixer* also known as a *cross-fader*. This device mixes two video sources together by doing a weighted average of their corresponding pixels according to a user-defined weight coefficient. This alpha (α) coefficient quantifies the weight value that each of the original pixels contributes to the final displayed pixel.

$$Pixel\_out = \alpha \ Pixel\_in\_A + (1 - \alpha) \ Pixel\_in\_B$$

Alpha can be viewed as the transparency factor of image B as it is superimposed onto image A. If $\alpha = 1$ the transparency of B is 100 percent and only the background image of A is visible; however, while if $\alpha = 0$ then B is totally opaque and no corresponding A pixels can be seen. It is important to note that

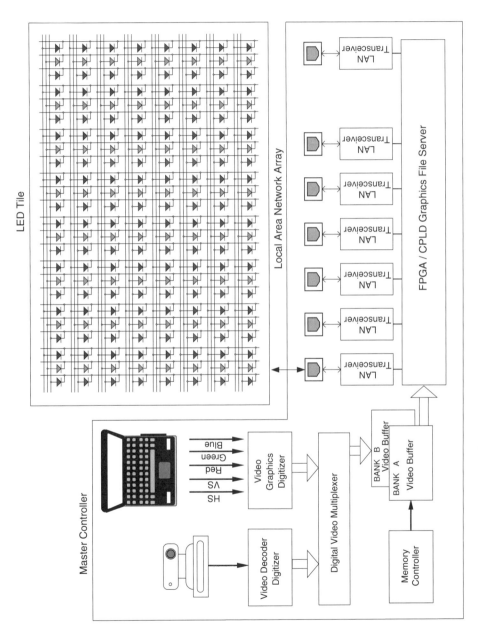

Figure 3.15 Video to LED Controller.

the above equation applies to all the data components of each pixel: YUV, YCbCr, YPbPr, or RGB depending on the color space we use.

We also must account for the zero value offsets of each color system component. So implementing a YCbCr alpha channel mixer, for example, requires calculating and rounding off the following equations:

Y out = α (YA in − 16) + (1 − α) (YB in − 16) + 16

Cb out = α (CbA in − 128) + (1 − α) (CbB in − 128) + 128

Cr out = α (CrA in − 128) + (1 − α) (CrB in − 128) + 128

Such sum-of-products calculations can be performed very effectively by specially designed *video DSPs* and *semi-custom FPGA cores*. For a compact and inexpensive solution we can also use a low-end FPGA or CPLD in conjunction with a *fast RAM look-up table* (Figure 3.16). In this arrangement the two product terms are calculated by supplying the values of α YAin, and YBin as the address inputs to a fast RAM and reading out of the RAM the value of Yout. If within the overlay area we work with YAin and YBin limited to 8 bits each, and allow for 32 different α values (5 bits), the total number of address lines sums to 21, thus requiring a 2M × 8 memory IC.

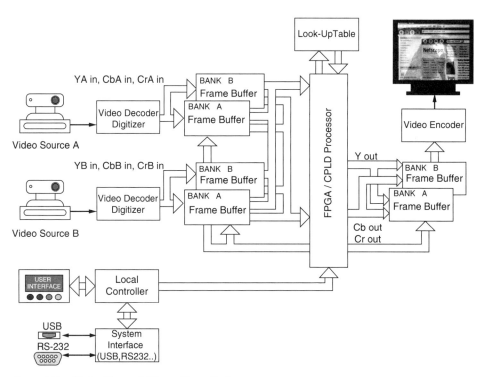

Figure 3.16   Alpha Channel Mixer–Cross Fader.

At this time, the access speed of a 2M × 8 bits asynchronous RAM is down to 15 nanoseconds (ns). It follows that within the 74 ns NTSC pixel period (13.5 MHz pixel frequency) we do have ample time to calculate Y out, Cb out, and Cr out using a single RAM IC.

3.11 Chroma Key Overlay

Among special effects, *chroma keying* is one of the most spectacular to see. We all marveled at yellow "first down" lines added to football telecasts and national flags seemingly located at the bottom of Olympic swimming pools. A more common sight (but technically no less impressive) is that of a weatherman waving in front of a computer-generated weather map.

At its essence, the process consists of the placement of a specially "keyed" area within the view of the camera (foreground), the consistent detection of this area even when partially occluded by foreground objects, and the pixel-by-pixel replacement of the foreground image within the "keyed" area with a background image (such as the weather map). The most commonly used "key" is a heavily saturated blue or green backdrop screen. Skin tones or any other color likely to be present in normal studio settings cannot be selected as a key color.

In practice, chroma keying is a rather complex process. Simple schemes—such as the raw sensing of a particular CbCr combination and switching accordingly to a foreground or background pixel—have a number of problems. Even if the foreground objects and people are carefully screened not to contain the key color, they will inevitably be illuminated by the reflected light from bright blue or green screen.

This creates various artifacts from edge halos and color contouring to background speckles showing through foreground objects. If the luminosity information of the backdrop screen is ignored, the final image will lack shadows, giving it an unreal, "computer game" look. The screen must also be absolutely perfect both in color content and in the way it is lit. A more tolerant method than simple keyed switching is depicted in Figure 3.17. To eliminate edge artifacts the foreground and the background images are processed separately. A *chroma processor* detects key areas and creates a coarse template, defining foreground and background image allocations. The template is not binary. Each of its pixels carry an alpha channel value α between 0 and 1, and the local key screen luminosity value Y.

The recording of Y and the "dimming" of the background image according to Y synthesizes the shadows that mark local lighting conditions, including those cast by foreground objects onto the backdrop screen. In keyed areas of the foreground $\alpha = 1$, in foreground object areas $\alpha = 0$, and in the edge regions it transitions gradually from 1 to 0 so it mixes the foreground and background in accordance with the alpha channel equations in section 3.10. The color components in the foreground image are reduced by values equal to the color components of the chroma key multiplied by α.

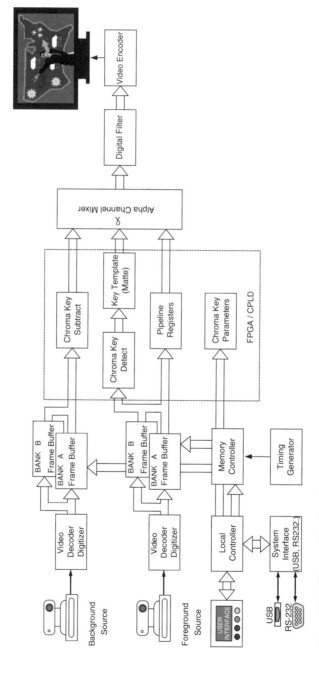

Figure 3.17 Chroma Key Overlay.

What we get is the foreground scene with chroma keyed regions replaced by black pixels, and with foreground object colors (and not Y values) compensated for backdrop color reflections and key "bleed." The final image is obtained by adding the luminosity-modified background to the chromacity-modified foreground image.

3.11.1 Simulated First Down Lines, Waving Pool Flags, and Ads

As for first down lines and flags in pools, things get more complicated—real-time intervention from a number of live operator is necessary. The stadium or the pool is first mapped from the vantage points of the cameras used by the broadcaster. Then, a *three-dimensional (3D) model* of the venue is created and stored in the computer. Using *pan-tilt-zoom* (PTZ) sensors that give the exact alignment of each camera, the 3-D computer model is refined during an extensive pre-broadcast calibration procedure. At the end of it, each set of PTZ coordinates received from a camera will prompt the computer to overlay an equivalent "virtual view" of the 3-D model onto the "live view" of the camera (Figure 3.18). A chroma key window having the shape of the first down line (or a flag or an ad), corrected for proper perspective, is inserted into the image at the location signaled by a field operator. Its relative screen position and perspective changes all the time, as the cameras follow the action.

If a player steps on the virtual first down line his image appears over the line, just as the weatherman appears to be in front of the weather map. The

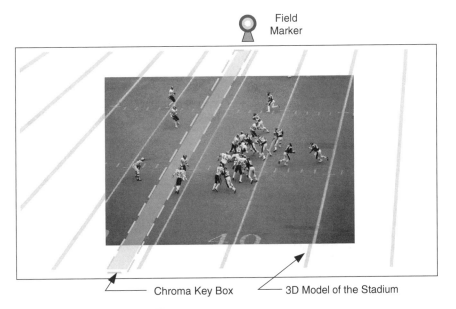

Figure 3.18 First Down Chroma Key.

color of the key depends on the application: turf green for football, blue for pools, white for ice hockey and skiing. Another operator continuously corrects the key color for unexpected interferences, such as blue swimming trunks or green football helmets.

3.12 360-Degrees Omnivision

The ability to capture a whole 360-degrees video-surround environment—instead of the customary 4:3 window—has been the holy grail for the photo and video industries since their beginnings. Although no *omnivision* product has reached the sophistication levels required by the consumer electronics market, two technologies are emerging as leading contenders. The first attains 360 degrees of coverage by synchronizing an array of cameras—eight, for example—that are placed in a circle and point outwards. For playback, an equal number of projectors with similar optical arrangements as the cameras are placed around a media room.

Each projector is located in a small window above the opposing screen or within the separation between two opposing screens. With a lot of computing power the images can be even stitched together into a continuous 360-degree image. Although not yet suitable for live applications, this technology is used in many amusement park attractions such as the Canadian pavilion at Epcot Center Orlando, Florida.

A second method is to use a *conical or parabolic mirror* and place the camera at the apex of the mirror (Figure 3.19). What the camera sees is an *image*

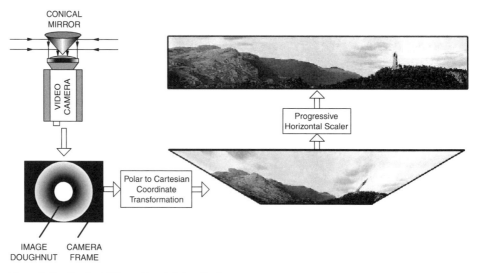

Figure 3.19   Conical Mirror Omnivision System.

doughnut at the center of the frame. To convert it into a linear image a Polar-to-Cartesian coordinate transformation is performed either by a DSP processor or a FPGA/CPLD block with a *sine/cosine look-up table* (Figure 3.20). The image obtained is still severely distorted: the bottom scan line (corresponding to the inner circumference of the doughnut) is much shorter than the top scan line (generated by the outer circumference).

For a conical mirror, the final correction consists in the *individual horizontal up-scaling* of each image line, with scaling factors progressively increasing from the top towards the bottom. In the future, with ever-increasing camera pixel densities, the algorithm will shift to progressive down-scaling of the lines, an inherently superior process. Once stored in the dual-port memory the output video is segmented into a number of subscreens, each driving a separate nonoverlapping projector image.

A low-cost implementation of this system is currently being used in surveillance applications. Instead of using a projector array it combines the various subscreens in a tiled image format (QUAD or 3×3), so it can be displayed on a single security monitor. In doing so it replaces four to nine cameras with a single camera of higher resolution and a mirror assembly. Another advantage is its highly simplified motion-detection algorithm, which searches the single doughnut image for frame-to-frame changes instead of analyzing a multitude of individual security camera outputs.

3.13 3-D Stereo Vision with Liquid Crystal Shutter Glasses

Not the exclusive domain of monster movies anymore, 3-D video has many civilian and military applications from aircraft simulators to surgeon training. The basic principles behind *stereo vision* are straightforward. In order to perceive a 3-D image each eye must look at an object from a slightly different angle. Therefore image capture must be done by two *separate and coplanar cameras* with views that converge on the object. The larger the distance between cameras the stronger the 3-D effect and the smaller the objects appear to the viewer.

The most common systems so far have been passive and consisted either of *red/green filtered movies* projected on *lambertian (white) screens* and viewed through red/green glasses, or *cross-polarized movies* projected on *silver screens* and viewed through *cross-polarized glasses*. However, systems based on newer *liquid crystal* (LC) *shutter glasses* are rapidly changing the 3D landscape (Figure 3.21). Making use of a PC monitor's ability to operate at refresh rates of 60 Hz and higher, this 3-D processor interleaves the left- and right-eye camera views at their natural 30 frames per second each.

Compatibility with PC monitors requires the videos to be deinterlaced and converted to RGB first.

The LC shutter glasses are controlled such that only one side is clear at any given time and the other side opaque. A frame synchronizer coordinates system operation by clearing the left side of the LC glasses when the left eye

66

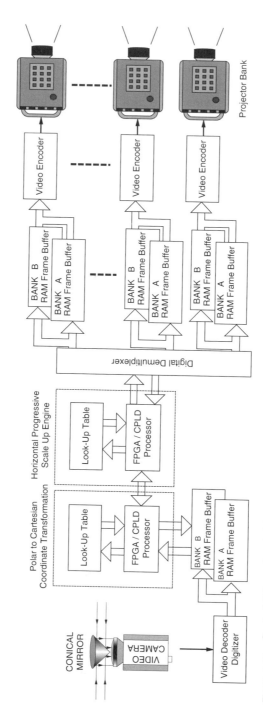

Figure 3.20 Conical Mirror Omnivision Block Diagram.

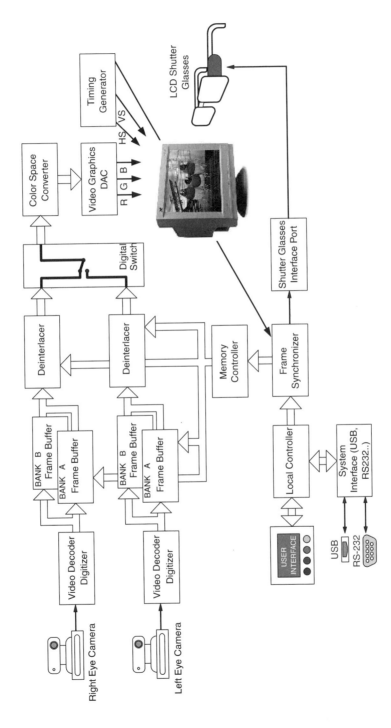

Figure 3.21 3D Stereo Video Using LCD Shutter Glasses.

67

image is displayed, then the right side of the glasses when the right eye image is shown. In theatre settings the LC glasses can be synchronized to the projectors via wireless links.

3.14 Telepresence Appliance for the Disabled

According to the United States General Accounting Office, an estimated 12.8 million Americans of all ages need assistance from others to carry out everyday activities. A highly integrated system that may reduce this dependence is shown in Figure 3.22. When used by a bedridden or wheelchair-bound person, this easy-to-use appliance facilitates interaction with appropriately equipped local and remote sites. Handicapped people can connect themselves to arrays of cameras and microphones located in museums, shops, or distant cities. They are able to interact with other disabled people or their caregivers, attend classes, or hold telecommuting jobs. If mobile, they can identify their current location and automatically find disabled-friendly paths and locations, or medical help. And of course, this system supports full Internet access.

The core capability of the system is to combine audio and video streams from a variety of sources and to provide local video and sound to selected remote destinations, all under the control of an extremely easy-to-use interface. The experience can be either passive or immersive, through a video monitor or lightweight electronic glasses (audio included). Special trackers read the position of the user's head and change the view accordingly.

However, as with any *virtual reality* application, synchronization between the head position and image must be as close to perfect as possible, otherwise "simulator sickness" (motion sickness) may result. Since current PTZ webcams have a significant accommodation delay, only direct connection camera clusters are used in this mode.

For a more realistic immersion experience, efforts are now being made to make the camera cluster output image seamless. Another important factor is compatibility with as many types of disabilities as possible. As a general rule, the needs of each disabled individual are different and so must be the services provided by the system as well as the user interface method.

3.15 Dual Energy X-Ray Security Scanner

With security concerns here to stay, *X-ray scanners* are becoming commonplace not only in airports but also in concert halls, stadiums, and even schools. One of the most capable current designs is the *dual energy scanner*.

To understand its operation we start with the *X-ray tube* (Figure 3.23). It consists of a *filament* (similar to that found inside light bulbs) and a pair of *electrodes* (cathode and anode) with a *high voltage* potential applied between them. When the filament is heated it releases electrons that are accelerated by the electric field between the electrodes; they are then projected onto the

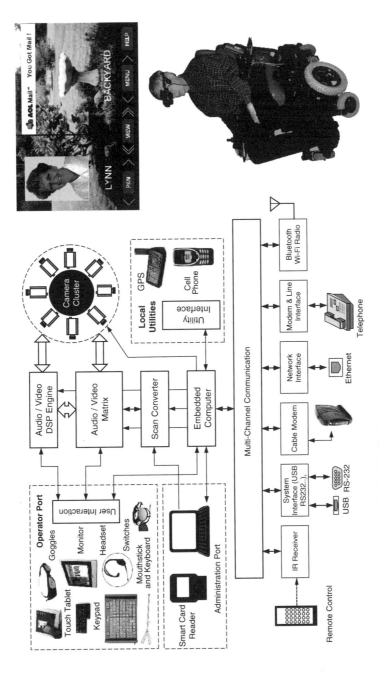

Figure 3.22 Telepresence Appliance for the Disabled.

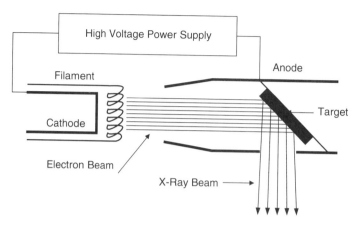

Figure 3.23   X-Ray Generator Tube.

tungsten anode. The resulting X-ray radiation carries about one percent of the energy of the incident electron beam; the rest is dissipated as heat.

X-ray detectors (Figure 3.24) use *cesium iodine scintillator* crystals to convert the *X-Ray photons* into *light photons*, which are then further converted into faint electrical currents by *low-noise photodiodes*. Detectable voltages are obtained by integrating these currents over the scan period of the *scanning analog multiplexer*.

Dual-view, dual-energy configurations use two pairs of X-ray generators positioned at 90 degrees with respect to each other (Figure 3.25). Each pair consists of two staggered generators, one *high-energy* operating at a higher voltage and producing more penetrating X-rays, and one *low-energy* producing softer X-rays.

Both beams are shaped into narrow fan patterns by collimators and projected through the target towards two opposing linear detector arrays (LDA). The fraction of the X-rays reaching the LDAs depends on the density (average atomic weight) and thickness of each of the objects placed in between. While high-energy X-rays are absorbed mostly by high-density materials like metals, soft X-rays are also absorbed by lower-density substances, such as plastic explosives, drugs, and currency (which has a higher density than general purpose paper). Less expensive scanners use a single X-ray generator pair located diagonally from the LDAs. The least expensive architecture uses a single pulsed X-rays generator and two LDAs; a low-energy X-rays LDA placed on top of a high-energy X-rays LDA, with a possible X-ray filter in between. The generator is pulsed alternately at 75 kV and 150 kV to generate the low-energy and high-energy X-rays, respectively.

The outputs of the LDAs are conditioned, digitized, and stored in the first layer of video buffers (Figure 3.26). The scanning process of the LDA combined with the movement of the conveyor belt in a direction perpendicular to the

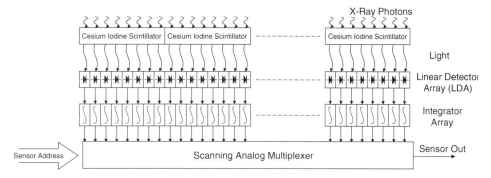

Figure 3.24 X-Ray Linear Detector Array.

LDA creates a de facto raster image of the target. However, the high-energy and low-energy X-ray images are offset by a distance equal to that which separates the generators.

A *machine vision* (MV) DSP layer compensates for this offset, enhances outlines, and detects contiguous areas of near equal density. The orthogonal *top* and *side* views are combined into 3-D objects by a *3-D synthesis engine* and then compared to a list of "threat" objects stored in a *data and image library*.

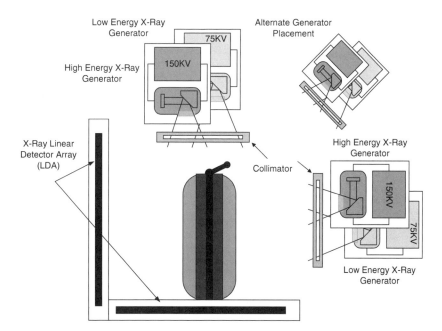

Figure 3.25 Dual Energy X-Ray Luggage Scanner.

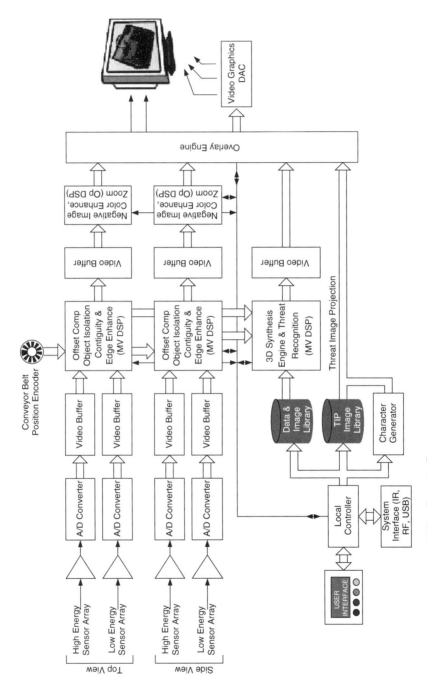

Figure 3.26 Dual-Energy X-Ray Security Scanner.

If a threat condition is detected, an alarm is triggered and the target object is closely inspected.

Even when alarms are few, a *threat image projection* (TIP) feature keeps the operators vigilant. It works by periodically injecting artificial threat objects from a TIP library into the output image. If the operator responds properly, the TIP image is removed without interrupting the conveyor flow. If it is ignored, the incident is recorded for supervisor evaluation at a later time.

A second DSP layer (Op DSP) offers a standard package of image enhancements. They include color selection and enhancements (X-rays are naturally colorless), contrast control, negative imaging, electronic zoom, and others. Top and side enhanced views, "threat" objects, TIP objects, and text are combined by an *overlay engine* and displayed on one or more high-resolution monitors.

3.16 High-Resolution/High-Definition Processors

How do our designs change as we move from traditional standard resolution to higher resolutions and HDTV formats? In most aspects the block diagrams remain the same. But the bandwidth, capacity, and required processing power of the blocks change. The increased data rates also change the interface techniques and devices used to connect different systems and assemblies together. Data compression becomes more important as it is employed to protect internal bandwidths and storage resources.

Let us take the example of a high-resolution *overlay processor* (Figure 3.27). The high-resolution inputs that need accommodation are analog component video (VGA, XGA, SXGA, UXGA...HDTV) and high-definition serial digital (HD-SDI). Back-compatibility with legacy video requires the acceptance of composite and analog RGB video as well as standard definition serial digital video (SDI).

After decoding, digitizing, or deserializing, the corresponding inputs are in parallel digital format. Some may need decompression, which is done by an MPEG-2 decoder resource.

Another processor resource is the *motion interpolator/deinterlacer/up-scaler:* It is used to convert standard resolutions to high-resolution or high-definition video. Image enhancements, video rotations, and image recognition are done by DSPs or FPGA/CPLDs with RAM support.

Graphics, animations, logos, and text are brought in by the local controller and stored in a local *image library*. The overlay engine combines the selected sources, adds graphics, and sends them to a down-scaler or directly to high-resolution output interfaces.

Since back-compatibility must also be maintained at the output, a number of different interfaces are used; *video graphics DAC* for analog PC monitors, HDTV *high-speed serializer* for HD-SDII, *Low-Voltage Differential Signaling (LVDS) driver* for flat panels, *video encoder* for composite video, and *low-speed serializer* for SDI. Most of the interfaces are rather simple and can be implemented with a minimal number of components.

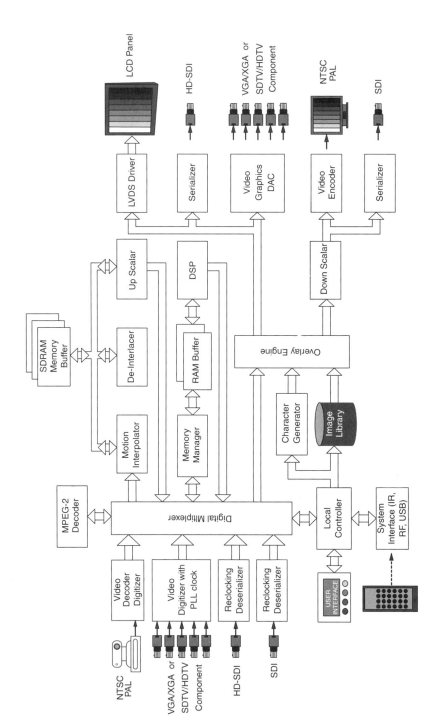

Figure 3.27 High-Resolution Processor.

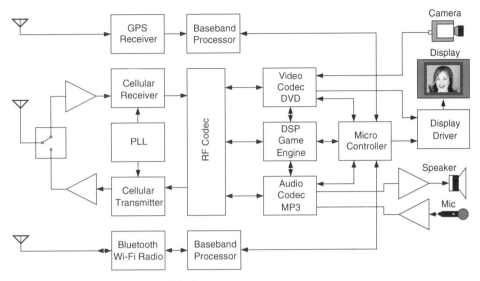

Figure 3.28   Portable Electronics Platform.

3.17 Portable Electronics

The dominant trend in consumer electronics over the past 30 years has been miniaturization and this will undoubtedly continue. Its latest directions are portable appliance convergence and wireless communications. Not only can we play videogames on our cell phones but we can also take pictures (in JPEG format) with them, make movies (MPEG-4), and find the closest Starbuck's using Global Positioning System (GPS). Soon our phones will help us check TV listings, pick up grocery lists directly from the refrigerator (Bluetooth, Wi-Fi) and probably take our pulse and blood pressure while displaying the NASDAQ ticker. A typical portable electronics platform is shown in Figure 3.28.

Note: So far the demands on portable device displays have been low cost, low power consumption, high brightness, high color saturation, and high contrast, at a relatively modest resolution and refresh rate. This will change when wearable computers and their accompanying high-resolution eyeglass displays take hold. Then many of the block diagrams covered in this chapter will need translation into the portable electronics domain. Block circuit implementations may be different, as small size and power consumption will take precedence over visual performance, at least in the short run.

4

Working with Function Blocks

4.1 Requirements Specification Document

The starting point for any development project is a product idea. The source of the idea may be a lone inventor, the research and development (R&D) department of a multinational corporation, or an unexpected offering by your competition. In all cases, the first task is the same. The idea is reduced to a product description, to a set of necessary and sufficient characteristics the product must have. This set of characteristics is embodied in a working document called the Requirements Specification Document (RSD).

Although there is no fixed set of rules to follow, an RSD may contain the following sections:

- *Introduction.* We should always start with the background for the project, the intended application, and the target markets (if any).

- *Function.* A single paragraph that describes the functions the product will perform, and the way it will meet the needs presented in the introduction.

- *Features.* In this section we describe in detail the key features of the product, starting with those that are essential to its success, followed by those that are desirable (and have only a minor impact on cost), and ending with those features that would be nice to have (at no additional cost, if possible). We should make sure we accommodate special users such as left-handed and colorblind people.

- *Product range.* What are the product variants? Are we developing high-resolution and standard-resolution models, PC-based cards and stand-alones, NTSC and PAL, portable and rack-mounted?

- *Physical characteristics.* If there are specific requirements concerning size, weight, power consumption, they must clearly stated.

- *Environmental.* Depending on where the product will be used there may be concerns regarding environmental temperature, humidity, vibration, shock, and so on. Even if the product will work in a nonchallenging setting such as a TV studio, this section still needs careful consideration.

- *Compatibility.* To what external equipment will the product be connected? What are the hardware and software compatibility issues? Do the menus need translation to foreign languages?

The RSD is not carved in stone. It will change quite a bit during the design process as what marketing wants is reconciled with what designers can do and with what suppliers can deliver. As a working example for this chapter, we use the development of a three-channel image mixer.

Introduction. In many videoconference applications, a local site (corporate headquarters, for example) needs to be connected to one or two remote locations such as a field offices or R&D facilities. The participants are seated at headquarters around a table occupying a relatively narrow horizontal strip of a standard 3:4 image. For the remote locations, experience dictates a minimum image size of one quarter screen (240 lines × 320 pixels per line). Our company needs to develop a video product that enables all participants in a three-way videoconference to see each other, all the time, during the course of a session (Figure 4.1.) This product is exclusively for internal use and the development effort needs to be completed within four to six months.

Function. We want to develop a product that combines a local, horizontal half-screen image with two remote images a quarter-screen size each. The local image is the center strip of the standard 3:4 video screen while the remote images are ¼ scaled versions of their corresponding video inputs. The product will be referred to as the *Image Stacker*.

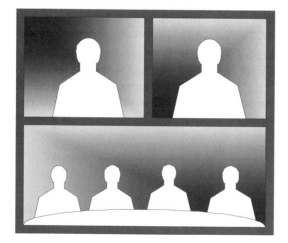

Figure 4.1   Image Stacker Screen.

Features. Essential to the Stacker is the basic image tiling function described in the previous section. Also essential is the ability to separately control the brightness, contrast, saturation, and hue for each input. All controls must be available through an RS-232 link to a host computer. Host computer software development is not part of this project. The manufacturing cost of the product, using American contractors, shall not exceed $300.00 per unit, in 100 unit lots.

Although the device will process analog composite inputs and outputs, it is desirable but not essential to also handle s-video input/output (I/O). Also desirable is the ability to bring any of the images to full screen format, and to freeze selected screens for download to the host computer. If there is almost no cost impact (less than one percent or $3.00), a minimal front panel interface should also be developed. This interface will allow us to cycle through the following screen sequence: tiled, full screen image 1, full screen image 2, full screen image 3. A separate button will "freeze-frame" the screen and initiate download.

Product range. Available in NTSC and as many PAL/SECAM variants as possible.

Physical characteristics. The final product will not be larger than half-rack size (1.75″ H × 8.75″ W × 9.5″ D). All input and output connectors must be BNC or miniDIN (for s-video) type.

Environmental. Temperature: 0–50°C; Humidity: 0–95 percent.

Compatibility. Interconnection with other NTSC/PAL video and control devices needs to be as simple and inexpensive as possible.

Although admittedly rudimentary, this RSD gives us a reasonable idea of what our finished product will look like. Everything included in the document must be addressed in the design. Things not included remain at the discretion of the designer, but consultation on such issues with the authors of the RSD (if not the designer) is essential.

4.2 Functional Block Diagram

The next step is the development of the system *functional block diagram*. Its purpose is to detail all the major functional blocks in the design (decoders, encoders, memories, and so on) as well as to highlight the requirements each block has in terms of block adaptors (to service unmatched connections to neighboring blocks) and system infrastructure (voltages, clocks, control lines, and so on).

From the RSD we know that the Stacker will have three inputs, two of which will be scaled to ¼ of screen size, and one which will be cropped by 50 percent at the top and bottom. While cropping can be done in the analog domain, scaling is always digital, so we immediately recognize the need for at least two video decoder-digitizers and image scaler blocks.

The function of a video decoder-digitizer is to first extract the analog luma and chroma information from an NTSC/PAL composite input (decoder) and then convert them separately into a parallel digital format (digitizer). To properly time its operation, the digitizing decoder also extracts the sync information embedded in the incoming video, recovering signals like VSYNC, HSYNC, and PXLCLK (pixel clock).

The image scaler receives the video data in digital format and, for a downscale function, averages two-dimensional (2-D) arrays of incoming pixels in order to create an accurate smaller image representation of the original image. To perform 2-D averaging it must buffer a number of input video lines, which requires either external or internal memory resources. For the time being we leave the third input alone and assume we can just pass through its "cropped" section at the right time and in the right screen position.

The output section will need an *encoder* block to convert the output luma and chroma information from the scalers back into NTSC/PAL format. Since we may keep the third channel in analog form we may need a fast analog SPDT switch to alternate between the analog output of the encoder and the third channel composite video.

This brings forth a possible synchronization issue. The timing of the three input signals is totally random. Each is driven by different cameras or CODECs, each with a different clock, vertical sync, and horizontal sync. Therefore at any given time the three input signals will reflect three different pixel ranks. If we give the output the same timing as the third input, this still will leave it out of sync with respect to the first two input channels. To solve the problem, we place a dual-ported *memory buffer* block after each scaler. Data is then written into the buffer in accordance to the timing details of the video input and the processing requirements of the scaler. Data readout is synchronous to the timing supplied by the third input. The extraction of third input timing information is done by a *sync separator* block.

The routing of the pixels from the two memory buffers to the encoder and the control of the *analog switch* is done by the *overlay processor* block. The precise time when each pixel is read, measured with respect to the HSYNC and VSYNC of the output, determines where each image window is opened on the screen. Other overlay processor functions are to implement the full screen and screen freeze commands, and also to generate all mandated image borders. Finally, since we are required to have an RS-232 control link, we need a local microprocessor controller with appropriate RS-232 drivers. A complete block diagram for the stacker is shown in Figure 4.2.

RSD Change. To keep the costs low we will not digitize and store the third input of the Stacker. This will save our using a decoder-digitizer, a scaler, and a memory buffer, but will not allow us to freeze frame a complete screen. We still can freeze the first two inputs.

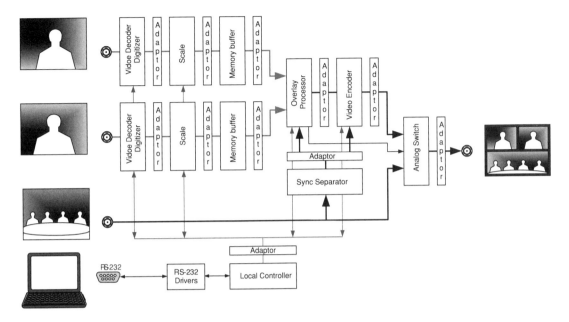

Figure 4.2   Stacker Functional Block Diagram.

4.3 Selecting the Core Integrated Circuits

Armed with some solid information on what each block has to do we can proceed to our component selection stage. Although there may be over 100 manufacturers of video ICs worldwide, most of them have a strong marketing presence in Asia, with American distribution either inexistent or minimal. This narrows considerably our choices to maybe a handful of possible ICs for each function block.

Looking at video decoder-digitizer chips we find that a few of them have a reasonable quality scaler built in. They also have composite and s-video inputs, and allow for the adjustment of brightness, contrast, saturation, and hue. These features accommodate a few of the desirable options listed in the RSD (s-video operation and control of channel video attributes), at least on the input side. We select the Conexant's Bt835 for our decoder-digitizer and scaler blocks.

Since manufacturers are "chip set" oriented we should also look at Conexant's video encoders. If any are suitable for our application they will tend to share the same design "philosophy" as the decoder, shortening the associated learning curve. Sure enough, the Conexant's Bt864 seems to fit all the requirements. It even has an s-video output. As it turns out, the Bt835 and Bt 864 are widely used in video security QUADs, which also makes them relatively inexpensive.

As for the memory buffers, we have to make an important decision. The required buffer size per image is:

$$240 \text{ lines} \times 320 \text{ pixels/line} \times 2 \text{ bytes/pixel}$$
$$(\text{luma, chroma in 4:2:2}) = 153.6 \text{ Kbytes}$$

To avoid image tearing we will need two banks of 153.6 Kbytes per channel. We could also do this using one bank but the algorithm becomes significantly more complex (see Chapter 10). Since, according to specifications, the data rate expected at the input of the encoder is 27 Mbytes/, the access time for the memory has to be better than 37 nanoseconds (nsec).

The decoders will output bursts of two pulses (Y and C in 8-bit mode) about 37 nsec apart, followed by inactive intervals proportional in length to the horizontal scaling factor. For a quarter screen output the data rate within a line will average to 13.5 Mbytes/sec, while on a per frame basis the average will drop to about 6.75 Mbytes/second. In conclusion, we will need four 256K $\times$ 8 memory buffers with a maximum access time of 37 nsec.

The best choice from the control point of view is the dual-ported first-in first-out (FIFO) memory buffer. Their connection to the decoders and the encoder would be almost glueless (logic "glue," of course!). But if we use an SDRAM type device a single IC could accommodate all the memory requirements of the Stacker, for a price less than half that of a single frame buffer FIFO. The 512 K $\times$ 16 bits $\times$ 2 banks KM416S1120D from Samsung, for example, has an access time under 10 nsec; this clearly allows us to service the decoders and the encoder, even when they both run at rates of 27 Mbytes/sec. The disadvantage is that the control of the SDRAM, when accessed in real time by three different clients, can be quite complicated. For this application we do not have a choice, however. To meet the cost requirements of the RSD we have to go with the SDRAM.

For the third channel *sync separator* we choose the EL4501 frontend IC from Elantec (now part of Intersil). Besides a sync separator function with an integral *data slicer* (see Chapter 8) it also offers *DC restoration*. For the first two channels the sync separation and DC restore functions are performed internally by the decoder digitizers.

There are very few options for the overlay processor. Aside from a couple of specialized ICs (like Averlogic's AL70X QUAD processor controller—www.averlogic.com), our only choice in implementing the overlay function is to use a *field programmable gate array* (FPGA) or a *complex programmable logic device* (CPLD). With a mandated four to six months in the development cycle, we have to select a device with which the designer has experience, or the one that is the easiest to learn. We select the Altera MAX3000A CPLD series. Its Max+plus II–Quartus II development software is easy to use and its entry level version, the Baseline, is free.

Within the MAX3000A family we start with the largest IC, the EPM3512A. If during development we find that we need fewer resources than anticipated, we can switch to the (almost) pin compatible and less expensive EPM3256A.

The only requirement for the analog switch is to be fast, with turn-on and turn-off times much smaller than the pixel period (about 70 nsec). Although we could use slower switches, the image quality will suffer somewhat if we allow the switchover to take multiple pixel periods. Besides, both Gennum and Maxim offer subpixel switching devices for reasonable prices. We select Gennum's GY4102A for its less noisy make-before-break architecture (no-load impedance discontuinty during switch-over).

Almost any microprocessor with a built-in Universal Asynchronous Receiver Transmitter (UART) seems to qualify for the *local controller* block. And we should keep in mind that with an internal timer even the UART can be easily bit-banged. Based on price performance alone, Atmel's AVRs and Microchip's PICs are the leading contenders. Their leftover I/O pins can be used to drive the optional front panel switches and LEDs. However, we do not know at this point what demands the other blocks will make on the microcontroller. On a preliminary basis we select the AVRTiny2313 with built-in UART, 2K FLASH, 128 Bytes EEPROM, 128 Bytes SRAM, 2 Timers, and built-in Watchdog.

Taking a look back at the RSD we conclude that we can probably achieve most of its goals, except for the freeze frame function of the third input channel. However, we cannot be certain until the design is completed. Selection of the core ICs is not an exact science—it is more of a navigation process. The designer has to find a set of parts that meet the RSD's needs, can be provided by suppliers, and match his or her level of experience.

4.4 Matching the Blocks

Each block should have a single major function IC at its core (with the notable exception of memory blocks that may contain multiple identical chips). If a block contains two or more major ICs, the block diagram may lack sufficient detail, and further decomposition into more sub-blocks may be needed. Once an IC is selected it will need a local infrastructure to make it work (clocks, decouplers, and so on).

Along its processing path each IC may receive input data, process it, and output processed or internally generated data of its own. The inputs and outputs of each block have to be matched to the I/O lines of the blocks feeding into it or being fed by it. In order to gain a better understanding of the I/Os and the infrastructure requirements of the ICs, we have to study their specification sheets.

A summary of our findings can be found in the redrawn block diagram of the Stacker (Figure 4.3). Some mismatches are immediately apparent. If we list the operating voltages for the core ICs we find 5 V, 3.3 V, and −5 V; some digital I/Os are TTL while others are LVTTL; the SDRAM has a single 16-bit bus input but it must store two asynchronous data streams, either 8- or 16-bit wide. We also have a total of four different clocks driving different digital circuits. Our next step is to eliminate mismatches and to provide a unified design infrastructure for the design.

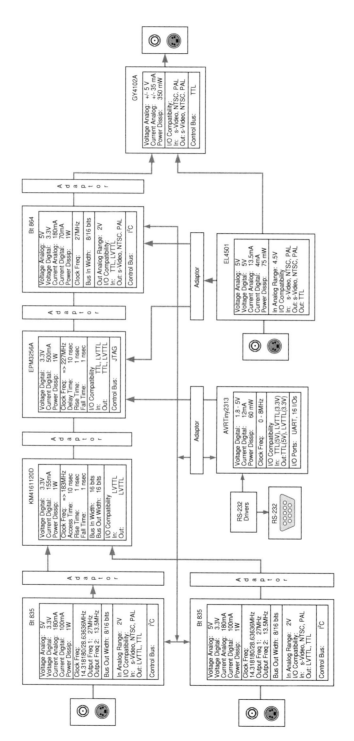

Figure 4.3 Stacker Interface and Resource Requirements.

4.4.1 Mixed Voltage Levels

One of the problems with which video designers must contend when dealing with current ICs is mixed voltage design. Most manufacturers are now producing 3.3 V ICs, with future products likely to be operating at levels of 2.7 V, 2.5 V, 1.8 V, or 1.5 V. However, some of the more common parts and processors continue to be available only in 5 V. There are significant advantages associated with the lowering of the supply voltage. For equivalent complexities, the lower-voltage ICs dissipate significantly less power. The reverse is also true. For the same power dissipation budget you can design faster and more complex ICs. At least for the foreseeable future though, our designs will have to accommodate ICs operating at a variety of voltage levels.

On a practical level, the difficulty of interfacing 3.3 V logic to 5 V logic depends on the direction of the I/O. To drive a 5 V input with a 3.3 V output we can use ACT/HCT logic. These IC families have TTL-compatible inputs (V_{IL} = 0.8 V, V_{IH} = 2.0 V) and CMOS outputs so no additional translation is necessary.

Connecting 5 V outputs to 3.3 V inputs is more difficult since standard LVTTL specifications require the voltage on any pin not to exceed Vcc + 0.5 V. Above this level the electrostatic discharge (ESD) protection diode built into standard 3.3 V logic (Figure 4.4) becomes forward biased and the 5 V output is shorted to the 3.3 V power supply. The answer is to use special 3.3 V logic with 5 V tolerant inputs such as the LVC, AHC, and LVT families from Texas Instruments Corporation (TI).

Even more complicated is the bidirectional translation of signals between 3.3 V and 5 V systems. One solution, effective in lower-frequency applications, is to use open-collector or open-drain logic. Then the driver can operate at its native voltage while its output is pulled up to the level of the driven logic. The disadvantage is that the use of pull-up resistors will introduce delays that may not be acceptable in high-speed applications.

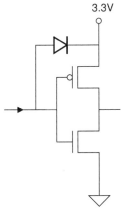

Figure 4.4 LVTTL ESD Diode Protection.

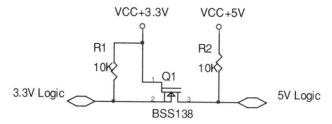

Figure 4.5 Bidirectional Translation 3.3 V↔5 V.

One example of such a circuit that translates the I²C bus between 3.3 V and 5 V levels is shown in Figure 4.5.

To bidirectionally interface more than one line we can turn to transceiver devices (74LV**245) in the LVT, LVC (TI) and LVX (Fairchild) families. Interfacing to lower voltages (2.5 V, 1.8 V, 1.5 V) poses similar kinds of problems: input and output voltage matching, ESD diode forward bias, and so on. However, because back-compatibility to 5 V and 3.3 V systems is a necessity, we can expect to see a wider variety of devices making their way to the market in the near future.

Figure 4.6 summarizes the voltage translation product offerings from TI and their range of applications. Almost all major suppliers of logic ICs provide some

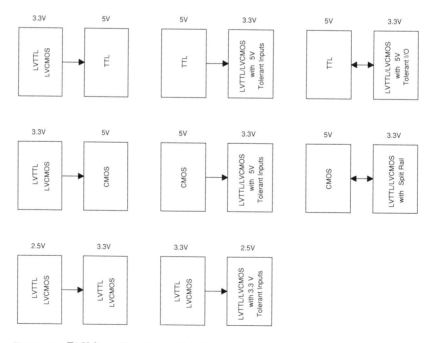

Figure 4.6 T1 Voltage Translation Solutions.

voltage translation solutions. To enhance their competitiveness some manufacturers of complex ICs include mixed-voltage compatibility in their products. The MAX7000A family from Altera Corporation, for example, will accept 2.5 V, 3.3 V, and 5 V inputs and will supply 3.3 V and 5 V compatible outputs.

4.4.2 Matching Bus Widths

One of the simpler examples of a bus width mismatch is that of a word size (16 bits) block output being stored in a byte-wide memory block. Aside from the obvious solution of using two-byte-wide memory blocks, we can also accommodate the mismatch by breaking down the words into *most significant bytes (MSB)* and *least significant bytes (LSB)* and by storing them separately in memory (Figure 4.7).

To extract a word we read the two bytes from memory and into two separate byte-wide registers. We then simultaneously enable the outputs of the registers, thus reconstructing the stored word.

For this process to work we need a memory device at least twice as fast as the speed of the buses it services. In other words, the worst case read/write cycle times have to be at most half the inverse of the bus data rate.

The memory controller must interpret the read and write commands from the local controller. When a write command is received, the memory controller generates the appropriate address where the incoming word is to be stored

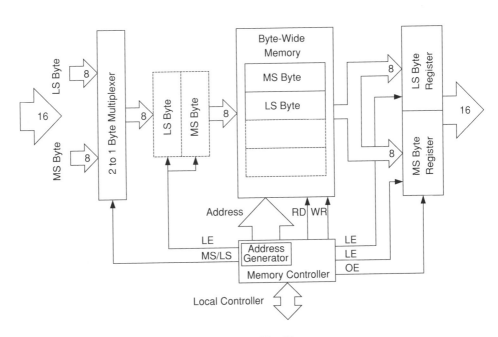

Figure 4.7   Wide Bus Narrow Memory Adaptor—Mux/Demux.

and steers the *byte multiplexer* to the MSB position (if not already there as idle state). It then pulses the write line (WR), increments the address, switches the byte multiplexer to LSB, and finally pulses the WR again.

For a read command, it generates (or forwards from the local controller) the MSB address of the word to be retrieved and strobes the memory read (RD) and *MSB register* latch enable (LE) lines; then increments the address and pulses the RD and LE of the *LSB register*. The speed of the memory controller has to be high enough to perform a complete read or write operation within the timing requirements of the input and output buses.

Going from a narrower to a wider bus is rarely a problem. The data from the narrow bus will just be placed on the wider bus either in the most or least significant positions, with the balance of the wide bus lines being given a default value.

Our Stacker has a different kind of data width problem. The SDRAM has a 16-bit data bus and both decoders can be programmed to operate either in 8-bit or 16-bit modes. But the data from the decoders comes in asynchronously with their own timing. So we cannot directly combine their outputs into 16 or 32 words.

What we have is two distinct devices that share the same memory but that are otherwise independent of each other. An ideal solution is one that allows each decoder to use the memory without any interference from the other decoder.

Such a solution is depicted in Figure 4.8. We start by assigning each device a different address block within the SDRAM memory: *block A* for bus A and *block B* for bus B. Since we do not know when data arrives on the two buses, we have to account for the possibility of data arriving simultaneously on both. We do so by storing the incoming data in FIFOs a few bytes deep. When a byte arrives on bus A, for example, it is stored in FIFO A and the memory controller is alerted via the WR A line (write request on Bus A).

The controller updates the *address generator* A and then proceeds to retrieve the data from FIFO A and store it in block A. A slightly different process takes place for a read request. It starts with a read request signal directed at the memory controller. Then, after the appropriate address is applied to the memory IC, a write strobe transfers the data from the memory to the corresponding output register and also informs the output bus that data is now available (RDY).

Address generators A and B are the address pointers for the A and B blocks. If bus A or the local controller requests access to block A, the output of the address generator A is placed on the memory address bus. Then the contents of address generator A are updated to the next likely address. Address generator B operates in a similar manner.

If two or more requests arrive simultaneously, the memory controller must queue the requests and service them sequentially within one external bus time cycle. It should be noted that if the memory controller is sufficiently fast the FIFOs can be replaced with registers.

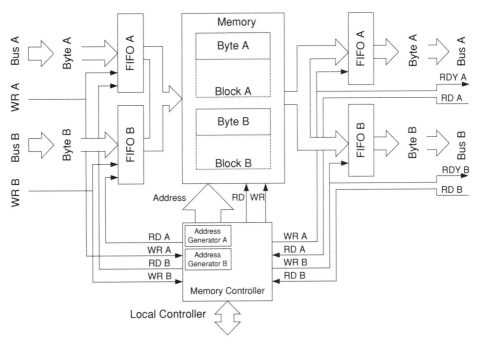

Figure 4.8   Wide Bus Narrow Memory Adaptor—Pixel Server.

4.4.3 Matching Data Speeds

Another condition common in video design occurs when a data stream is so fast that no single memory device can store it in real time. In the case of UXGA, for instance, the pixel rate is over 200 Mp/sec. The total time available to read a pixel from memory is then under 5 nsec, clearly a tall order at least for any reasonably priced commercial application.

A solution is to split the fast bus into a number of slower buses connected in parallel, and then sequentially distribute the fast data onto the slower buses. A memory buffer architecture that uses this technique is shown in Figure 4.9.

At the beginning of an input stream an initial address is placed on the common input address bus of the A, B, and C memory ICs. The front 1-to-3 demultiplexer then divides the incoming traffic three ways and directs it towards three input registers; these hold the data long enough for the memory controller to write it in the corresponding memories. In this way each memory chip stores every third input data unit (byte, word...). The address remains constant for each data triplet, and the individual write time is increased threefold.

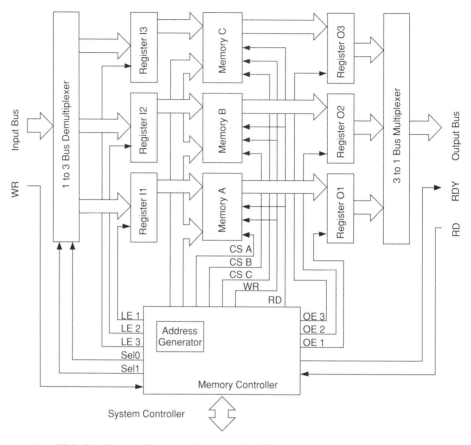

Figure 4.9   Wide Bus Narrow Memory Adaptor—Pixel Server.

Reading of the data is done by first setting up a common read address for all three memories, transferring a data triplet from the three memories to their respective output registers, and by sequentially strobing the output enable lines of each register.

Because of speed requirements, buffering live UXGA resolution signals uses at least twice as much memory as the UXGA pixel count would indicate. The memory buffer will also have to accommodate simultaneous read and write cycles as the video data stream to the output DACs (and therefore the computer screen) can never be interrupted.

Figure 4.10 shows a ping-pong memory structure that accomplishes simultaneous read and write operations by continuously swapping the input and output memory banks. Its operation is simple. Input data is written in bank 1 and read from bank 2 until a full frame is read on the output bus. At that time the two banks are swapped and data is written in bank 2 and read from

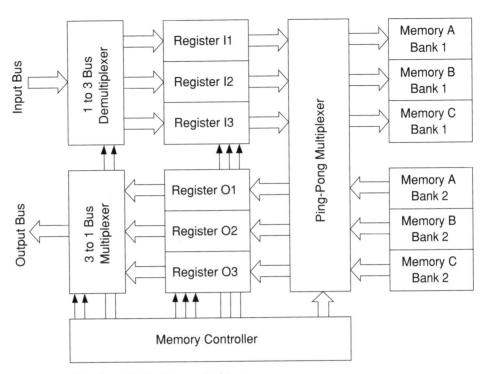

Figure 4.10 Ping-Pong Buffer Memory Architecture.

bank 1. The swapping is done by the ping-pong multiplexer which switches the address, data, and control lines of the two banks. The ping-pong memory architecture works well when the refresh rate is high but the rate of information upgrade is low. This is the case with computer graphics applications where the screen data rate is high but the image content changes little from frame to frame. Fast moving images will require a different approach (see Chapter 10).

4.5 Clock Generation

Ideally, each video system should have a single clock. However, since each manufacturer has a different set of design priorities, their integrated circuits will tend to use different types of support services and different clock frequencies. If we design a general purpose microcontroller with a built-in UART we want a clock frequency that can be easily divided down to generate RS-232 baud rates. 11.0592 MHz sounds like a good number in this case. Just divide it by $2^7 \times 3^2$ and you get exactly 9600 baud.

If our instruction cycle is 12 clocks long and we want our product to be designer friendly, we choose 12 MHz. This way each instruction cycle takes

1 microsecond, which makes software timing considerably easier. If we work with NTSC video the 3.579545 MHz color subcarrier frequency (or a multiple of it) is likely to be someplace in our design already. Of course, the NTSC pixel rate of 13.5 Mp/sec is not a multiple of any of the above. In the high-definition realm, frequency variations proliferate even more as we start from 74.25 MHz, 148.5 MHz, and 270 MHz for standard definition and parallel digital video, and go as high as 1.485 GHz for serial high-definition formats.

The main reason for consolidating our clocks is to minimize the number of possible noise sources. For example, even as most decoders have built-in oscillators we should always use the output from one to drive the others. This reduces color interference between decoders and simplifies the PCB layout.

4.5.1 PLL with Integer Frequency Multiplier

The key circuit for generating and controlling clock frequencies is the phase locked loop or PLL. As Figure 4.11 shows the PLL consists of a *phase detector*, a *loop filter*, a *voltage controlled oscillator* (VCO), and a *divide by M* circuit connected as a feedback element between the VCO output and the phase detector.

The simplest form of a phase detector is an XOR gate. Its output is of a train of pulses, each pulse located exactly where the two input signals differ. If we reference the output of the XOR gate to the midpoint of its swing (between 0 and 1), we see that the only time we obtain a zero average (50 percent duty cycle) is when the two inputs are precisely 90 degrees out of phase (Figure 4.12).

If the phase difference between the input and feedback signals increases, the average becomes positive; if it decreases it becomes negative. The purpose of the loop filter is to integrate (average) the output of the phase detector and apply the resulting value to the VCO. The time it takes the loop filter to average the pulses from the phase detector has to be much smaller than the "beat" period between the input and feedback signals. If it is not, its output will just oscillate and so will the output frequency of the VCO. The PLL will not "lock

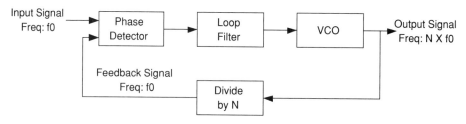

Figure 4.11   Phase Locked Loop.

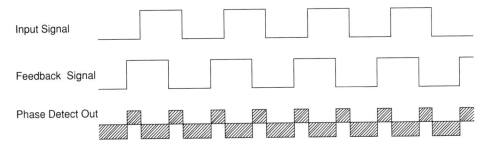

Figure 4.12   PLL Phase Detector Signals.

on" the input signal. When the PLL is locked on the input signal its output has a frequency exactly M times that of its input, where M is the division factor in the feedback loop. In essence, the PLL is a frequency multiplier.

4.5.2 PLL with Rational Frequency Multiplier

With a slight modification, PLLs can produce outputs with frequencies that are rational multiples of the input frequency, that is, $(N/M)f_0$. To generate them we must first divide the input frequency by the desired denominator factor M (Figure 4.13).

4.5.3 Clock Tree

Returning to the Stacker requirements, our clock system has to provide the following frequencies: 14.318180 MHz (2 decoder-digitizers), 27 MHz (encoder), about 8 MHz (local controller), and a multiple of the 13.5 MHz (overlay processor). Adding to our complications is the fact that the HSYNC,

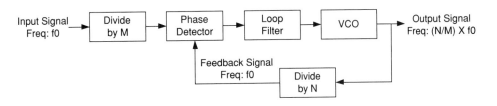

Figure 4.13   Rational Multiplier Frequency Synthesis.

VSYNC, and PXLCLK signals applied to the encoder have to be the same as the ones embedded in the third video input. If they are not, the color burst generated by the encoder will be different than the one of the third video input. Then the two inputs to the analog switch will not have the same color reference and pixel-by-pixel overlay will no longer be possible. The solution rests with the use of multiple PLLs (Figure 4.14).

To the first PLL we apply the 15.625 KHz HSYNC signal extracted by the sync separator from the third video input. If we set the multiplication factor to 3,456 the output of the PLL will be 54 MHz, ideal for the counters and state machines of the overlay processor (4 times the pixel rate). If we divide the output of the first PLL by 2 we obtain 27 MHz, the frequency required for the operation of the encoder. The 27 MHz is then fed to a second PLL with a 35/66 rational multiplication factor in order to generate the 14.31818 MHz clocks for the video decoder-digitizers.

A reasonable local controller clock can be obtained if we divide the 27 MHz by 4. This produces a frequency of 6.75 MHz. If we program the controller baud rate generator to further divide the clock by 704 we come within 0.12 percent of 9600 baud, well within the tolerance margin of the RS-232 standard.

The *flip-flops* (FF) are used to register the clocks with the highest frequency available in the system, a measure that eliminates unwanted glitches and minimizes phase jitter. Examples of PLL ICs well suited for this design are the MK2712 and MK1575-01, both from Integrated Circuits Systems, Inc.

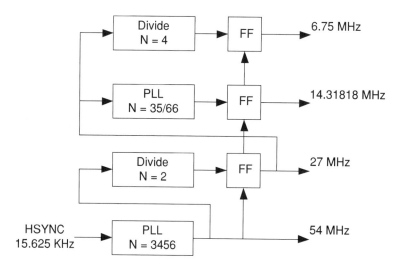

Figure 4.14   Stacker PLL Clock Generation Tree.

4.6 Clock Distribution

If a system operates at very high frequencies, as is the case with serial HDTV, each branch of the clock trace on the printed circuit board will experience a different delay. Looping the trace would make the delays cumulative, while wiring the clock in a star configuration makes termination of the clock line impossible (Chapter 14).

4.6.1 Zero-Delay Clock Buffer

An effective solution to clock distribution problems is to have a separate clock line for each IC that uses the clock. We can achieve this with zero-delay clock buffers (Figure 4.15). The operation of the buffer is based on using a PLL circuit to close the feedback loop between the output and the phase detector of the PLL. Then the output clock will be in phase with the input (zero delay). Since such buffers generally have multiple outputs we can use them to drive different ICs each with its own line terminator.

If the impedance of the clock traces to different ICs is different, then the clock will be delayed for different amounts of time along each line, even if we use a zero-delay buffer.

4.6.2 Clock Delay Equalizer Buffer

What we need is the ability to individually equalize the lines by introducing a different user programmable delay on each. This is done by adding delay lines to the individual outputs of the zero-delay buffer (Figure 4.16).

Clock distribution ICs are produced by a number of manufacturers including Cypress Semiconductor Corporation and Integrated Circuit Systems, Inc.

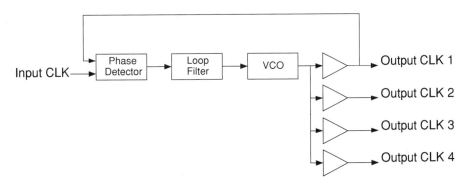

Figure 4.15   Zero-Delay Clock Buffer.

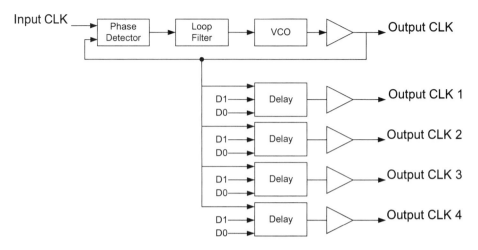

Figure 4.16   Clock Buffer with Programmable Delays.

4.7 Multiple Voltage Power Supplies

4.7.1 Multiple Voltage Ladders

The power supply structure for a mixed voltage circuit has to take into account the heat dissipation of the on-board regulators and the separation of the analog and digital power rails. In a linear design we can minimize heat dissipation by using the lowest possible input voltage in conjunction with linear *low drop-off* (LDO) regulators. If we use switching power supplies we must also use heavy filtering or employ additional linear regulators on the output supply lines.

Multiple on-board regulators can be connected as one of the ladders shown in Figure 4.17 or as a combination thereof. The first ladder concentrates power dissipation on the first regulator column, since the +/–5 V regulators carry the total load currents for the board. This is convenient when the enclosure or the board allows us to heat-sink only one or two of the regulators. If the power dissipation is more evenly distributed, we can use the second ladder where all regulators are connected directly to the input power line although, in this case, they may all need heat-sinking.

4.7.2 Mixed Voltage Supplies

The best way of protecting sensitive analog sections from digital supply noise is to power them with separate analog supplies and grounds (Figure 4.18). The

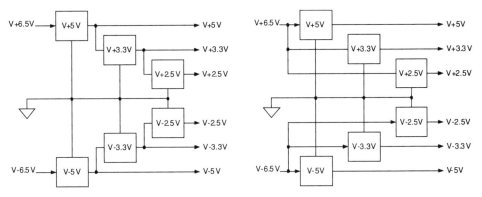

Figure 4.17 Multiple Voltage Power Ladder.

digital and analog grounds should meet at the input power connector and the cables bringing the DC power to the board should have the smallest possible impedance (Chapter 14). Otherwise any voltage drop on the ground line due to digital power noise will also be seen by the analog ground regardless of the on-board precautions we may take.

Another fact we must take into consideration is that in mixed voltage design on power-up, different voltages may have to come on at different times.

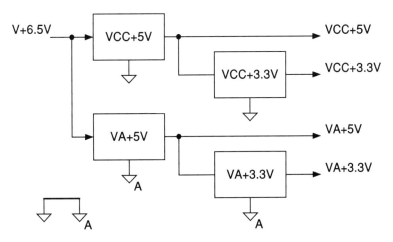

Figure 4.18 Alternate Multiple Voltage Arrangement.

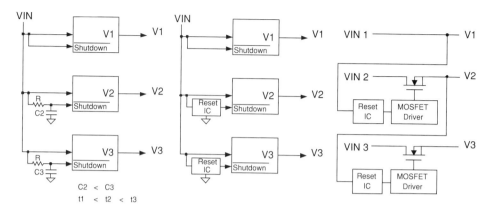

Figure 4.19   Power Supply Sequencing Methods.

4.7.3 Power Sequencing

Our voltage sources may have to be sequenced (Figure 4.19). Many multi-voltage ICs such as FPGA/CPLD chips with different core and I/O supplies have specific requirements and these requirements must be satisfied for proper operation. Not sequencing may also create latch-up conditions in older chips.

The sequencing method for regulators with integral shutdown is to use a power-up delay circuit to retard the shutdown input. If the input provides a hysterisis function a simple RC circuit is sufficient; if it does not, the same delay function can be performed by *reset ICs*. If the regulators have no shutdown inputs we need to use pass MOSFETs, which in turn are controlled by reset ICs. However we should keep in mind that MOSFETs have a non-zero channel resistance; this may become a factor at high currents.

Video Input and Output Circuits

5.1 Video Transmission Lines

If we place a number of positive and negative electrical charges within a given workspace and then proceed to move them around, we generate an electromagnetic (EM) field which propagates outward with a speed determined by the properties of the medium. In free space this velocity is given by

$$c = \frac{1}{\sqrt{\varepsilon_0 \mu_0}}$$

where ε_0 and μ_0 are the electric permittivity and magnetic permeability of vacuum.

Since light consists of electromagnetic waves within a certain wavelength bracket (380 nanometers to 780 nanometers, to be precise), c is nothing else but the speed of light in vacuum, approximately 300,000,000 m/sec or 186,000 miles/sec.

If we look at EM propagation from a slightly different perspective, it takes 84.7 picoseconds (psec) for the waves to travel through one inch of free space. In a material with permittivity $\varepsilon_r \varepsilon_0$ and permeability $\mu_r \mu_0$, the propagation time increases to:

$$\frac{\delta t}{\delta l} = 84.7 \sqrt{\varepsilon_r \mu_r} \ \text{psec/inch}$$

For a PCB trace this delay is about 180 psec/inch, while for a standard coaxial (coax) cable it measures roughly 128 psec/inch.

The same process takes place when we apply a potential difference between two points of a circuit. The propagation velocity in this case depends on the detailed characteristics of the materials we use and the values of any lumped or distributed impedances the EM wave finds in its path.

When the signals we use are of relatively low frequency we can assume the propagation from the input of the circuit to the output to be practically instantaneous. A composite video signal, for example, has a bandwidth of 4.2 MHz. When traveling through a Belden #8221 coax wire with a nominal velocity of propagation 66 percent of c (EM propagation velocity in vacuum), the NTSC signal is delayed 128.33 picoseconds per inch of cable.

Since the shortest signal period in the NTSC spectrum (the inverse of maximum frequency) is about 238,000 picoseconds, it is clear that the phase difference between the input and output signals is practically imperceptible until the cable length approaches 25 feet. We conclude that "EM wave" or "transmission line" effects may be important when wiring composite video from room to room or house to house; they can be safely ignored for short haul routes such as equipment rack connections or PCB traces.

Things change when we work with high-frequency systems such as a 71.4 MHz UXGA scaler, or a 1.485 Gbps HD serial interface. In the first case the wavelength is down in the 100-inch range, and any cable or PCB trace longer than seven inches (1/16 of wavelength) should be given special attention. For the second example, the critical length is down to 0.5 inches.

What is the nature of these "electromagnetic wave" effects? Why are they so detrimental to system performance? The most important effect is the partial reflection of the incident signal both on the load and at the source side.

Let us model our video transmitter as a current source I in parallel with a source impedance Z_s, and deliver the signal to an arbitrary load impedance Z_L by means of a transmission line with a characteristic impedance Z_0 (Figure 5.1).

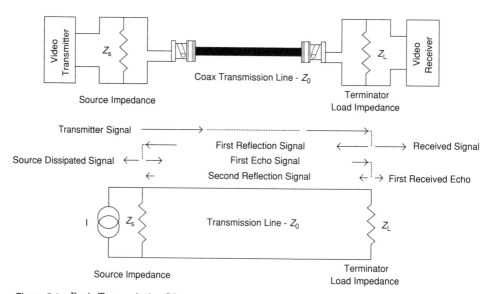

Figure 5.1   Basic Transmission Line.

The transmitter signal upon arriving at the load side is partially absorbed by the load and partially reflected back, just as light is partially absorbed and reflected when incident on an arbitrary surface.

The reflected signal will split again when reaching the source impedance Z_s where it gives rise to an echo component that travels once more towards the load. The net signal across the load is a fraction of the current incident video signal plus a number of time delayed echoes.

Fortunately, there is a transmission line configuration that is nonreflecting to EM waves. To achieve it we must match the signal source impedance to the transmission line characteristic impedance and the load impedance. In short, we eliminate unwanted EM reflections if $Z_s = Z_0 = Z_L$.

The transmission line most commonly used in video is the coax cable. Its high-frequency characteristic impedance is given by:

$$Z_0 = \frac{1}{2\pi} \sqrt{\frac{\mu}{\varepsilon}} \ln \frac{D}{d}$$

where μ = magnetic permeability of the insulation material
ε = electric permittivity of the insulation material
D = diameter of the outer shield
d = diameter of the center conductor

If we switch to Thevenin's equivalent and analyze the transmission line under matched conditions (Figure 5.2), we find that the voltage at the receiver end is not altered by echoes, although it is attenuated by the Z_s–Z_0–Z_L voltage divider and it is delayed by a phase angle:

$$\theta = \frac{2\pi}{\lambda} z$$

where ω = signal angular frequency
α = line attenuation coefficient
λ = signal wavelength
Vt_0 = source signal amplitude
Vr_0 = load (receiver) signal amplitude
z = line length

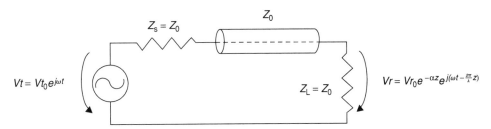

Figure 5.2 Transmission Line—Thevenin's Equivalent Circuit.

A number of problems now become apparent. The load (receiver) signal amplitude Vr_0 is smaller than its corresponding source (transmitter) value Vt_0 and is dependent on the length z of the line. Also the phase angle delay is dependent on the signal wavelength and therefore frequency. This tends to distort the received signal by delaying different source frequency components by different amounts. Such frequency spreading can be readily observed when high-bandwidth analog signals are transmitted over long cables.

Finally α, although relatively small in value, does increase with frequency. As a result, high-frequency components will be attenuated faster (over shorter distances) than the lower-frequency ones, despite the apparent frequency independence of Z_0 (for coax cables). One consequence is the distortion we see when transmitting high-frequency digital pulses, such as serial HDTV signals, even over moderate lengths of uncompensated transmission lines. Most transmission lines currently used in the video industry have a 75 Ω or a 50 Ω characteristic impedance. This is deemed to be an acceptable compromise between reasonable cable design and good line performance.

The impedance elements placed at the source and load sides for the exclusive purpose of matching the source and load to the interconnecting transmission line are known as "terminators." In most instances terminator impedances are purely resistive. Most analog video equipment has 75 Ω terminators connected in parallel to their inputs and 75 Ω terminators connected in series with their outputs, thus matching them at both ends to standard 75 Ω coax cables.

Terminators are even more extensively used in digital video design where, as we saw earlier, PCB traces as short as half an inch can cause problems. In this case they are necessary not only at the inputs and outputs of the system, but also when interconnecting high-speed integrated circuits on the boards themselves. Since not every single high-speed PCB copper trace can be perfectly matched, some tolerance to reflections must be built in. Terminators provide this degree of tolerance by quickly dissipating the minor echoes and oscillations that inevitably occur in such systems. The most common values for terminator resistors R_t are in the 50 Ω to 100 Ω range (Figure 5.3). Although these low values require higher driving currents, they are necessary because of the non-negligible input capacitance of the ICs. Larger values would result in larger RC time constants and in the deterioration of the signal edges when operating at 100 Mbps or higher.

While any good quality low-reactance resistor (of appropriate power rating) can be used as a terminator, you may also consider terminator IC arrays especially since some come with ESD protection on each line (TI, Maxim). We will revisit the topic of terminators in Chapter 14.

5.2 Input/Output ESD Circuit Protection

Any electronic system open to the outside can be subjected to an electrostatic discharge or ESD. Protection against such events is particularly important in

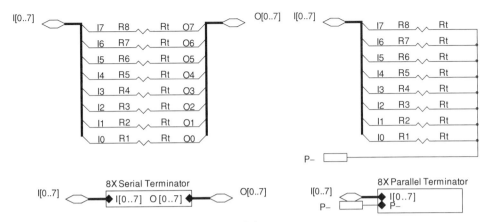

Figure 5.3   Multiline Resistive Terminator Modules.

video equipment where components may be interconnected via hundreds of feet of coax or twisted pair cables. Buildup of electrostatic charges may be due to weather phenomena or everyday activities like handling a wool sweater or walking on a linoleum floor. Some ESD disturbances may even reach our equipment over the power lines.

The destructive effects of ESD on electronics are easy to understand. An electrostatic discharge or spark creates a high-amplitude electromagnetic pulse with a very steep rise time. In turn, this generates very high voltages across any inductive elements on its path, PCB traces included. If the discharge makes contact with, or is in the proximity of, an electronic circuit a number of failures—from transient to catastrophic—may take place. They include generation of false bits, memory erasure, and puncture of dielectric layers and p-n junctions.

ESD protection for video circuits is, by necessity, light to moderate. We cannot use high-power Zener diodes directly on video lines since their considerable capacitance would degrade the signal.

The only "in box" high-power protection we can generally provide consists of galvanic isolation from the power line, filtering, and solid grounding. The use of coax cables and shielded twisted pairs can reduce the risks considerably, but only if their shields have low-impedance paths to the system ground. Still, unless ESD protection is designed in, low-power, high-frequency transients will go through.

One arrangement for video line input protection is shown in Figure 5.4(a). The video input signal is coupled to the protected side through a DC blocking capacitor Ci. The value of Ci depends on the input impedance of the protected circuit but its voltage rating should be as high as possible.

ESD protection is provided by a dual low-capacitance Schottky diode connected to the VA+ and VA−power rails. Further clamping may be provided by

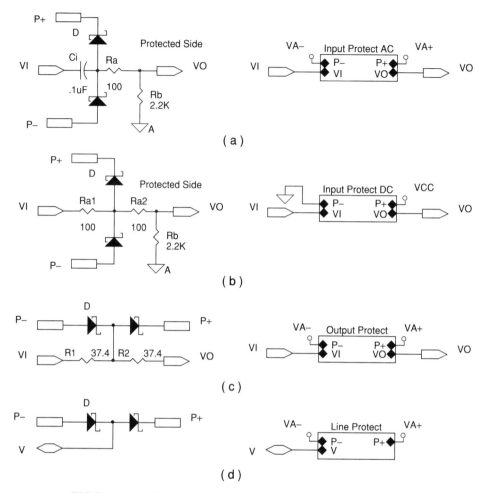

Figure 5.4 ESD Protection Circuits.

the protected IC itself through its substrate junction, with the series resistor Ra limiting the backflow current. Finally, Rb references the video to ground. If the protected circuit is DC coupled, the blocking capacitor is removed and replaced by the current limiting resistor Roa[Figure 5.4(b)].

Output protection is less demanding since the relatively low impedance and high power output stages are inherently more robust. The simplest approach is to use a dual Schottky diode for clamping the line to the power rails [Figure 5.4(c)]. In this circuit the output terminators R_1 and R_2 also act as a current limiter for the substrate junction. Many times, when the protection is against local ESD, R_1 and R_2 are combined in a single 75 Ω resistor. If the protected circuit has built-in current limiting, even this resistor may be removed [Figure 5.4(d)].

In the case of differential lines the protection circuit must clamp the line-to-line voltage to safe values, while still allowing each line to float to common mode levels.

This can be done by using two columns of Schottky diodes in a series-parallel connection (Figure 5.5). The differential signal can now pass unobstructed, as long as its amplitude is smaller than the sum of the threshold voltage drops of the Schottkys. Higher voltages will open the diodes and safely clamp the signal.

Depending on the tolerance of the protected circuit, each line common mode voltage can also be individually clamped by diodes going to VA+ and VA−.

ESD protection for bussed or multiline signals can be just a multiple version of the circuits we have seen so far [Figure 5.6(a)] or may have additional components for other functions, such as filtering [Figure 5.6(b)]. We could also use an integrated ESD protection IC such as those available from STMicroelectronics, Semiconwell, and others.

All our discrete protection circuits use Schottky diodes because they offer a lower forward voltage drop and a lower capacitance than silicon diodes. The lower forward voltage drop gives us an additional margin when protecting silicon based ICs, while their low capacitance shunts less signal current at high frequency. Their inherent higher leakage current is not a serious drawback when attached to 75 Ω video lines. On video inputs there should always be a resistor between the diodes and the pins of the ICs in order to prevent shifting of the input DC bias levels.

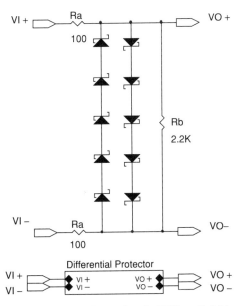

Figure 5.5 ESD Protection for Differential Lines.

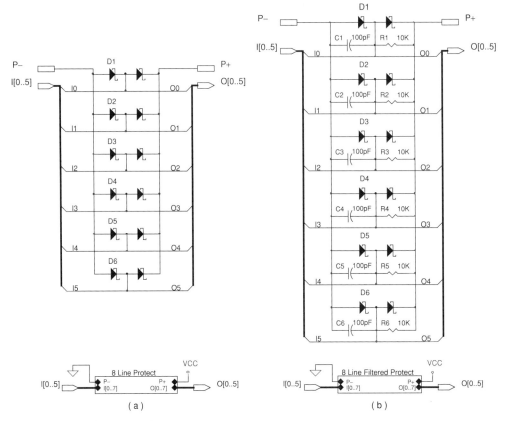

Figure 5.6 Multiline Protection.

To make our schematics easier to read we occasionally use the convention in Figure 5.7 to illustrate multiple protection circuits, where N is the number of independent identical circuits in the block.

We use a similar convention where multiple decoupler or filter capacitors would otherwise clutter the schematics. In Figure 5.8 the nXCa, mXCb block consists of n tantalums of Ca value and m ceramic caps with a value of Cb, where n, m, Ca, and Cb are different for different applications. These caps are to be placed on the board according to the PCB layout rules outlined in

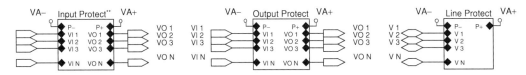

Figure 5.7 Protection Module Convention.

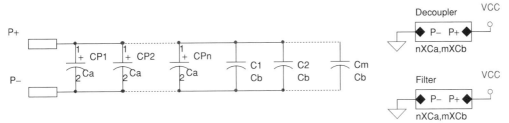

Figure 5.8   Decoupler and Filter Module Convention.

Chapter 14 or according to the specification documentation of the IC to which they are connected.

5.3 Analog Inputs and Outputs

Transporting an analog video signal across a low-impedance transmission line can be a challenge for any electronics designer. The circuits need to have a gain-bandwidth of 200 MHz or more, gain-flatness less than 0.5 dB at 10 MHz, low differential gain error (.1 percent or better), low differential phase error (.1 percent or better), and the ability to drive 75 Ω video loads.

5.3.1 Single-Sided Analog Drivers and Receivers

Despite these difficulties, designers have now available to them a number of viable video driver and receiver solutions in a form we are all familiar with; the video operational amplifier. And just like our standard opamp, the video opamp can be used either in an inverting or noninverting configuration (Figure 5.9).

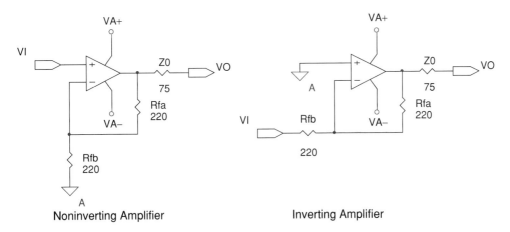

Figure 5.9   Standard Operational Amplifier Configurations.

The gain G for the noninverting amplifier circuit is given by:

$$G = 1 + \frac{Rfa}{Rfb}$$

In the inverting configuration the gain is negative:

$$G = -\frac{Rfa}{Rfb}$$

Video opamps can be used both as drivers and receivers as shown in Figure 5.10. On the driver side we use a gain 2 noninverting amplifier to compensate for the voltage drop across the Rs/Rl voltage divider. Both Rs and Rl are 75 Ω terminators in anticipation of their use with 75 Ω coax lines. Since the driver and receiver may be separated by long cable runs, we must reference them to their local DC bias points. This necessitates the use of the DC blocking capacitor Ct.

On the receiver side the coupling capacitor Cr passes the AC portion of the video signal to the high-input impedance receiver opamp or buffer. Cr allows the receiver to create its own optimal DC bias point independent of the state of the driver or coax line.

The rather large value of Ct is necessary in order to pass through the low frequencies at the bottom of the NTSC/PAL spectrum. However, if we increase the gain in the low-frequency region we can significantly reduce the values and physical size of the DC blocking components (although their number will increase). Figure 5.11 shows such an arrangement.

In this circuit the high-frequency gain of the driver is determined mainly by the ratio of Rfa2 + Rfa3 to Rfb, since Cta and Ctb have negligible impedances and the parallel combination of Rfa1 and Rfa2 is practically equal to Rfa2. At low frequencies the contribution of Rfa1 increasingly becomes a factor. This

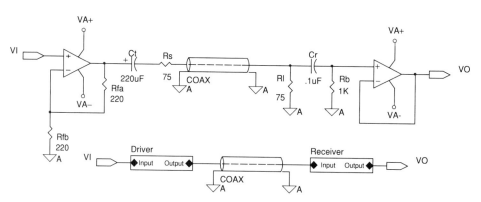

Figure 5.10 Simple Opamp Video Link.

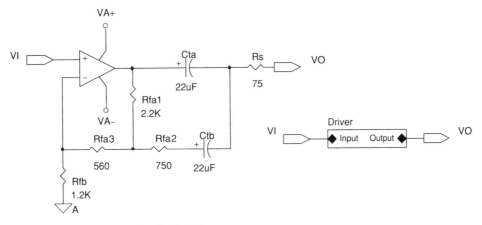

Figure 5.11   Bootstrap Video Cable Driver.

boosts the gain of the driver and compensates for the increased attenuation brought on by lower capacitor values.

Video opamps come in a wide variety of performance classes, prices, and packages. In their quest to simplify video design as much as possible, many manufacturers even offer fixed-gain (G = 1 or 2) opamps that do not use any external components and come in packages as small as SOT23 (MAX4200 from Maxim).

5.3.2 S-Video Drivers and Receivers

If we want to move beyond composite to s-video or RGB, both our cables and driver/receiver circuits must change but in predictable ways. The cables need to incorporate multiple coax lines within a common shield and jacket, while the number and performance of the drivers and receivers also change to meet the demands of the s-video (Figure 5.12) and RGB (Figure 5.13) formats.

Since the individual bandwidths of the luma and chroma signals of s-video are comparable to the bandwidth of composite video we can use the same amplifier circuits and ICs (MAX4389-MAX4396, for example) as before.

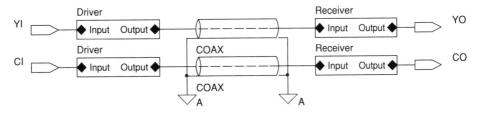

Figure 5.12   Simple Opamp S-Video Link.

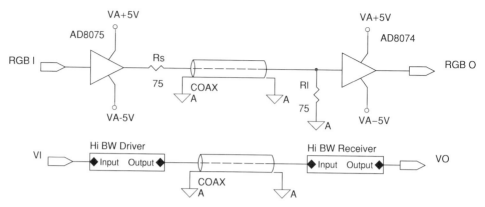

Figure 5.13   Simple High Speed Video Link.

5.3.3 VGA, XGA, UXGA Video Drivers and Receivers

To operate in the frequency range associated with computer monitors (XGA, UXGA, WXGA), our drivers and receivers need to handle bandwidths in excess of 70 MHz.

A pair of buffers offered by Analog Devices, Inc. can handily satisfy this requirement. The AD8075 has a gain $G = 2$ and a bandwidth of 500 MHz making it suitable for the driver section while its unity gain version, the AD8074, makes for a good receiver circuit (Figure 5.13). Three pairs of such buffers and a few logic drivers and receivers complete a minimal VGA link (Figure 5.14).

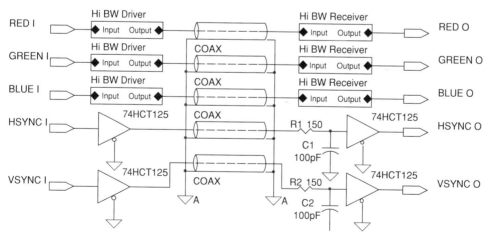

Figure 5.14   VGA, UGA, XGA Video Link.

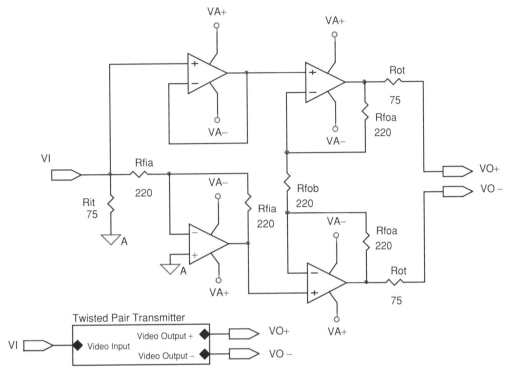

Figure 5.15   Differential Line Transmitter.

5.3.4 Opamp Differential Drivers

Drivers and receivers for twisted pair cables and other differential lines are more complex than their single-sided equivalents. Let's start with a typical design using generic video opamps (Figure 5.15).

The two input amplifiers buffer (top) and invert (bottom) the incoming video thus converting the single-sided input into a differential signal. The two output opamps form a differential in–differential out amplifier with a gain:

$$G = 1 + \frac{2Rfoa}{Rfob}$$

Since twisted pair lines have relatively high ohmic losses, we can compensate by pre-amplifying the transmitter signal using a smaller Rfob value.

5.3.5 Opamp Differential Receivers

The receiver (Figure 5.16) is a standard differential amplifier with a single sided output and a gain of

$$G = -\frac{Rfoa}{Rfob}\left(1+\frac{2\,Rfia}{Rfib}\right)$$

Building differential drivers and receivers with individual opamps offer the utmost control over the parameters of the design. It also makes it more dependent on circuit layout, component precision, environmental noise, and so forth.

5.3.6 Twisted Pair Differential Link

If detailed control is not required you may choose to use a driver/receiver chip set such as the MAX4147/MAX4144 from Maxim Integrated Products (Figure 5.17). A twisted-pair cable and two BNC connectors complete the link (Figure 5.18).

Important note: All driver and receiver designs must include appropriate ESD protection circuits and power-supply filtering capacitors. Their omission

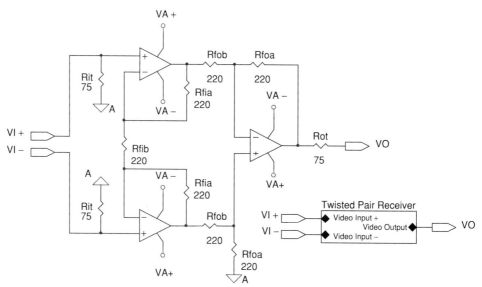

Figure 5.16   Differential Line Receiver.

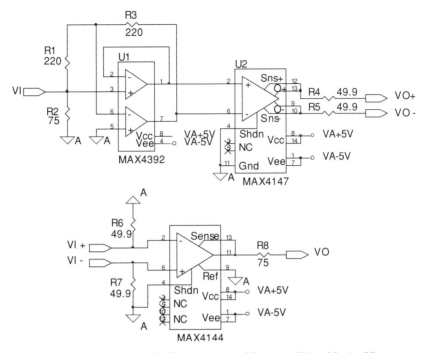

Figure 5.17   Twisted Pair Cable Transmitter and Receiver Using Maxim ICs.

from our figures reflects a desire to maintain clarity of presentation and does not imply that their use is only optional.

5.4 Digital Inputs and Outputs

If when designing analog video systems we need to occasionally treat communication links and interconnection wires as transmission lines, HD video makes such considerations mandatory. At 1.485 Mbps, all PCB traces half an inch or longer must have their impedances accounted for in the design of the circuit (controlled impedance). All off-board lines must be shielded and

Figure 5.18   Twisted Pair Video Link.

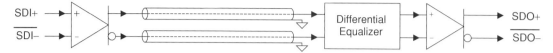

Figure 5.19   SDI Link Arrangement.

matched to their termination impedances, and even a few feet of cable needs frequency compensation (equalization).

5.4.1 Digital Video Drivers

Serial digital interface (SDI) streams are normally transported from device to device in differential formats using two 75 Ω coax lines (Figure 5.19). The cable driver is a high-speed, high-input impedance differential in–differential out amplifier. The logic circuitry driving its inputs is by necessity ECL or single-voltage ECL, also known as PECL (Pseudo-ECL). The main concern when designing SDI drivers is the impedance matching of the two outputs to their respective transmission lines.

5.4.2 Digital Driver Circuit

A design using the GS1528 driver from Gennum Corporation is shown in Figure 5.20. The circuit itself is easy to use with most critical design issues

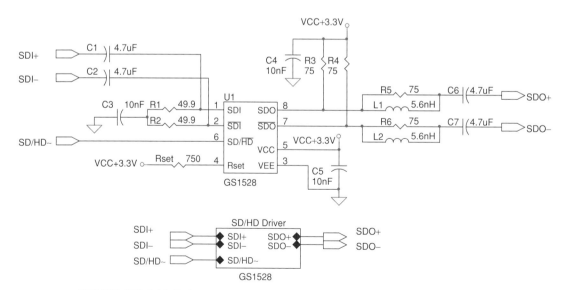

Figure 5.20   SDI, HD-SDI Cable Driver.

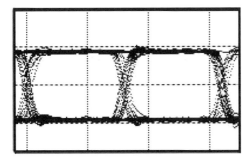

Figure 5.21 Eye Pattern.

being handled internally by the IC. Still, since the circuit operates at rates up to 1.485 Gbps, the input lines need proper termination, which are done via the R1, R2, and C3 network.

The input and output signals are coupled through 4.7 uF or larger tantalum capacitor. Line compensation inductors L1 and L2 are needed to minimize trace impedance effects, but their exact value is found through experimental adjustment. The desired result of the adjustment is a clean "eye pattern" output on the video waveform monitor (Figure 5.21), which corresponds to minimal signal distortion. Finally, Rset controls the amplitude of the output signal and the SD/HD~ input adapts the rise and fall times of the device to the requirements of standard definition (SD)/high definition (HD) conventions.

5.4.3 Eye Pattern Test

Eye diagrams are multivalued displays, with each point along the time axis corresponding to multiple voltage values. In order to generate the "eye pattern" for a particular data link (Figure 5.22), the clock signal of the driving

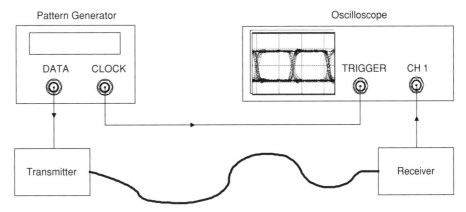

Figure 5.22 Eye Pattern Generation.

pattern generator is connected to the trigger input of the oscilloscope, and the data output is routed through the link into one of the input channels. If the oscilloscope is set on infinite persistence the individual signal traces will overlap, creating a distinct eye pattern. The less distortion the signal accumulates while in transit, the cleaner the eye pattern.

Because of their cumulative nature, eye patterns are stable and can be used to measure noisy signals where single traces are unreadable.

5.4.4 Digital Receiver/Equalizer

As noted earlier in this chapter, the attenuation coefficient for coax lines is a function of frequency, with higher frequency components being attenuated more than the lower-frequency ones. To counterbalance this effect, the first function performed at the receiving end is equalization, or the preferential amplification of high-frequency components. The transfer function of the *equalizer* circuit is a vertically flipped mirror image of the coax transfer function (Figure 5.23).

In practice, the equalizer uses a *high-pass filter* to separate the higher frequency range of the input spectrum, then amplifies it and *adds* it to the input signal (Figure 5.24). Since we are transporting digital pulses the main effect of the coax frequency dependence is to increase the rise and fall times of the signal, or to "round off" the pulses. By boosting the high-frequency part of the spectrum we decrease the rise and fall times; we "square up" the pulses.

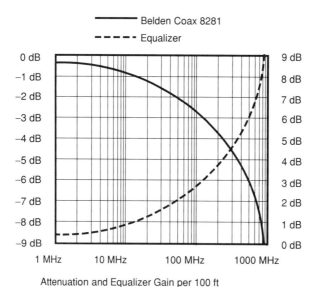

Figure 5.23   Equalizer Frequency Transfer Function.

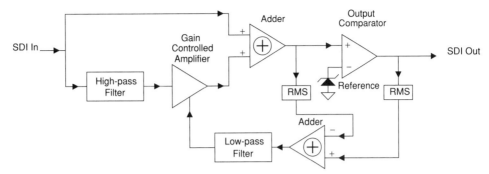

Figure 5.24   Equalizer Block Diagram.

The gain applied to the high-frequency range is determined by a feedback loop driven by the difference in *RMS* (root-mean-square) between the compensated signal and its local pulse equivalent obtained by comparing it to a *reference voltage*.

If the compensated signal has good edges, running it through the *comparator* yields a waveform very similar to the compensated signal itself. The widths of the difference pulses generated by the *feedback adder* will then be small, the amplitude of the *low-pass filter* output will be small, and the gain of the high-frequency range amplifier will also be small.

By contrast, if the edges of the compensated signal are poor, there will be larger delays between the start of a rise or fall slope and the point where the waveforms reach the reference level of the comparator. The difference pulses will grow wider, and the output signal of the low-pass filter will reach higher amplitudes. This results in a higher amplifier gain and a stronger compensation being applied to the input signal.

5.4.5 Receiver/Equalizer Circuit

The application circuit in Figure 5.25 depicts a straightforward implementation of an Equalizer/Receiver using Gennum's GS1524 IC. The only critical area is the input interface which must account both for the coax cable behavior and for the PCB traces leading to the IC. Just as for the drivers, we can use the inductors L1 and L2 for fine tuning. For short cable lengths they may be omitted.

Although dual coax cables increase system performance, a single coax is sufficient for many applications. Then the SDO- line of the driver IC is left open and the R4 of the equalizer is shorted. For all digital video design we strongly recommend the use of ground and power planes (see Chapter 14).

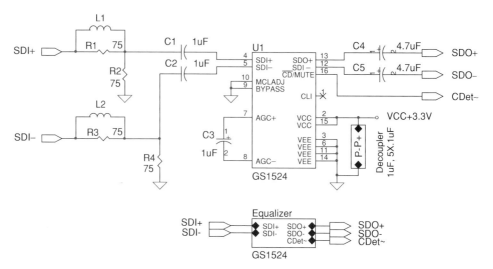

Figure 5.25   SDI, HD-SDI Equalizer Circuit.

5.4.6 Clock and Data Recovery, Reclocking

A standard SDI stream contains both data and clock information (NRZI-encoded), which must be separated at the receiver side. In addition, some timing degradation (jitter) may also occur during the equalization and switching processes. The jobs of extracting the clock and data from the SDI stream and of restoring their proper relationship belong to the *reclocker* circuit. At the core of a reclocker is a PLL feedback system comprised of a *voltage controlled oscillator* (VCO), a *frequency divider*, a *phase detector*, and a *charge pump* (Figure 5.26).

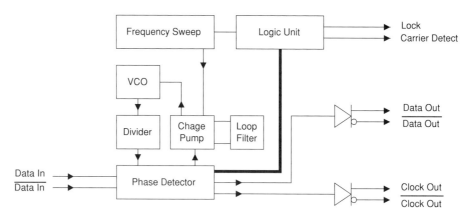

Figure 5.26   Reclocker Block Diagram.

The controlling voltage of the VCO is produced by a charge pump driven by the phase error between the incoming signal and the VCO output. Since the VCO will only provide signals within a relatively narrow band around its center frequency, a programmable divider is needed to bring the output of the VCO within the range of the input. Phase error correction pulses are generated by the phase detector and integrated by the charge pump and its loop filter.

Some reclockers use a frequency sweeper to widen the PLL's acquisition range (which is about the same as the loop bandwidth) to match the range of the VCO. The sweeper generates a triangular or trapezoidal waveform that also controls the VCO via the charge pump. This forces the VCO to ramp its output frequency up and down in search of input signals to lock on. When found, the sweeper stops scanning and maintains its value until the lock is lost. Once the PLL is locked, the phase detector realigns the data and clock signals and passes them on to differential line drivers. Finally, lock and carrier detect logic flags validate the status of the device to the system controller.

5.4.7 Reclocker Circuit

The reclocker circuit is only of moderate complexity and is usually found integrated together with other functions such as deserializers or switchers. The implementation in Figure 5.27, for example, uses a Gennum GS1535 IC that also includes a 4:1 input multiplexer.

This application circuit is straightforward, with the usual PCB design considerations to be taken into account: the use of ground and power planes, short and properly terminated traces (on the IC side the terminations are internal), and proper power decoupling. The control signals select the input source (DI Sel0, DI Sel1), mute the output (SDO Mute~), and suppress a rarely used input frequency (ASI/177~ see datasheet for details).

5.5 LVDS Drivers and Receivers

The transport of large amounts of parallel digital data even across moderate distances can be a difficult task. Consider the link between a PC graphics controller and an LCD panel. The controller outputs 18 or 24 bits of RGB data (for 256,000 or 16 million colors), plus line and frame signals (equivalent to the HSYNC and VSYNC in analog video). This requires 26 digital cables a few feet long, each carrying 3 V, 20 mA signals with a bandwidth of 65 MHz. The noise and power consumption problems are obvious, not to mention the inconvenience of stringing 26 coax or twisted pair cables from the controller to the LCD.

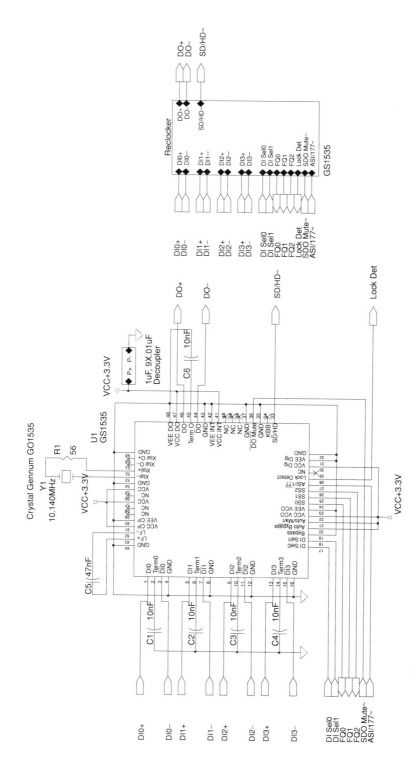

Figure 5.27 SDI, HD-SDI Reclocker Circuit.

120

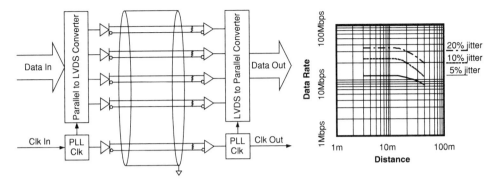

Figure 5.28   LVDS Link.

A practical solution to the problem is provided by the LVDS (Low Voltage Differential Signaling) technology. On the source side LVDS transmitters serialize the parallel information using a built-in PLL clock, and drive it on five low-voltage differential lines. On the receiver side the signals are converted from differential to single-sided standard logic and deserialized, thus completing the data recovery process (Figure 5.28).

The use of LVDS transmitters and receivers simplifies considerably the architecture of the PC-controller-to-LCD-panel link (Figure 5.29). A single, flexible, shielded cable containing five individually shielded twisted pairs is all that is required.

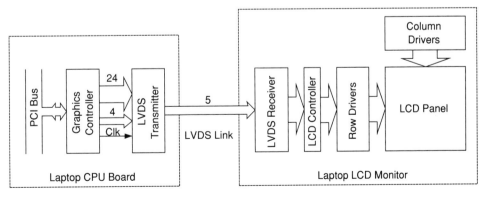

Figure 5.29   Laptop CPU–LCD Monitor Link.

The use of LVDS links, however is not limited to driving display panels. Any application requiring the transport of parallel data over distances up to 15 meters is a good candidate for LVDS. Figures 5.30 and 5.31 show a practical implementation of a LVDS link using the DS90CR287 transmitter and the DS90CR288 receiver from National Semiconductor. All termination resistors are 50 Ω.

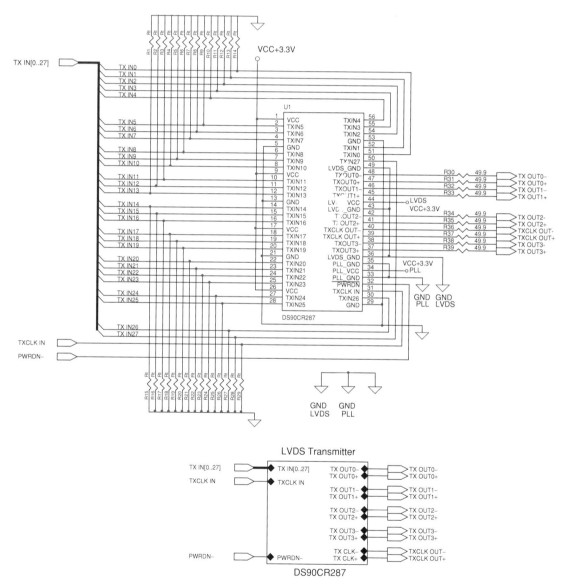

Figure 5.30   LVDS Transmitter.

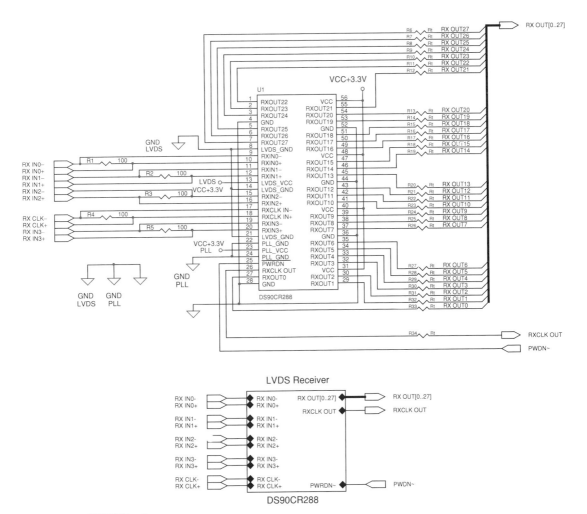

Figure 5.31   LVDS Receiver.

Video Distribution

6.1 Analog Distribution Amplifiers (DAs)

As we have seen in Chapter 5, the most common form of video transmission line is the 75 Ω coax cable. Since nearly all video systems must be matched to it, their input impedance is also equal to 75 Ω. It follows that wiring multiple devices to the same cable feed is not simply a matter of using a number of coax T adaptors, for example. If we do, all the input impedances will be connected in parallel and present to the cable junction an equivalent impedance of 75 Ω divided by the number of such devices. With the source and line impedances remaining the same, the signal voltage across their inputs will drop below acceptable levels. The coax transmission line will no longer be matched.

6.1.1 Composite DA

A simple solution to this problem is to use a distribution amplifier or DA. This is a single input/multiple outputs amplifier with all input and output impedances equal to 75 Ω.

The 1-to-4 composite video DA circuit in Figure 6.1 uses five discrete opamps. The input noninverting amplifier has a gain of two and feeds four video buffers, which in turn drive 75 Ω video lines. The gain is necessary to compensate for the voltage drop across the individual Rot resistors when driving 75 Ω loads at the far side of the lines.

6.1.2 S-Video DA

Distribution amplifiers for s-video signals (Figure 6.2) consist essentially of two composite DAs in parallel, one handling the luma (Y) and the other the chroma (C) components. For best performance the opamps should be as well matched as possible. A good approach is to use a dual amplifier for the input stages on the Y and C paths, and a quad unity buffer for the output stages.

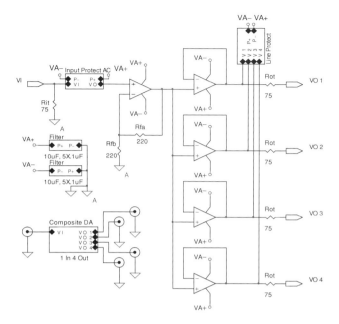

Figure 6.1   Composite Distribution Amplifier.

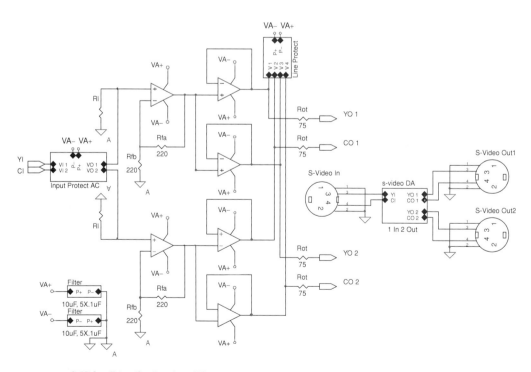

Figure 6.2   S-Video Distribution Amplifier.

When we lay out the printed circuit board (PCB) for this application we must be particularly careful in the areas where the luma and chroma signals cross each other. With output currents in excess of 10 mA, the chroma input lines can easily pick up interference from the output traces.

6.1.3 VGA, XGA, UXGA DA

For the design of a VGA, SVGA, UVGA, or WXGA distribution amplifier we use three separate high-bandwidth DAs, one for each RGB color component, and digital buffers for the HSYNC and VSYNC (Figure 6.3).

Individual DAs can be architecturally similar to the one in Figure 6.1, but the opamps or buffers we use must be high-bandwidth (AD8074/8075, MAX4178/4278). In this particular circuit we make use of the fact that some ICs can deliver a high enough output current to perform the distribution function without the need of further buffering. A single MAX4278, for example, can drive two or three terminated video lines at full spec.

As a general reminder the circuits in Figure 6.1 through Figure 6.3 include I/O line protection circuitry, power supply filters, and decouplers. We will not explicitly show them in all our diagrams but they are assumed to be present wherever they are needed. It is the prerogative of the designer to decide when, where, and how they are used.

6.2 Analog Switches and Multiplexers

A key capability needed by most video distribution systems is to select an input signal from a number of available sources and to deliver it to a particular device, under user or software control. Video switches and multiplexers both perform this job but they do it differently.

A multiplexer is a device that outputs a buffered or amplified version of a selected input [Figure 6.4(a)], while a switch will directly connect its output to the selected input with only a small channel resistance in between [Figure 6.4(b)]. Therefore a switch is a bidirectional device while a multiplexer is not. In both cases a logic control signal (Source Select) determines which input source makes its way to the output.

6.3 Analog Matrix Switches

Most flexible of all switching systems is the matrix switch. Its function is to connect any one (or more) of its outputs to any of its inputs. Due to its architecture, multiple parallel paths can be easily accommodated as seen from the block diagram in Figure 6.5.

At the core of a matrix switch is an array of SPST analog switches. Each of them can be opened or closed by a logic control signal generated and stored by a decoder circuit. Its function is to translate an input address into switch commands and insure that illegal conditions, such as connecting multiple inputs together, are avoided.

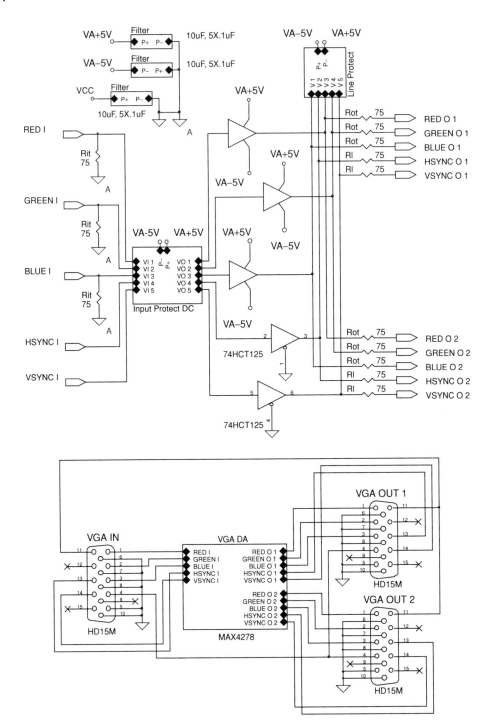

Figure 6.3 VGA Distribution Amplifier.

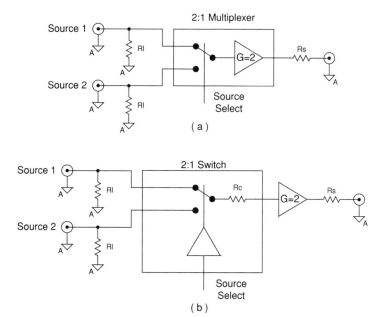

Figure 6.4   Video Multiplexer and Switch.

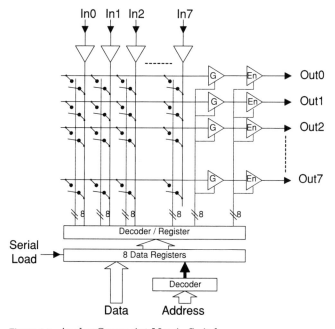

Figure 6.5   Analog Crosspoint Matrix Switch.

Driving the decoder is a set of registers (one for each output) that holds the address of the source input and the desired status of the output (active, tri-stated, and so on). The registers are loaded either through a serial or parallel port. Depending on the application the outputs of the matrix can be buffered (G = 1) or amplified (G = 2) before being passed on to the tri-state output drivers.

6.3.1 Expanding Matrix Switches

Matrix switches can be easily expanded to form arrays as large as 512×512. Their tri-state outputs hold the key to the expansion process. To design a 2N $\times$ 2N matrix using N $\times$ N arrays, for example, we apply the bottom N input lines to two N $\times$ N matrices and the top N to two others. Then we connect the outputs of the matrices together as shown in Figure 6.6.

To route an input in the [1 ... N] range to an output in the [1 ... N] range, we enable the corresponding output of array A and tri-state that output of array C. For an [N ... 2N] input to be connected to a [1 ... N] output we tri-state A and enable C, and so on.

6.3.2 Analog Matrix Switch Circuit

A high-performance IC that can be used to build crosspoint matrix blocks for composite, s-video, and even UVGA applications is the AD8108 from Analog Devices, Inc (Figure 6.7). It boasts a –3 dB bandwidth of 325 MHz and a –0.1 dB bandwidth of 70 MHz with very good channel crosstalk performance. Its parallel user interface enables the updating of individual switches without globally reloading the entire array configuration. Larger arrays can use the serial interface which allows us to daisy-chain multiple ICs without the help of any additional logic.

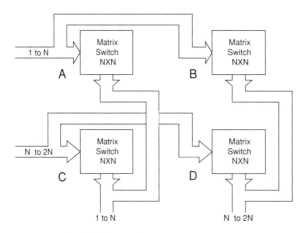

Figure 6.6 Switch Expansion.

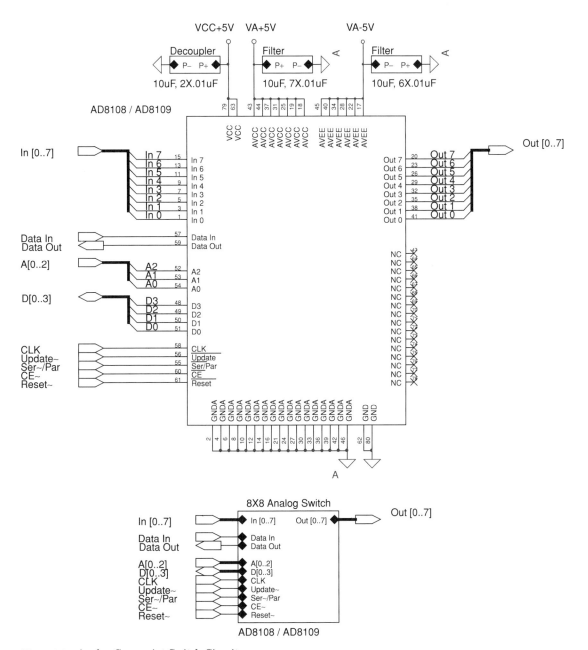

Figure 6.7   Analog Crosspoint Switch Circuit.

6.4 SDI, HD-SDI Distribution Amplifiers

The elementary building block for digital distribution amplifiers is the cable driver.

In the input stage of a typical digital DA the outputs of a local cable driver are used to feed the input signal to an internal DA transmission line. Connected to this line we find the inputs of a number of other cable drivers, which collectively form the DA's output stage (Figure 6.8). The internal transmission line has to be properly terminated at both ends and has to form a continuous open loop without any branches.

Furthermore, the input impedance of the output drivers needs to be much larger than the termination impedances on the line; in this way their presence does not disturb the local voltage and current distribution.

6.4.1 SDI, HD-SDI 1 → 2 DA Circuit

A design for a one input–two outputs digital DA, used both for standard and high-definition video, is shown in Figure 6.9. The driver we selected is the GS1528 from Gennum, one of the few drivers on the market capable of operating at 1.485Mbps.

Although the circuit appears to be quite simple, its main challenge rests in the design of the board. Even with CAD systems that calculate the trace impedance while the designer enters the layout in the computer, the value measured on the fabricated board is invariably different. Therefore all termination resistor and inductor values on the schematics are only tentative. The final values are arrived at by experimentation with the actual printed circuit board.

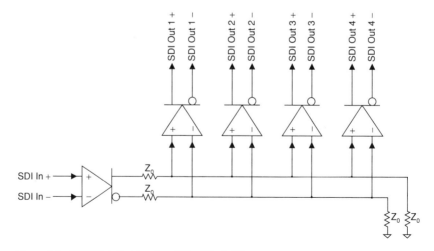

Figure 6.8   SDI Distribution Amplifier Block Diagram.

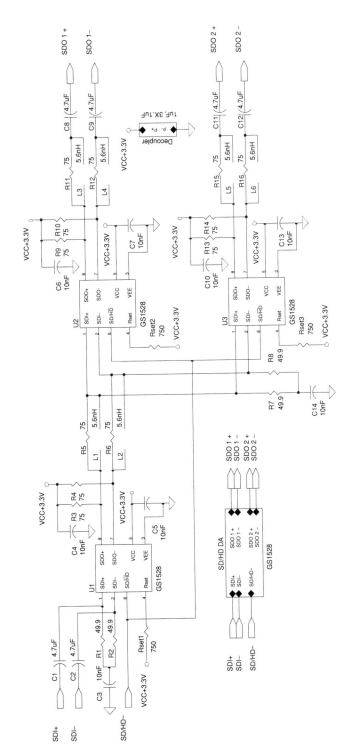

Figure 6.9 SDI Distribution Amplifier Circuit.

6.5 SDI, HD-SDI Matrix Switches

The general structure of a *digital crosspoint switch* (Figure 6.10) is similar to the one of its analog counterpart. Each output is assigned a data register that holds the address of the input it is connected to and the desired output status (active or tri-stated). This information is provided by the user via a parallel or serial interface. A set of configuration registers then converts it to a logic connection pattern for the switch matrix.

6.5.1 SDI, HD-SDI 8 × 8 Matrix Switch Circuit

In practice, 8×8 array blocks are implemented using single chip ICs such as National Semiconductor's CLC018. As with any circuit that operates in the Mbps range, the focus of the design effort is the board layout. The circuit schematic is quite simple showing the usual bypass and termination passives (Figure 6.11).

The switch configuration is controlled by the I/O address bus (OA2, OA1, OA0, IA2, IA1, IA0, Tristate) and the Load and Configure lines. To change a connection the three bits of the output address are placed on the OA2, OA1, OA0 lines, the three bits of the input address on the IA2, IA1, and IA0 lines, and the desired output status (tri-state or active) on the Tristate line. Then Load is strobed, and finally Configure is strobed. To change all connections at once the data for all eight outputs must be loaded first, after which the Configure line is pulsed.

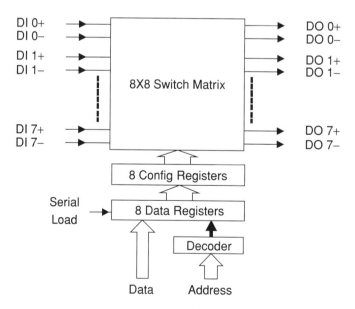

Figure 6.10   SDI Crosspoint Switch Matrix.

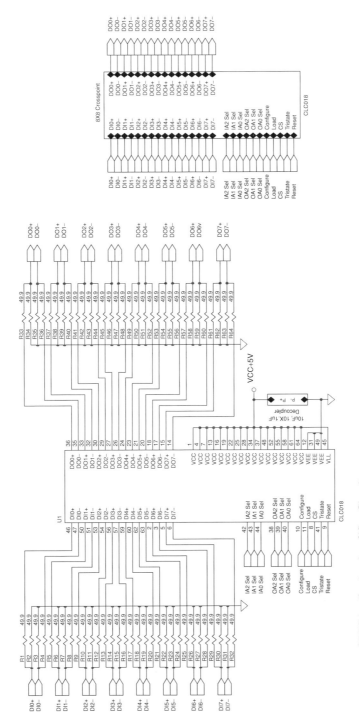

Figure 6.11 SDI Distribution Amplifier Circuit.

Larger arrays can be obtained using the same methodology we used for analog matrix switches. Of course, the layout becomes that much more complex, with termination resistors required for most of the chip-to-chip interconnection lines as well.

6.6 Parallel Digital Switches

On occasion we may need to switch parallel video buses, especially when the video routing is internal to the board. This happens, for example, when we overlay multiple 24/30-bit RGB sources, each of them already digitized and stored in video memory.

From the circuit architecture perspective we have three options. The first is to use bus switches, multiplexers, or demultiplexer ICs. They are available in widths ranging from 8 bits to 24 bits, and provide an easy to use modular solution to the problem (Texas Instruments, IDT).

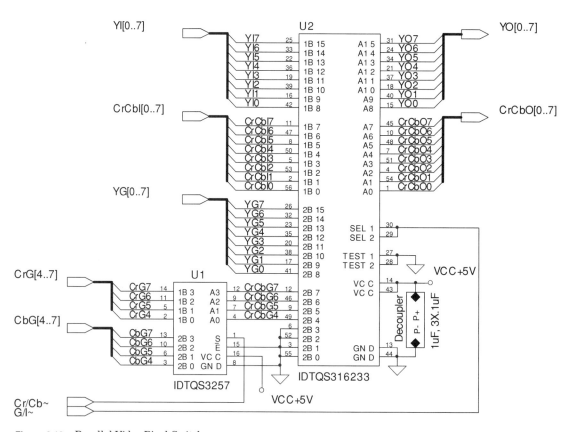

Figure 6.12   Parallel Video Pixel Switch.

A second option comes from Fairchild Corp. in the form of the MSX and OCX families of massively parallel switches. They can handle up to 512 I/O lines at data rates up to 1.6 Gbps (OCX), and can operate in TTL, LVTTL, LVDS, and LVPECL environments. They require controller programming.

Finally we can switch parallel video buses using FPGAs or CPLDs. Despite its more limited capacity and bandwidth, this approach is more flexible since, once loaded in the IC, the data can be locally processed before being transferred to the next block.

6.6.1 Parallel Video—Background Color YCbCr Switch

Figure 6.12 illustrates a video mixing application utilizing two ultrafast digital multiplexers from IDT. On input channel 1B of the 32×16 mux-demux (IDTQS316233) we have a parallel video stream that consists of 8 bits of luma (YI) and 8 bits of interleaved chroma information (CrCbI). This stream could originate, for example, from a decoder-digitizer connected to a video camera.

On the 2B input channel we connect a parallel graphics source. The YG bus carries 8 bit of graphics luma data with the LSB of 2B split into two nibbles. The lower nibble is grounded while the upper nibble is connected, through a fast 8×4 mux-demux, either to 4 bits of graphics red chroma (CrG) or to 4 bits of blue chroma (CbG), depending on the state of Cr/Cb~. Line G/I~ may be controlled by an overlay processor and selects which type of pixel, input or graphics, is allowed on the output bus. Line Cr/Cb~ controls which component of the 4:2:2 format, Cr or Cb, is processed at a given time.

The function of this circuit is to mix, on a pixel-by-pixel basis, a 4:2:2 video stream with a 256 colors graphics stream. Such a circuit can then be used in a logo or frame overlay application. The graphics information can be dynamic (when retrieved from a graphics memory buffer) or static (as generated by DIP switches or static registers).

7

Controller Infrastructure

7.1 Controller Video IC Interfaces; I²C, SPI

Many peripheral video chips require internal register programming both on startup and during normal operation. Information such as desired contrast and luminosity, scale factors, video format, and many others needs to be communicated by the local controller to the video ICs. On occasion, the peripherals will pass on to the controller, and therefore the user, status information, or captured data (camera images, for example).

The most widely used interface standards for communication between controllers and peripherals are the I²C (Inter-IC) developed by Philips and the SPI (Serial Peripheral Interface) introduced by Motorola. Both are serial in order to minimize the number of pins that the peripheral ICs have to dedicate to the communications task.

7.1.1 I²C Bus

The I²C bus is a two-wire interface that can be used for data rates up to 100 Kbps, although newer peripherals support high-speed modes up to 3.4 Mbps (Figure 7.1). The maximum load capacitance allowed on the bus is 400 pF, which limits the number of peripheral devices that can be placed on a given I²C loop to about 16. Both the data line (SDA) and the clock line (SCL) are bidirectional, open-collector (open-drain) lines and therefore need connection to the local power supply through pull-up resistors or current sources. The I²C bus can be used in single master or multiple master configurations. However, in most practical video applications we have a single master controller driving a number of peripheral slave devices.

To initiate a data transfer on an I²C bus the *master* will create a so called *start condition* (S) on the bus by holding the clock line SCL high while gener-

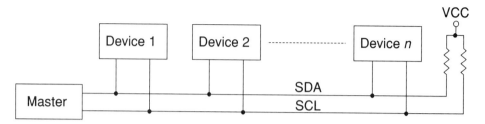

Figure 7.1 I²C Bus.

ating a high-to-low transition on the SDA data line. Following *start* data placed on the bus either by the *master* or any of the *slaves* will be sampled by the SCL clock [Figure 7.2(a)].

The data must remain stable while the clock is high. If the addressed device is not ready to accept or generate another data bit, it can force a *wait state* by pulling the clock low. Since all I²C devices have open-collector or open-drain interfaces to the bus, their outputs create an equivalent wired AND gate, and therefore any of them can force SCL low.

Sensing the wait state, the master suspends further bus transactions until the condition is cleared. After the successful transfer of a full byte the addressed device must acknowledge (ACK) the operation by holding the data line SDA low during the ninth clock cycle.

If the transfer is not successful (NACK), SDA will be kept high and the master is flagged. At the end of a data transfer process the master will generate a *stop condition* (P) by holding SCL high and generating a low-to-high transition on SDA.

To work with multiple slave devices the I²C Master must be able to address them individually. Each slave comes with a unique I²C address or with the ability to be assigned an address by strapping a few programming pins. To initiate a data transfer with a particular slave device the master issues an 8-bit packet header, with the first seven bits of the packet being the slave address and the eighth a read or write instruction (R/W~).

The header is then followed by data or control bytes put on the bus either by the master or the addressed slave, depending on the value of the R/W~ bit [Figure 7.2(b)].

Even though some microprocessors do have a built-in I²C interface, the most common ones do not. However, implementation of the I²C protocol using almost any microcontroller is fairly straightforward. All that is needed is two I/O lines and the appropriate software driver (see Catalyst App. Note AN2). But it does require overhead. The sampling frequency for the SDA and SCL lines must be at least twice the SCL rate.

If you cannot afford the overhead of an I²C software driver and still want to use a generic, low-cost processor such as the 8031/8051, you can use the PCF8584 or PCA9564 I²C controllers from Philips. They are designed mainly

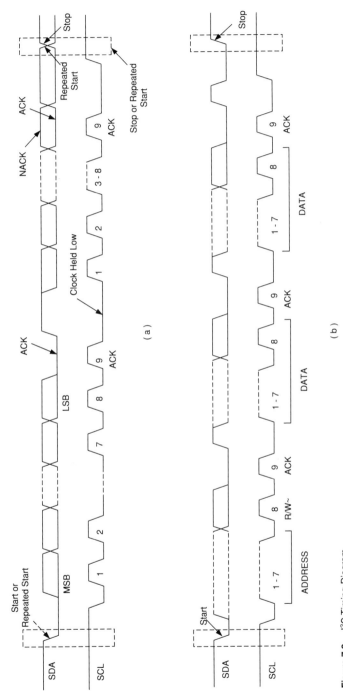

Figure 7.2 I²C Timing Diagram.

141

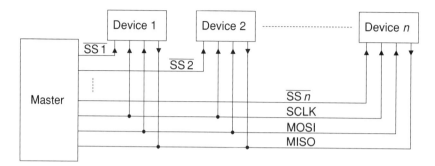

Figure 7.3   SPI Bus.

for the Intel type bus but can be easily adapted for other configurations. Dedicated I²C controllers take care of all master functions and interface to the local controller through a simple, interrupt-driven parallel port.

7.1.2 SPI Bus

The SPI hardware differs from I²C in a number of ways. SPI is a four-wire interface with separate input, output, and clock lines. The fourth line, the Slave Select (SS~), is separate for each peripheral device and, when low, it enables the SPI port of that device. While simpler, this scheme requires more dedicated interface lines both for the master and the slaves (Figure 7.3).

Without the pull-up resistors to slow things down the SPI is faster than the I²C, with its maximum data rate determined only by the characteristics of the ICs themselves. Depending on your choice of processor and peripherals, SPI data rates can go as high as 10 Mbps.

As it can be seen in Figure 7.4, the internal structure of an SPI link consists of two back-to-back shift registers, one each for the master and the slave, and an SPI clock generator controlled by the master. Each register is connected to an internal bus so it can be locally loaded and unloaded. The *master out slave*

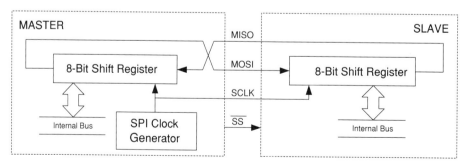

Figure 7.4   SPI Register Loops.

in (MOSI) line connects the output of the master shift register to the input of the slave shift register. In a circular arrangement the *master in slave out* (MISO) line connects the opposite ends of the same shift registers.

The transfer of data between the master and a slave device starts with setting the corresponding SS~ line low (Figure 7.5). Then for each rising edge (or falling edge if so programmed) of the clock, one data bit is shifted from the master to the slave on the MOSI line and one bit is shifted in reverse direction on the MISO line. Upon the completion of the data swap the master pulls the SS~ line high.

Both can then proceed to unload the shift registers, load the next data byte, check for integrity (Parity, CRC . . .), and so on. If your microcontroller does not have an SPI port it can be easily synthesized using general purpose I/O ports. The speed of SPI bus operation then depends on the processor speed, but will still be significant.

7.2 System Controllers

System controllers or local controllers perform a number of functions critical to the operation of the design. Therefore a great deal of attention must be paid to the selection of the appropriate CPU for the job, and to the selection of the necessary and sufficient on-board resources.

On power-up the system controller must initialize the board and all its ICs. The initializing sequence must account for the various communications standards individual ICs may use, such as I²C, SPI, or custom. If no hardware facilities are available for a particular standard, the CPU needs to simulate or "bit-bang" it using its I/O lines. Once the system is initialized the controller has to retrieve the last operational settings from local nonvolatile memory and load them into the appropriate programmable ICs.

The controller has to maintain a real-time user interface which may include LCD and LED displays, switches and keyboards, speakers, and so on. When the user interaction is mediated by a host computer, the controller must service the host interface, an RS-232 or a USB port most of the time. Any change

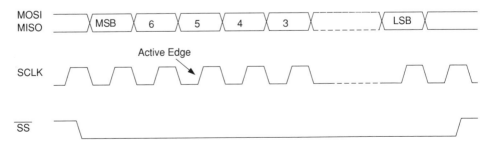

Figure 7.5   SPI Timing Diagram.

in the operating settings of the system, once entered by the user through the user interface or host interface, must be translated into individual IC commands and the commands must be then ferried to the proper peripherals. Status information from the ICs, if available, are also read, stored in memory, and possibly displayed.

If the system is involved in data collection (image capture), the controller has to manage the transfer of information from the data collection ICs (camera chip, video digitizer) to the host computer interface (USB, Firewire...).

The controller must service the on-board resources that require real-time interaction. For example, if we use an RTC (real-time clock) for time stamping the video output, the controller will frequently poll the RTC and update the user interface or graphic display to show current time and date information. In case of system malfunction the controller has to alert the user and take protective action, if such capability is provided for in the design. This may include immediate settings and data backup or switching to an alternate power source.

Although not an encouraged practice, in some designs the controller is placed directly in the video data path. We may find this in a cost-sensitive security QUAD application where the system is required to detect image changes indicative of unauthorized intrusions. In this case, the controller compares successive luminosity values for a discrete number of points on the screen (the sensor grid). If there are changes an alarm condition is triggered.

In a system with controlled shut-down, the controller first saves all important information in nonvolatile memory, then turns off the main power relay; it does this while possibly keeping itself in a minimum power sleeper mode until the *on* switch is activated.

The first step in selecting a microprocessor is to list all the jobs it must perform in the context of a particular design. This should be done as a time progression from startup to shutdown, with clearly identified worst-case conditions when the micro is under most stress. Each job needs to be described in terms of required hardware resources and software overhead. In doing so, one has to consider the initialization, service, and organized shutdown requirements of each IC (I/Os, interrupts...). The more detailed the analysis, the better the micro will fit the design. Ideally the operating system and all service routines should be flowcharted and each flowchart block assigned a conservative number of instruction cycles. At the end of this process a requirements specification list is drafted and then compared with the data sheets of available microprocessors. Once a "short list" is in place, the final choice is made based on the match between available development tools for a particular processor and in-house software skills and experience.

The application notes and projects in the remainder of this book will use the 8031/8051 processor family, although many others would perform as well. Our choice is primarily based on its availability from a great variety of vendors, and in a variety of configurations (Philips, Intel, Atmel, STMicroelectronics, Dallas-Maxim). We can also take advantage of the wide range of development tools and application notes, available free or at reasonable cost from manufacturers or third-party suppliers.

7.2.1 Minimal Controller Circuit

If the demands on the system are fairly moderate a single chip controller built around the 87C51/87C52 may prove adequate (Figure 7.6). This basic microcontroller has up to 8KB of OTP/ROM program memory and up to 256 bytes of scratchpad RAM on board. In can operate at power supply voltages between 2.7 V and 5 V (although it does not support mixed-voltage I/O), with clock speeds up to 33 MHz (at 5 V). Its peripheral interfaces consist of four 8-bit parallel ports, a duplex UART, three 16-bit timers, and a four-priority level interrupt structure.

A more powerful version of this chip is the pin-compatible 87C251 from Intel. It boasts 16 KB of OTP/ROM, 1 KB of RAM, and 2 UARTs, for a fifteenfold increase in performance. Another enhanced version, the T89C51IC2 from Atmel has 32 KB of flash memory, an I²C interface, but no UART. Finally, there is DS89C420 from Dallas Semiconductor (now part of Maxim) with 16 KB flash and twelve times faster processing than a standard 80C51 operating at the same clock frequency.

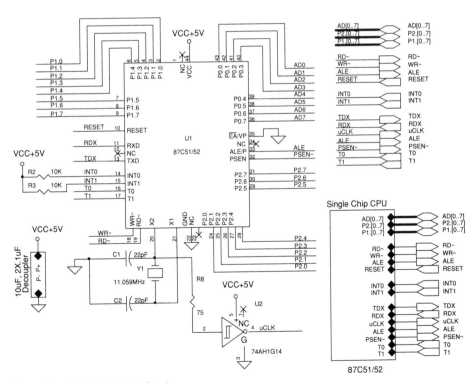

Figure 7.6 Minimal Controller Circuit.

With a 33 MHz clock the DS89C420 can achieve processing rates up to 33 MIPS (millions of instructions per second), enough for any conceivable non-DSP application. The interface between the controller module in Figure 7.6 and the rest of the system consists of 16 parallel I/O lines, 8 address/data lines (AD), read (RD~), write (WR~), and address latch enable (ALE) control signals.

Communication with the host is facilitated by a built-in UART (TXD, RXD), which leaves the SPI or I^{2}C networks to be implemented in software or to be handled by a stand-alone controller. The circuit also generates a buffered oscillator output to be used as a synchronous clock reference for peripherals that need a timing source in phase with the controller clock.

7.2.2 Expanded Controller Circuit

If our software requires amounts of program and data memory beyond the internal resources of the 87C51/87C52, we can use its ROM-less version, the 80C31/80C32 with external OTP/ROM and RAM (Figure 7.7). The total program and data memory spaces can go to 64 KB each, although for our circuit we chose only a 32 KB RAM leaving 32 KB for I/O interfaces. We should note that by banking the ICs we can add a practically unlimited amount of memory to our system. This is important if picture and video storage are part of the design.

Needless to say the designs we present here are only two of the possible implementations of microcontroller modules. There is a practically unlimited number of variations to choose from, starting with different processors and memory architectures, and ending with different operating voltages and packaging options. You can use, for example, the highly successful Stamp (Parallax Inc.) or Rabbit (Rabbit Semiconductor) integrated modules. They both come with a powerful processor and an extensive set of peripherals surface-mounted on a DIP packaged mini-board.

7.3 System Resources

Working in support of the local system controller we have a number of utility ICs that form the *system resource package* (Figure 7.8). In a bid to simplify the board layout and minimize microprocessor overhead, we chose to interconnect the ICs using the two-wire I^{2}C bus.

7.3.1 I^{2}C Controller

Since neither the single chip nor the expanded processor modules discussed earlier have an I^{2}C port, the management of the I^{2}C bus is done by a dedicated controller U2 (PCA9564, Philips). On the processor side the PCA9564 has an 8051-type interface with data lines D[0 . . 7] and RD~/WR~ signals controlling the read and write operations. The chip and its registers are addressed using the chip select line (I2CCS~) and two address lines (A0, A1). Internally the

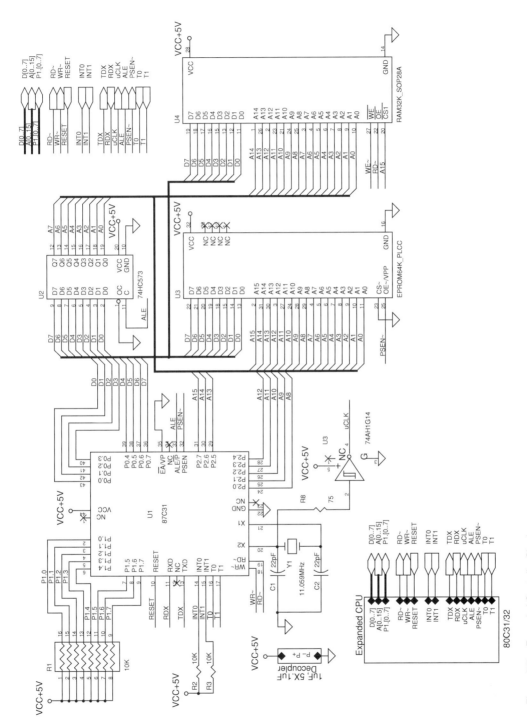

Figure 7.7 Expanded Controller Circuit.

147

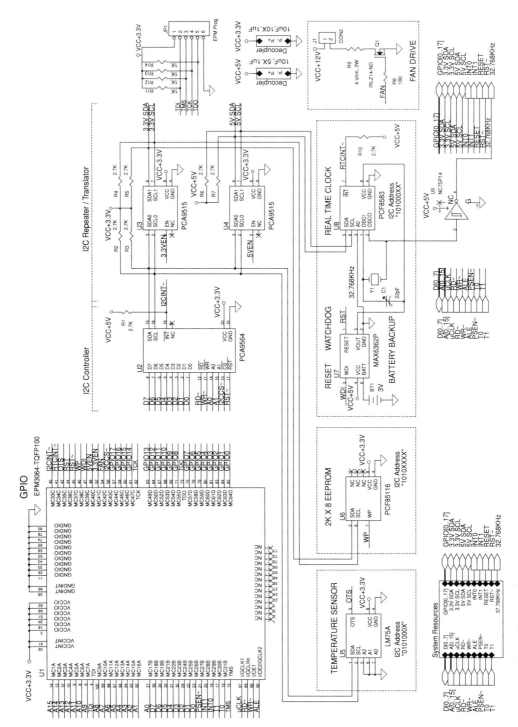

Figure 7.8 System Resource Package.

SDA and SCL lines are generated (transmitted) or processed (received) by a state machine timed by a local clock.

In master mode the bit rate is programmable between 36 Kbps to 330 Kbps, while in slave mode it automatically synchronizes to the master rate up to its 400 Kbps limit. Although powered from a 3.3 V source, the I/O lines of the PCA9585 are 5 V tolerant which, in our case, enables us to use the chip with a 5 V microprocessor module. The operating voltage for the I^{2}C bus is 3.3 V as determined by the voltage source to which the pull-up resistors are connected.

7.3.2 I^{2}C Repeater/Translator

When designing a general purpose utility package we must assume there will be other devices, external to the *system resource package*, connected to the I^{2}C bus. Some of them will be 3.3 V and some 5 V. To accommodate them we create two additional external I^{2}C loops using the U3 and U4 repeaters (PCA9515, Philips).

Besides extending the range of the I^{2}C bus to another 400 pF worth of devices, the PCA9515 can also translate the 3.3 V SDA and SCL signals to other voltage levels. We use this capability of U4 to drive the 5V external I^{2}C network. A convenient enable (EN) input also allows us to use I^{2}C devices with overlapping or identical addresses. We simply place them on different loops that are then individually enabled.

7.3.3 Nonvolatile Memory (EEPROM)

Most systems capture, generate, or otherwise use data that is important for their proper operation or their desired function. User settings are one example of such data; we always want to have the same resolution, luminosity, color, and contrast on our computer screen every time we turn it on. We therefore need a moderate capacity and speed memory device, immune to power fluctuations or power loss.

We chose for this function a 2KX8 I^{2}C EEPROM device (PCF85116, Philips). The chip generates its own timing and programming voltages, making it very easy to use. Data protection measures include the use of redundant EEPROM cells, and an external write protect (WP) input.

7.3.4 Reset/Watchdog/Battery Backup

Insuring the proper startup and operation of the controller software is the task of the microprocessor supervisory IC, U7 (MAX6362P, Maxim). On power-up it holds the reset (RST) line high for about 150 ms, enough for all the internal CPU registers to stabilize. Then it monitors the activity on its watchdog input line (WDI) and if it does not sense at least one logic transition within the watchdog period (1 to 2.25 seconds), it starts a new reset cycle (RST high).

For its part, the controller software must ensure that there is at least one logic transition on the WDI line within each watchdog period. If the software gets stuck in a loop or a transient nudges the program counter, the WDI line will remain low or high for longer than the watchdog period, and the watchdog circuit will reset the system.

The MAX6362P also has a voltage monitoring function. If the supply voltage VCC + 5V falls below a factory-preset threshold value (4.5 V, for example), the chip will generate a new reset sequence. When the power voltage drops below the battery voltage VBATT an internal analog switch connects VOUT, normally tied to VCC + 5V, to the battery. Therefore any circuit powered by VOUT will continue to operate. Such circuits may be the system RAM or a *real-time clock*.

7.3.5 Real-Time Clock (RTC)

Time and date stamping of video are almost universal. All camcorders, digital cameras, and VCRs have built-in real-time clocks. So do most studio and security monitoring equipment. Implementation of the time and date function involves the use of an RTC chip U8 (PCF8583, Philips) with an external clock grade crystal Y1 (32.768 KHz) and, if accuracy is very important, a tuning capacitor C1.

The PCF8583 provides 12- and 24-hour formats, a four-year calendar, and an alarm function. With minimal additional logic you can use the alarm signal (INT~ line) to control the shutdown input of a voltage regulator (LT1129, Linear Technology). This will wake up the system connected to the voltage regulator, which can be a battery-powered nature cam or a telemetry device. The most important auxiliary feature the PCF8583 has to offer is its 240 bytes of battery-backed RAM. It has the advantage of allowing us continuous storage of important system information, while if we update the EEPROM once every second we would exhaust its 1,000,000 write cycles limit in about 278 hours.

If you look at the address range of the RTC (U8) and compare it with that of the EEPROM you will notice that they overlap. To resolve this conflict we connected the RTC on the external 5 V I²C loop and the EEPROM on the external 3.3 V I²C loop. Since the corresponding repeaters can be individually enabled, we can access the RTC and EEPROM separately without any address conflict.

7.3.6 Cooling Fan with Temperature Sensor

Video circuit designs feature many wide buses operating at high speeds. With so many lines working at relatively high frequencies, video equipment tends to run hot. As a result, temperature sensors and cooling fans are becoming a necessity for almost all video products.

Besides acting as a thermometer, the temperature sensor has to convert the temperature data to a digital format. U5 (LM75A, Philips) is such a digitizing sensor. Over the –55°C to 125°C range, the 11-bit analog-to-digital converter of the LM75A delivers an impressive 0.125°C resolution.

When the measured temperature exceeds a user-programmable threshold the LM75A issues an interrupt (OTS) to the local controller. Then the controller software turns on the FAN I/O bit and the associated FET driver powers an external exhaust fan, thus completing a "thermal watchdog" feedback loop.

7.3.7 General Purpose Decoder and IO

To tie all loose ends we need an address decoder, IO, and miscellaneous logic processor chip. We found the EPM3064 (U1) from Altera to be ideal for the task. The functions that the EPM3064 performs within system resources are address decoding for U2, implementation of parallel ports WDI, WP, 5VEN, and FAN, and bit inversion of RTS and 5VEN to obtain RTS~ and 3.3VEN, respectively.

Externally, the EPM3064 can be used as a general purpose resource to implement any logic function that the overall system may require. A total of 18 lines GPIO[0..17] are available for this purpose. Programming of the EPM3064 is done using the JP1 header, a ByteBlasterMV cable from Altera, and their free Max+Plus II BASELINE design software. If this is the only circuit the ByteBlasterMV is connected to, then TDI, TDO, TCK, and TMS need to be pulled up to VCC +3.3 V with 1K resistors.

7.4 Host and Slave Interfaces

7.4.1 RS-232/422/485 Driver Circuits

The most common interface between the local controller and a host PC or a slave device is the serial RS-232. The protocol consists of a start bit (0), seven data bits, a parity check bit, and a stop bit (1) (Figure 7.9). Signal levels are +12 V (actually +5 V to +12 V) for logic 1 and –12 V (–5V to –15V) for logic 0.

Besides transmit (TDX) and receive (RDX) lines, a complete RS-232 interface also has a number of handshake lines. Request to Send (RTS) instructs the receiver to prepare to receive data, Clear to Send (CTS) tells the transmitter that the receiver is ready to receive, Data Terminal Ready (DTR) informs peripherals that the transmitter is active, Data Set Ready (DSR) tells the controller the receiver is active, and Data Carrier Detect (DCD) tells the controller that a peripheral modem detects the carrier signal of a remote transmitter.

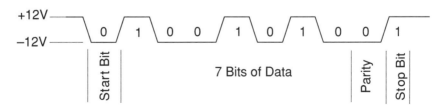

Figure 7.9   RS-232 Protocol.

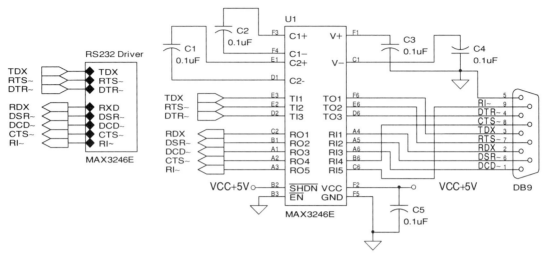

Figure 7.10   Full RS-232 Interface.

In most applications, the RS-232 protocol is implemented by a stand-alone UART device or a UART functional block inside a microcontroller. Voltage translation from the logic levels of the UART to +/–12 V is done by dedicated RS-232 interface chips, such as the MAX3246E from Maxim (Figure 7.10). The MAX3246E operates from a single voltage source and uses internal DC-to-DC converters (charge-pumps) to generate the +/–12 V levels.

With the important exception of modem interfaces, hardware handshakes are rarely used. Information is simply sent by the transmitter on the RS-232 link and when a complete data unit is transmitted the UART of the receiver proceeds to interrupt the local controller, which then retrieves the data. In this case the handshake lines are wired in loop-back mode with the RST~ tied to CTS~ and DTR~, DSR~, and DCD~ connected together (Figure 7.11).

For higher data rates or longer communication lines we can keep the RS-232 protocol but change the cable driver from the +/–12V single-sided arrangement to a 5 V differential line driver. Using drivers, receivers, and cables compliant with the RS-422/485 standard we can attain data rates of up to 10 Mbps at 1.2 meters or a range of 1200 meters at 100 Kbps. The RS-422/485 transceiver shown in Figure 7.12 uses the MAX1481 IC from Maxim which, if connected to a UART and two controller I/O lines, can implement both the RS-422 and RS-485 links.

The only difference between the two is that the RS-422 drivers and receivers are always *on,* making it a point-to-point standard, while the RS-485 has tri-state control over the drivers, which makes it suitable for multidrop networks. Most current microprocessors have at least one built-in UART, with some having two or more. Basic UARTs generate the necessary RS-232 clock rates (baud rates), implement the serial RS-232 protocol, and interrupt the

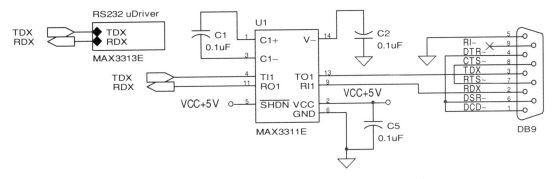

Figure 7.11   Minimal RS-232 Interface.

controller when a data unit is transmitted or received. More advanced UARTs offer much more. The SC16C554 from Philips, for example, is a quad UART that can operate at data rates as high as 5 Mbps.

Internal 16-byte transmit and receive FIFO buffers give the controller the resources needed to read and write serial data in fast bursts instead of servicing single-byte interrupts. A built-in modem manager frees the local controller from handshaking duties while the IrDA encoder and decoder interface directly with most infrared transceivers.

7.4.2 IrDA Interface and Controller Circuit

One method of doing away with cumbersome cables and dongles between your equipment and the host is to use an infrared short range link. The IrDA (Infrared Data Association) standard offers such an option. Although the range of IrDA-compliant equipment is only about one meter, it supports speeds from 115.2 Kbps to 4 Mbps. Such rates are sufficient to download picture files, connect to live audio, and even download MPEG video.

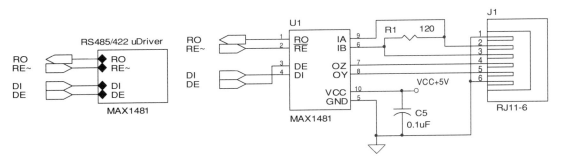

Figure 7.12   RS-422/RS-485 Interface.

The timing of the IrDA encoder or decoder is controlled by an IrDA clock with a frequency 16 times the data rate. IrDA encoding consists in generating a pulse at the center of each "0" data bit, while staying at 0 for all the "1" bits. Each output pulse starts with the seventh IrDA clock after the falling edge of the input, and lasts for three clock pulses [see Figure 7.13(a)]. To decode IrDA data we generate a pulse 16 clocks in length for each input pulse three clock pulses long [see Figure 7.13(b)]. The algorithm remains the same for positive polarity IrDA inputs, as can be seen in see Figure 7.13(c).

A stand-alone encoder/decoder IC is the TIR1000 from Texas Instruments (Figure 7.14). The Sharp IrDA transceiver GP2W1001YP consists of a high-frequency LED IR source and a matched IR receiver. Together they form an IrDA interface that can be connected to most standard UARTs.

As we can see from the timing diagrams in Figure 7.13, the IrDA protocol is simple enough to be easily embedded in CPLD/FPGA or microprocessor firmware. A dedicated 8-pin PIC or Atmel microprocessor is ideal for this task, although not at a 4 Mbps throughput.

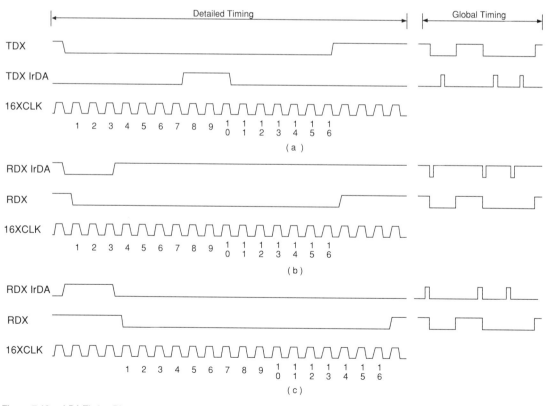

Figure 7.13   IrDA Timing Diagram.

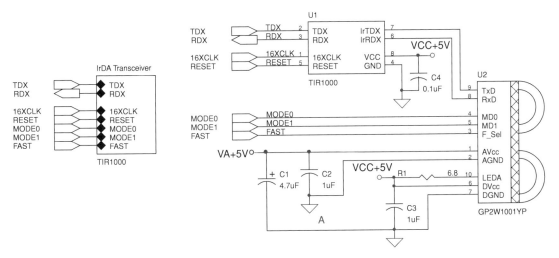

Figure 7.14 IrDA Transceiver.

Some more recent UARTs incorporate the IrDA encoder/decoder as an internal resource. In the SC16C554, for example, any of the four UART channels can be connected to the IrDA processor.

A communication interface package built around the SC16C554 is shown in Figure 7.15. In this circuit the first serial port (A) is wired in full handshake mode to accommodate an anticipated external modem. If port A is connected to a two-line RS-232 device, the IC will check for handshake protocols, and if none present it will ignore the corresponding lines. Ports B and C are operated as two-line serial ports with the handshake lines in loop-back mode.

The A and B channels are connected to RS-232 drivers; A to a five inputs–three outputs MAX3246E and B to a one input–one output MAX3311E.

For compatibility with industrial equipment channel C, supplemented by two I/O lines, RE~ and DE, implements an RS-422/485 port. The interface IC is the MAX1481 differential driver/receiver, with the RE~ and DE inputs independently enabling the receiver and transmitter, respectively. The last channel, channel D, is internally connected to the IrDA encoder/decoder and externally drives the Sharp GP2W1001YP IrDA transceiver.

Different SC16C554 variants support different microprocessor interface standards. To maintain compatibility with 80C51 family we use the SC16C554IB80. Operationally, the controller can manage the UARTs either by polling the *transmit register ready* (TXRDY~) and *receive register ready* (RXRDY~) outputs, or by servicing UART interrupts.

Note that each UART issues its own active high interrupt, which requires a bit of logic to convert to the single, active low interrupt used by the 8051. This can be easily accommodated by the MAX3064 in the system resources pack.

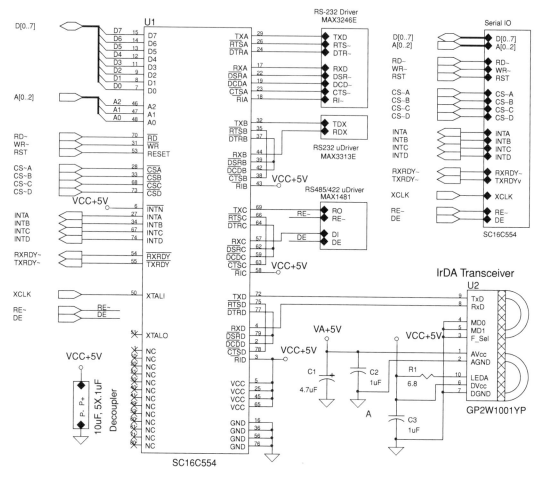

Figure 7.15   Serial Communication Package IrDA, RS-232, RS-422/485.

7.4.3 USB Interface and Controller Circuit

None of the standards covered so far can accommodate the bandwidth needed to transport real-time video, even when the video is MPEG compressed. The required step-up in transfer rates is provided by the *Universal Serial Bus* (USB) interface. Although the USB 2.0, also known as High-Speed USB, supports data rates up to 480 Mbps, controllers that operate at that speed are hard to find and even harder to use. However, designs in the 100 Mbps range can readily be implemented with currently available single-chip USB controllers.

First, let us take a brief look at the standard. The physical layer of USB consists of a 3.3V bidirectional differential line with three states: J, K, and SE0 (Single Ended Zero). The "J" state corresponds to a logic 1 and, for a high-

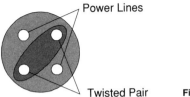

Power Lines

Twisted Pair

Figure 7.16   USB Cable.

speed (480 Mbps) or full-speed device (12 Mbps), is characterized by a differential voltage between D+ and D– larger than 200 mV. For a low-speed USB device (1.5 Mbps) the roles of U+ and U– reverse and the J state is associated with a differential voltage between U– and U+ larger than 200 mV. Repeated J states also indicate an "idle" condition when no transactions take place on the USB bus. The "K" state corresponds to a logic 0 and is the reverse of the J state; for high-speed and full-speed devices $U_+ - U_- < -200$ mV, for low-speed ones $U_+ - U_- > 200$ mV. The SE0 state, when both U_+ and U_- are close to ground, is used to signal the end of a packet (EOP).

The maximum length for a USB 2.0 cable is five meters and the maxim number of peripherals you can run on a single bus is 127, although the overall data rate is split accordingly. Mechanically, the USB cable consists of a shielded, twisted pair that carries the D_+ and D_- signals, and two power lines for +5V and ground (Figure 7.16).

The USB data stream is NRZI (non-return to zero inverted). This allows for easier clock extraction and synchronization. The encoding scheme is straightforward [Figure 7.17(a)]. For a 0-input bit the output executes a logic transi-

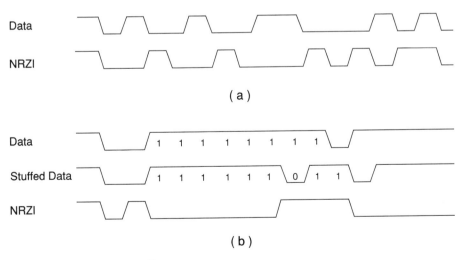

Figure 7.17   USB–NRZI Timing Diagram.

tion from the current state to the opposite state. There are no transitions if the input is logic 1.

If there are seven or more consecutive 1's at the input, the encoder is required to force a 0 bit into the sequence (after the sixth logic 1 bit), a process known as bit "stuffing" [Figure 7.17(b)]. This guarantees a transition in the NRZI output at least every seventh input period, which is essential for reliable clock synchronization. We should reiterate that in this instance an output 1 corresponds to a USB J state and a 0 to a K state.

Under the USB, standard information is grouped in packets, each packet preceded by a *sync pattern* (SP) and followed by an *end of packet* (EOP) marker (Figure 7.18). The sync SP consists of seven 0 bits followed by a 1 bit (all NRZI encoded) while two SE0 states followed by an idle forms the EOP marker.

Within each packet we find an 8-bit packet identifier (PID) and, depending on the nature of the packet, a data payload and a *cyclic redundancy code* (CRC) error check segment. The first four bits of the packet identifier form the *PID* code which describes the type of the packet being transmitted or requested. The last four bits are the 1's complement of the PID code and are used for error detection.

USB uses four types of packets: *tokens, data packets, handshake packets*, and *preamble*. Their functions and PID codes are summarized in Table 7.1. All tokens are issued by the host to set up USB devices for impending bus operations. The *start of frame* (SOF) token attaches a marker and an ID number to the subsequent frame (sequence of packets).

A *setup token* contains the *address, endpoint*, and *control command* for a target USB device. An *endpoint* is the USB equivalent for an internal control or status register. The *IN* and *OUT tokens* identify the source or recipient USB device for the next *data packet*.

The Data 0 and Data 1 packets are identical in all respects except for their PID codes. Checking the codes (odd or even) before and after data transmissions is used as another error-protection measure. Data packets can be either

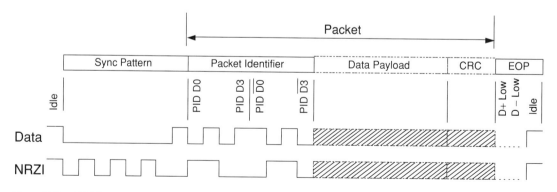

Figure 7.18   USB Packet.

TABLE 7.1   USB Packet Types

Type	Name	PID	Payload Content
Tokens	Start of Frame	0101	Frame Number
	Setup	1101	Sets up USB Device w Address / Endpoint
	OUT Packet	0001	USB Device Address / Endpoint
	IN Packet	1001	USB Device Address / Endpoint
Data	Data0	0011	Data Payload Even Parity PID
	Data1	1011	Data Payload Odd Parity PID
Handshake	Acknowledge	0010	
	No Acknowledge	1010	
	Stall	1110	
	Preamble	1100	Idle Mode Delay for Low Speed USB

up to 64 bytes for standard transfers or up to 1023 bytes for periodic (isochronous) transfers.

The success or failure of a transmission is reflected in the *handshake packet* issued at the end of a standard transaction. An *acknowledge packet* confirms an error-free transaction; a *no acknowledge* flags a transmission error or a "device busy" condition, and a *stall packet* indicates a target device problem.

The way that different packets can combine into complete transactions is shown in Table 7.2. How does a device know it is engaging in an isochronous data transfer? The OUT token addresses its isochronous endpoint instead of its standard endpoint.

Finally, multiple transactions are combined by the host into frames and are placed on the USB bus, from where individual USB devices retrieve and execute their allocated commands (Figure 7.19).

Although its performance is impressive, the key to the success of the USB bus is its affordability. To make it inexpensive, most of the functions on the device are performed by as single USB IC. This makes the USB bus a practical choice for consumer electronics devices like mice, keyboards, and MP3 players.

TABLE 7.2   USB Transactions

		Host			USB Device			Host	
IN, No Errors	Sync	IN Token	EOP	Sync	Data up to 64 bytes	EOP	Sync	ACK	EOP
IN, Data Errors	Sync	IN Token	EOP	Sync	Data up to 64 bytes	EOP			
IN, Device Busy	Sync	IN Token	EOP				Sync	NACK	EOP
IN, Device Error	Sync	IN Token	EOP				Sync	STALL	EOP
IN, Isochronous	Sync	IN Token	EOP	Sync	Data up to 1023 bytes				EOP

		Host			USB Device			USB Device	
OUT, No Errors	Sync	OUT Token	EOP	Sync	Data up to 64 bytes	EOP	Sync	ACK	EOP
OUT, Data Errors	Sync	OUT Token	EOP	Sync	Data up to 64 bytes	EOP			
OUT, Device Busy	Sync	OUT Token	EOP				Sync	NACK	EOP
OUT, Device Error	Sync	OUT Token	EOP				Sync	STALL	EOP
OUT, Isochronous	Sync	OUT Token	EOP	Sync	Data up to 1023 bytes				EOP

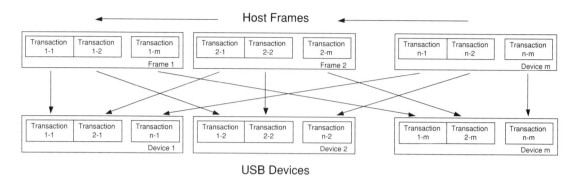

Figure 7.19   USB Frame Sequence.

Figure 7.20 illustrates the structure of a full- or high-speed USB controller. It is obvious that when dealing with 100 Mbps serial streams we cannot rely on a microprocessor for any meaningful mediation function. Even when deserialized in word-width data (16 bits), the required processing bandwidth is beyond the capabilities of most DSPs, let alone microprocessors.

The function of a USB controller is to retrieve information from a local data source (memory, ADC, and so on) and convert it to a USB-compatible

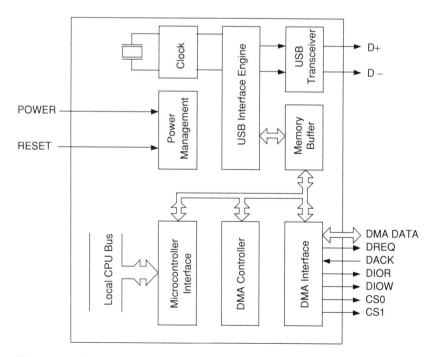

Figure 7.20   USB Controller Block Diagram.

data stream, under the supervision of a local system controller. It also must do the reverse, to convert USB data from the host to a standard logic format and deliver it to local memory storage, DACs, and so on.

For video applications, the ability to transfer large amounts of data from a local buffer memory (cameras, VCRs . . .) to the host is very important. An efficient way to achieve it is to use a *direct memory access controller* (DMA) to read bursts of information from the external memory and place them in an internal *memory buffer*. From there the *USB interface engine* converts it to USB format and outputs it to USB transceivers. To use it in products powered by the USB bus, the USB controller has a *power management unit* that steps down the +5 V level found on the bus to 3.3 V, the level at which the controller itself operates. A limit of 500 mA is allowed for the total current that can be drawn from the bus.

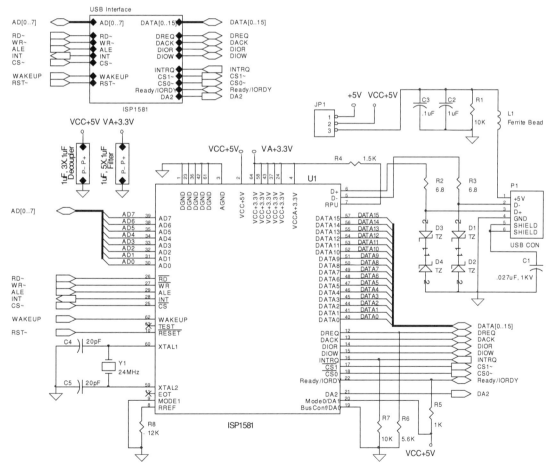

Figure 7.21   USB Controller Circuit.

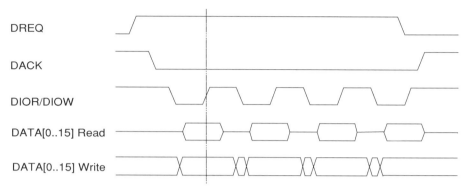

Figure 7.22   USB DMA Timing.

An example of a USB 2.0 controller module is shown in Figure 7.21. The module is built around an ISP1521 single chip USB controller from Philips operating in split bus mode; a multiplexed address/data bus interfacing the IC to the local controller, and a separate DMA data bus interfacing it to the data source. The microcontroller bus type is selected by the *mode* pin, with 0 configuring it for an Intel bus. Maximum data rates are 12.8 Mwords per second (204.8 Mbps) on the DMA bus and 12.5 Mbytes per second (100 Mbps) on the microcontroller bus. Internally the ISP1521 maintains eight IN and eight OUT endpoints, including the control endpoints, and supports automatic detection of the *high-speed* mode.

The ISP1581 can be programmed to enter a minimal power *suspend state* from which it recovers when it detects a low-to-high transition on the *wakeup* input. This is essential if the unit is powered by the USB bus (JP1 setting). The DMA interface supports a number of different modes of operation, one of which is shown in Figure 7.22. Using it to access standard video memory requires a minimal but fast external *DMA manager* or CPLD/FPGA state machine.

In normal operation, when the USB Controller requests a data transfer transaction it raises the DREQ line, which is followed by the DMA device acknowledging it on the DACK line. Next the DMA device outputs a burst of data in sync with the DIOR (read) output of the internal DMA Controller. The sequence for writing data to the DMA device is similar but uses the DIOW output instead of the DIOR. The transfer ends when the DMA Controller drops its DREQ line. The IORDY and DA2 signals can be used to access ATA/ATAPI devices like ZIP drives and memory cards.

7.5 User Interface

The operation of a stand-alone device can be controlled either through a host computer and its GUI (Graphical User Interface) software, or by means of its front panel. In this section we will turn our attention to the menu of options one might consider when designing a front-panel user interface for video equipment.

Front panels are relatively simple assemblies. They consist mostly of displays and switches, but they do have to accommodate the idiosyncrasies of the respective application. Video equipment, for example, is used mostly in the dark and in many instances as part of large racks chock full of similar boxes. It then makes sense to use active displays such as LEDs or LCDs with strong backlights. As for keypads and switches they all must be well-lit and provide tactile feedback.

7.5.1 Segment Displays

The simplest character display you can use is the seven-segment LED. It requires a direct drive which can be easily implemented with a 74XX374 register/driver. However, in Figure 7.23 we selected the PCF8574 I^2C LED driver instead. Since the display is not changing very often, the slower two-wire I^2C interface should not be a problem. The PCF8574 can also be programmed to "blink" the LED segments, each at a different rate. This distinctive feature can be used to signal error conditions or other operational codes.

7.5.2 Matrix LED Displays

If a seven-segment LED is not sufficient, the matrix LED represents the next step up. 5×7 LED alphanumeric displays have long been associated with high-overhead microprocessor multiplexing. With the MAX6953 this problem disappears. Not only does it multiplex up to four 5×7 common cathode LED matrices without any controller assistance, but it also has a built-in character generator (104 predefined Arial and 24 user-programmable), can automatically blink between two messages, and allows the user to separately program the brightness of each matrix. You can use multiple MAX6953 ICs to drive larger, single-color arrays or to drive dual-color or white (RGB) LED matrices, all under I^2C control.

7.5.3 LCD Displays

Alphanumeric LED displays are not always the best choice. They are rather large, relatively expensive, and use a significant amount of power. If your design requires higher-density graphics or multiline menus, a better and more economical choice is an LCD module. With all the difficult electronics handled inside the module, the interface is a simple parallel scheme that only needs some I/O bit-banging.

A few smaller companies offer I^2C-driven LCD panels, both alphanumeric and graphic, and some even include a keypad encoder on the LCD controller board (Matrix Orbital Corporation).

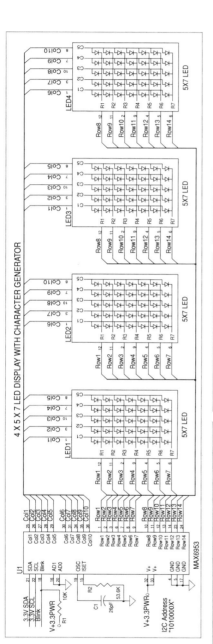

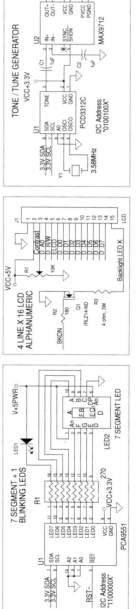

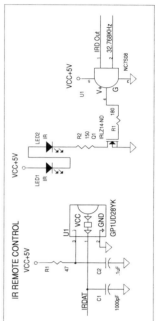

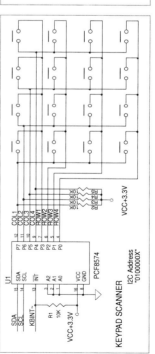

Figure 7.23 General Purpose User Interface Panel.

7.5.4 Keypad Encoder

If you need a separate keypad encoder that operates on the I^2C bus, the PCF8574 8-bit I/O expander offers a partial solution. You have to scan the rows but you will get an interrupt if there is a change in inputs. This reduces somewhat the software overhead by simplifying the scanning algorithm. A better approach is to implement the I^2C keypad protocol using one of a number of inexpensive microcontrollers. As a matter of fact, since the algorithms to encode keypad switch closures and to control LCD modules are quite simple, both drivers can be implemented by a single-chip 8051 type device, programmed as an I^2C slave peripheral. Not only is the unit cost of this solution much lower, but you can customize it much closer to your panel (LCD type, keypad size, and so on).

7.5.5 Tone/Tune Generator

For audible signaling in a dark environment or anyplace where the operator cannot readily see the panel, we can use a tone generator. The I^2C compatibility of the PCD3312C makes it a logical choice. The PCD3312C can synthesize two octaves of musical scales in semitone steps, and can also generate DTMF tones for a bonus "dial-out" capability. Distinctive sounds can be used to identify the source and nature of a problem (loss of signal, for example) without posting warning text on the output video. In Figure 7.23, the output of the PCD3312C drives a 0.5 W D-class amplifier from Maxim (MAX9712), which we selected for its good performance and minimal heat dissipation characteristics.

Video Decoders, Graphic Digitizers, and Deserializers

Before any video processing can take place, the input signals have to be translated into a digital format compatible with the internal structure of the system. Most video processors operate only with parallel digital data. If the input is analog, then, we have to convert it to digital; if the input is serial digital, we must deserialize it, and so on. Converting an analog signal to digital is governed by a set of mathematical principles that we need to understand in order to make full use of the IC building blocks that implement such conversions.

8.1 Theory of Signal Sampling

The cornerstone of electronic engineering is the duality between the behavior in time of electrical signals and their frequency spectrum. The mathematical connection between the two is embodied in the definition and properties of the *Fourier series*. If $x(t)$ is a periodic but otherwise arbitrary function of time t, with a period T, it can be shown that $x(t)$ can be reconstructed by the superposition of an infinite number of sinusoidal waveforms. If we describe a sinusoid using Euler's formula

$$e^{j\omega t} = \cos(\omega t) + j\,\sin(\omega t)$$

then the expansion of the function $x(t)$ in sinusoidal (also referred to as *harmonic*) components is given by

$$x(t) = \sum_{m=-\infty}^{m=+\infty} c_m e^{jm\omega_s t}$$

where the coefficients c_m are given by

$$c_m = \frac{1}{T} \int_{\frac{T}{2}}^{\frac{T}{2}} x(t)e^{-jm\omega_s t} dt$$

The above pair of mathematical expressions defines the Fourier series.

If we calculate, for example, the Fourier series of an infinite periodic sequence of pulses with amplitude A, period T, and width τ (Figur. 8.1), we find

$$c_m = \frac{1}{T} \int_{-\frac{T}{2}}^{\frac{T}{2}} x(t)e^{-jm\omega_s t} dt = \frac{A\tau}{T} \frac{\sin\left[\dfrac{m\pi\tau}{T}\right]}{\dfrac{m\pi\tau}{T}} = \frac{A\tau}{T} \mathrm{sinc}\left[\frac{m\pi\tau}{T}\right]$$

The c_m coefficients are the amplitudes of the sinusoids that comprise the $x(t)$ waveform. The c_m set of values forms the frequency spectrum of $x(t)$ (Figure 8.2). In the above expressions, f is the frequency of $x(t)$ while $\omega_s = 2\pi f$ is defined as its angular frequency.

Very important to our analysis is the *unit impulse* or Dirac function. By definition, the unit impulse function is infinite at the origin and 0 elsewhere (Figure 8.3).

In principle the $\delta(t)$ function can be thought of as a square pulse of unit area with a forever narrowing width. If we build up a pulse function $P(t)$ as a periodic sequence of δ functions

$$P(t) = \sum_{m=-\infty}^{m=+\infty} \delta(t - nT)$$

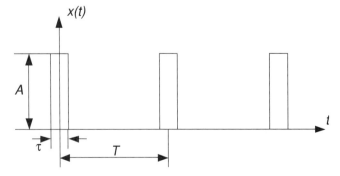

Figure 8.1 $x(t)$ Pulse Sequence.

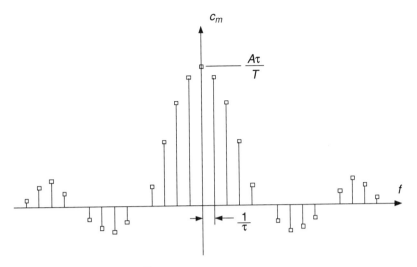

Figure 8.2 *X(f)* Frequency Spectrum.

its associated Fourier coefficients are calculated to be

$$c_{mp} = \frac{1}{T} \int_{\frac{T}{2}}^{\frac{T}{2}} \delta(t)e^{-jm\omega_s t} dt = \frac{1}{T}$$

Then *P(t)* can be written as

$$P(t) = \frac{1}{T} \sum_{m=-\infty}^{m=+\infty} e^{jm\omega_s t}$$

As you may already have guessed, *P(t)* will serve as a sampling function in our image digitizing process. However, the video images themselves are not periodic, and therefore cannot be described in terms of a Fourier series.

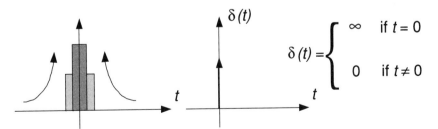

Figure 8.3 *x(t)* The Dirac Function.

The equivalent mathematical tool for nonperiodic functions is the *Fourier Transform* (FT), which is arrived at from the Fourier series equations by letting the period T go to infinity. Then the frequency spectrum of a nonperiodic function $x(t)$ is found by calculating the integral

$$F[x(t)] = X(f) = \int_{-\infty}^{+\infty} x(t)e^{-j2\pi ft}dt = \int_{-\infty}^{+\infty} x(t)e^{-j\omega t}dt$$

The *Inverse Fourier Transform* (IFT) recovers the time function $x(t)$ if its associated spectrum or Fourier Transform function is known.

$$x(t) = \int_{-\infty}^{+\infty} X(f)e^{j2\pi ft}df = \frac{1}{2\pi}\int_{-\infty}^{+\infty} X(\omega)e^{j\omega t}d\omega$$

Some of the FT properties with important consequences in signal analysis are the following:

- If $X(f)$ is the Fourier transform of $x(t)$ then the transform of a time-shifted function $x(t - \tau)$ is given by $e^{-\omega\tau}X(f)$

- $F\left[\dfrac{d^n}{dt^n}x(t)\right] = (j\omega)^n X(f)$

- $F\left[\displaystyle\int_{-\infty}^{t} x(\tau)d\tau\right] = \dfrac{X(f)}{(j\omega)^n}$

The Fourier transform $|X(f)|$ of a finite energy signal $x(t)$ is also finite. The boundaries $-f_B$ to f_B of $|X(f)|$ define the absolute bandwidth of x (Figure 8.4).

Let us introduce the sampled function $x_s(t)$; we obtain it by multiplying the arbitrary function $x(t)$ by a sampling sequence of δ functions $P(t)$ (Figure 8.5). For our purposes, $x(t)$ represents an arbitrary analog video signal, $P(t)$ the

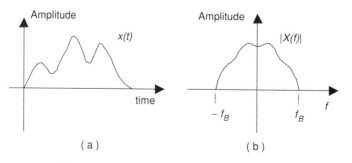

(a)   (b)

Figure 8.4   Spectrum of a Nonperiodic Function.

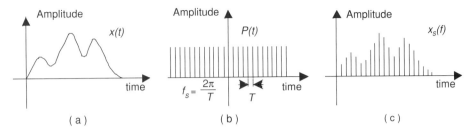

Figure 8.5   Signal Sampling Process.

sampling process by which digitizers acquire measurements of $x(t)$, and $x_s(t)$ the resulting output sequence of video samples.

$$x_s(t) = x(t)P(t) = \sum_{n=-\infty}^{n=+\infty} x(nT)\delta(t - nT)$$

If we replace $P(t)$ by its Fourier series expansion we get

$$x_s(t) = x(t)P(t) = \frac{1}{T} x(t) \sum_{m=-\infty}^{m=+\infty} e^{jm2\pi f_s t} = \frac{1}{T} \sum_{m=-\infty}^{m=+\infty} x(t) e^{jm2\pi f_s t}$$

8.1.1 Frequency Spectrum of Sampled Signals

Now let us apply the Fourier transform to both sides of this equation and make use of the first FT property

$$F[x(t)P(t)] = \frac{1}{T} \sum_{m=-\infty}^{m=+\infty} F[x(t)e^{jm2\pi f_s t}] = \frac{1}{T} \sum_{m=-\infty}^{m=+\infty} X(f - mf_s)$$

where f_s is the sampling frequency, the inverse of the sampling period T.

What this tells us is that $X_s(f)$, the spectrum of $x_s(t)$, consists of a scaled copy of the original spectrum $X(f)$ plus a multitude of scaled copies of the $X(f)$ shifted by multiples of the sampling frequency f_s (Figure 8.6).

If we want to recover $x(t)$ from $x_s(t)$ all we need to do, aside from a gain factor, is to remove the so called spectral "images" of $X(f)$ from $X_s(f)$. We can do it by using a low-pass filter with a cut-off frequency between f_B and $f_s - f_B$. We can also see that for the signal recovery to be unique the $X(f)$ spectrum must not overlap any of its spectral images. Therefore $f_s - f_B$ must be equal to or bigger than f_B; in other words, the sampling frequency must be at least twice as large as the largest frequency in the spectrum. This condition is nothing else but a restatement of the well known *Sampling Theorem*. We recognize the critical frequency value $f_N = 2f_B$ as the *Nyquist frequency* limit.

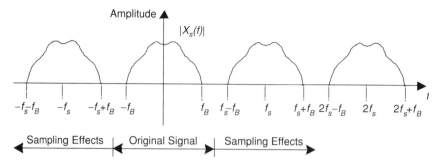

Figure 8.6 Spectrum of a Sampled Function.

8.2 Front-End Functions

8.2.1 Anti-aliasing Filters for Composite, S-Video, RGB, and YPbPr

To accommodate the 4.2 MHz bandwidth of analog video, ADC section for a typical video digitizer uses a sampling rate of 14.31818 MHz, which is four times the frequency of the color subcarrier. The ADC also digitizes any high-frequency signal it finds above 4.2 MHz, something which may result, as we saw in the previous section, in unwanted spectral images. If the spectral images are not removed their time domain components, known as *aliases*, distort the digitized image. A low-pass filter placed close to the input of the ADC that confines the input bandwidth below the Nyquist limit is called an *anti-aliasing filter*. A *reconstruction filter*, by contrast, relates to the conversion of digital signals to analog signals by DACs. They are also low-pass filters and they are placed after the DAC to remove the high-frequency harmonics of the sampling clock (see Chapter 9).

For low-cost applications such as security systems, a simple passive π network can adequately do either job (Figure 8.7). In higher-end applications

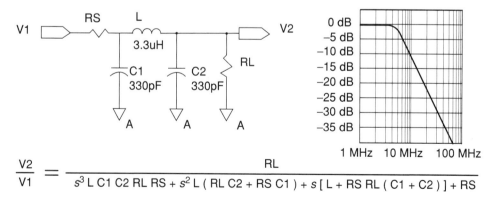

$$\frac{V2}{V1} = \frac{RL}{s^3\,L\,C1\,C2\,RL\,RS + s^2\,L\,(\,RL\,C2 + RS\,C1\,) + s\,[\,L + RS\,RL\,(\,C1 + C2\,)\,] + RS}$$

Figure 8.7 π (Pi) Lowpass Filter (LPF).

where more aggressive filtering is required, active filters, such as the third-order multiple feedback design in Figure 8.8, are needed. As can be seen by comparing their attenuation graphs, the improvement over the passive π network is significant.

For high-performance, standard-definition systems, Maxim offers a sixth order, fully integrated filter (MAX7428) with some noteworthy features. Besides its impressive filter function (see Figure 8.10), this IC offers programmable high-frequency boost to compensate for cable losses, output disable for parallel connection, programmable clamp voltage for use with Y (1 V) or C (1.5 V), bypass function, and a two-channel input multiplexer (Figure 8.9).

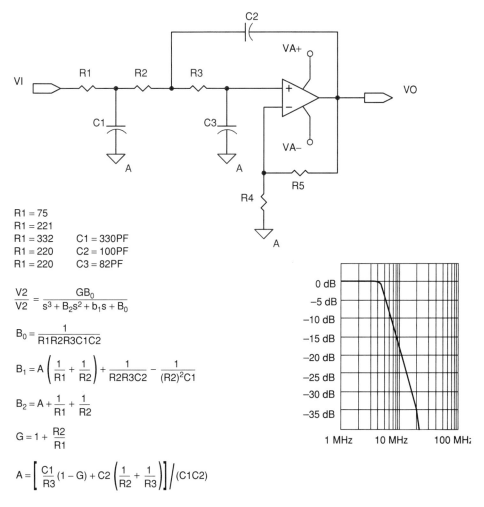

R1 = 75
R1 = 221
R1 = 332   C1 = 330PF
R1 = 220   C2 = 100PF
R1 = 220   C3 = 82PF

$$\frac{V2}{V2} = \frac{GB_0}{s^3 + B_2 s^2 + b_1 s + B_0}$$

$$B_0 = \frac{1}{R1 R2 R3 C1 C2}$$

$$B_1 = A\left(\frac{1}{R1} + \frac{1}{R2}\right) + \frac{1}{R2 R3 C2} - \frac{1}{(R2)^2 C1}$$

$$B_2 = A + \frac{1}{R1} + \frac{1}{R2}$$

$$G = 1 + \frac{R2}{R1}$$

$$A = \left[\frac{C1}{R3}(1 - G) + C2\left(\frac{1}{R2} + \frac{1}{R3}\right)\right] / (C1 C2)$$

Figure 8.8 Third-Order Multiple Feedback LPF.

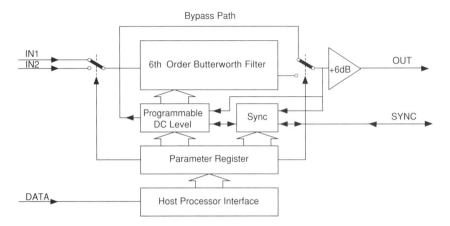

Figure 8.9   MAX7428 Block Diagram.

The MAX7428 is programmed via a single wire interface that, although cumbersome, allows for up to 16 devices to be controlled on a single loop. In color component applications a sync circuit detects the Y-embedded sync pulse and provides it as a reference to the other filters. For proper operation the Y filter is then set as master with its sync line programmed as output, while the other filters are programmed as slaves with their sync lines set as inputs.

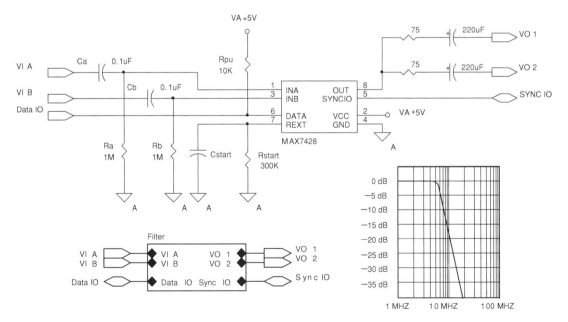

Figure 8.10   LPF Using MAX7428.

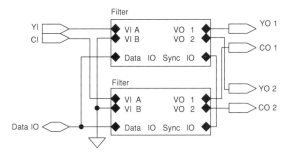

Figure 8.11   S-Video Anti-Aliasing and Reconstruction LPF.

This allows multiple MAX7428s to be used in s-video (Figure 8.11), RGB, and YPbPr designs (Figure 8.12).

8.2.2 Automatic Gain Control

When using analog inputs we must ensure their levels are within the specifications of the standards to which they belong. After going through tens of feet of cabling, the amplitude of the video signal occasionally decreases below acceptable values. To restore it we use an automatic gain control amplifier or AGC (Figure 8.13).

The key component of the AGC is a *variable gain amplifier* (VGA). As its name suggests, the gain of the VGA is proportional to the voltage applied to its gain control pin. In the AGC configuration the output of the VGA drives the input of an envelope detector circuit. The detector output is subtracted from a reference voltage Vref, with the resulting "error" signal being applied to the input of an integrator block. The loop is closed by connecting the output of the integrator to the gain control input of the VGA.

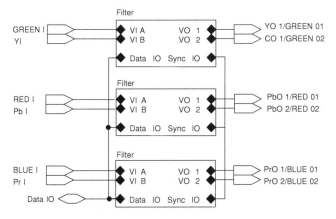

Figure 8.12   RGB, YPbPr Anti-Aliasing and Reconstruction LPF.

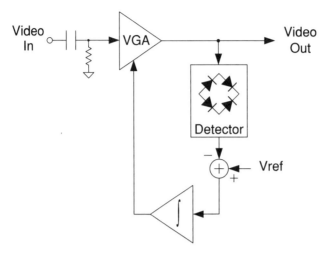

Figure 8.13   AGC Block Diagram.

If the input level drops, the output of the detector becomes lower than the reference voltage and a positive error signal is applied to the input of the integrator. The output of the integrator then ramps up and therefore increases the gain of the amplifier. As the gain increases, the output of the detector raises until it balances the reference voltage. At this point the AGC is in equilibrium. If the signal level increases the process is reversed, with a negative error voltage driving the integrator output and the VGA gain lower until we get back to equilibrium.

8.2.3 DC Restoration

The input AC coupling capacitors produce a different kind of problem. Since the DC level of the video signal varies from frame to frame and line to line, so should the DC bias of the capacitors. However, when we combine the 75 Ω terminator resistance with a commonly used 22 µF coupling capacitor, the time constant of the resulting RC network is 1.65 msec. This is much larger than the 63.5 µsec duration of an NTSC line. The result is that the capacitor DC bias cannot follow the line-to-line changes in luminosity and the scan lines tilt, as shown in Figure 8.14. We can bring the RC time constant in line with the horizontal sync period by using much smaller capacitor values, but this reduces video bandwidth.

A better solution is offered by the DC restore circuit in Figure 8.15. Its function is to reestablish the DC bias of the video signal to a value given by a reference source, in this case the Zener diode Vref.

In operation, during the back porch period of the video signal the sample and hold (S/H) analog switch is closed. This closes the feedback loop of U1 which

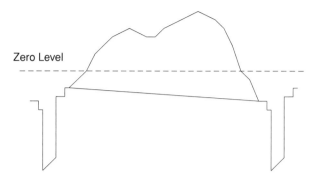

Figure 8.14   Effect of Coupling Capacitor.

acts to equalize the voltages on its + and − inputs. Consequently, this forces the output of U2 to equal Vref. The mechanism that balances the voltages in the U1 feedback loop is the charging of capacitor Chold. When the S/H switch opens Chold is left with no discharge path. Therefore it maintains its voltage which offsets the U2 input signal and biases the U2 output to the Vref level.

All we need is a way to detect the back porch period of the video signal. The circuit that extracts the back porch along with the rest of the timing information from a video stream is the *sync separator* (Figure 8.16).

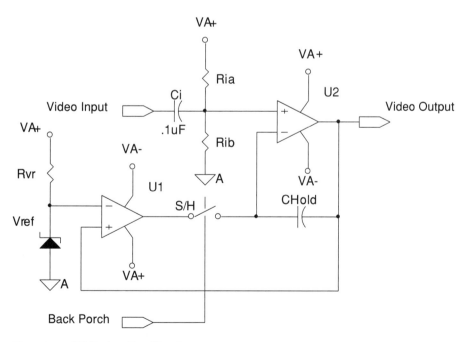

Figure 8.15   DC Restoration Circuit.

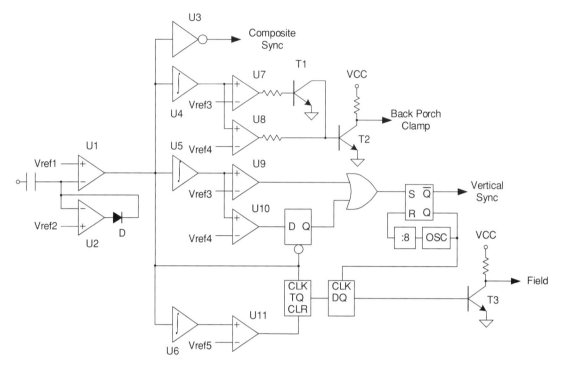

Figure 8.16   Sync Separator Block Diagram.

8.2.4 Sync Signal Separation

To reliably extract the composite sync, the video input is AC coupled and then clamped by amplifier U2, diode D, and reference source Vref2. The level of the clamp is one diode voltage drop below Vref2.

If the input voltage drops below the clamp voltage, the diode opens and the coupling cap is charged to a higher bias point. This effectively fixes the tips of the syncs to the clamp voltage. The composite sync is then extracted by comparing the input voltage to the voltage reference Vref1. U3 converts the inverted sync pulses emerging from U1 to normal sync pulse polarity.

The extraction of the vertical sync and back porch signals relies on using leaky integrators as timing elements. During normal horizontal lines the length of composite sync pulses applied to the input of U5 is too short to bring the output of the integrator within the range of the Vref4 and Vref3 reference sources.

During vertical sync, however, right before the first serration pulse, we have the first negative pulse wide enough for the output of the integrator to go past Vref4 (negative sync pulses become positive after U1). Then the output of the U10 comparator is registered into the D flip-flop by the following serration pulse. The output of the D flip-flop through the OR gate sets the SR flip-flop and drives its output Q high and the vertical sync output low. In turn, Q enables the oscillator OSC which, after eight cycles, brings R high and terminates the vertical sync pulse. If the input vertical sync has no serration pulses or is too long, the integrator passes the Vref2 level and sets the SR flip-flop directly. This triggers a duplicate vertical sync pulse.

To generate the back porch output the integrator U4 starts a new voltage ramp at the trailing edge of each horizontal sync pulse. When the ramp passes the Vref4 level the output of the comparator U8 goes high, turning on transistor T2 and bringing the back porch output low. As the voltage ramp passes Vref3, transistor T1 is turned on. This shunts the base of T2 to ground and terminates the back porch pulse.

The timing for the field output is controlled by the U6 integrator. It is designed not to reach Vref5 if the integration time is half-line or less; however, it drives U11 high if the integration time approaches one full scan line.

The T flip-flop is clocked by the composite sync which includes horizontal sync pulses, equalizing pulses, and serration pulses. However, for each normal line, the T toggle generated by the horizontal sync is cleared before the end of each line by U11. During the vertical interval the higher frequency of the equalizing and serration pulses do not give U6 enough time to drive U11 high and therefore clear the T flip-flop. Q toggles with each pulse and, since we have an even number of such pulses, the T flip-flop status at the end of the vertical interval is the same as at the beginning. What also toggles the T flip-flop is the half- line at the end of the odd field (or field 1). This toggle, registered by the Q output of the SR (the beginning of the vertical sync), determines the output field signal.

Outside the vertical interval, the horizontal sync is the same as the composite sync so it can be generated by gating the composite sync with the complemented VSYNC signal. Inside the vertical interval, HSYNC can be obtained by counting and decoding the composite sync transitions.

8.2.4.1 Sync Separation Circuit with DC Restore

A practical sync separator circuit with built-in DC level restoration is shown in Figure 8.17. It uses Elantec's EL4501 *video front-end* IC, a remarkable sync separator with a 100 MHz bandwidth and an integral data slicer that can be used to retrieve codes embedded both in the active video and video-blanking regions. The data slicer is a comparator that "slices" the video out at a programmable threshold and provides a digital signal corresponding to the segments of the video above (or below) the threshold voltage.

8.3 Video Decoder-Digitizer

The *video decoder-digitizer* is a multifunction, integrated circuit that converts a composite or S-video input into a standard parallel digital format. Far more

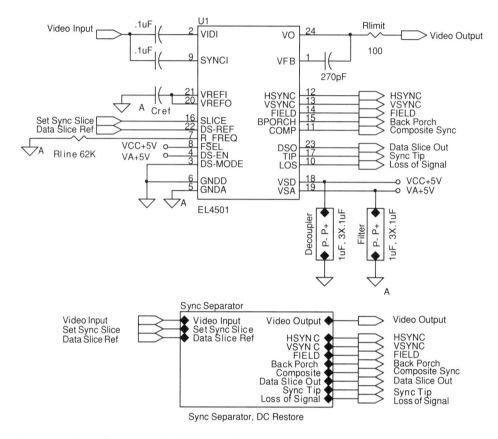

Figure 8.17   Sync Separator with DC Restoration.

than an AD converter, the decoder-digitizer incorporates the complete chain of functional blocks required for the conversion (Figure 8.18).

To start with, most newer decoders include a front *video multiplexer* that enables one unit to digitize video from a number of sources. Some multiplexers are fast enough to capture a single frame from one source, then switch to the next, capture one frame, then switch again. This can be used to build very inexpensive, PC-based security multiplexer cards. They work by sequentially capturing single frames from a number of different cameras and then tiling them together on the PC monitor.

Past the multiplexer, the luminance and chrominance signals are processed by built-in *clamping circuits* and *automatic gain control* amplifiers. Then the Y component is applied to an internal *sync separator* circuit that extracts the horizontal sync and provides it as the reference to the *phase locked loop clock* (PLL) of the decoder. The PLL then generates a clock that is an exact multiple of the line frequency of the input signal. This clock drives not only the

ADCs but also all the internal digital subsystems of the decoder including the state machine of the output timing generator.

The frequencies generated by the PLL vary according to the standard being decoded, selected resolution, pixel format, and so on. Most decoders sample the input luma and chroma at $8f_c$ (28.63636 MHz), where f_c is the frequency of the color subcarrier. The ADCs can be 8 to 12 bits and generally have integrated band-gap references. Since they dissipate significant power, the chroma ADC can be turned off during composite processing. Once in the digital domain the luma and chroma are passed through two consecutive filters. The first is a low-pass anti-aliasing filter designed to limit the bandwidths of the signals below the Nyquist limit: the second is a digital comb filter.

8.3.1 Y/C Separation

The function of the Y/C comb filter is to separate, to the degree possible, the narrow luma and chroma frequency bands from the common portion of the composite spectrum. These bands are due to the regular structures inherent to composite timing. They are caused by the highly periodic horizontal sync pulses

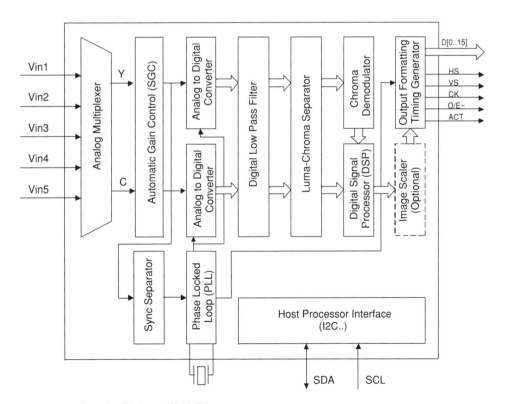

Figure 8.18 Decoder-Digitizer Block Diagram.

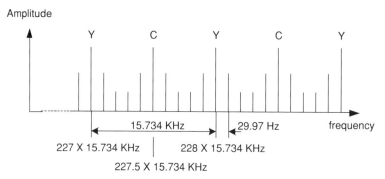

Figure 8.19   NTSC Spectrum—Detail.

and the vertical interval complex of pulses (equalizing, serration) that act as modulators of the analog luma signal.

If f_H and f_V are the frequencies of the horizontal sync and vertical interval pulses, the spectrum will consist of a number of main bands separated by f_H *(15.73425 KHz)* each main band flanked by sidebands spaced at f_v (59.94 Hz) and $f_v/2$ (29.97 Hz). In a 4.2 MHz luma bandwidth there are approximately 267 main bands.

When designing the NTSC standard, the logical place for the color carrier was found to be at the high end of the luma spectrum but in between the luma main bands. The color subcarrier frequency was chosen to be $f_c = 227.5 \times f_H =$ 3.579541875 MHz.

Since the color information is subjected to the same type of sync modulation as the luma, its spectrum exhibits the same type of band structure (Figure 8.19). However, it is shifted by half f_H.

From the filter point of view, the separation of Y and C requires a frequency domain transfer function that is periodic. Such filters are known as "comb filters" (see filters in Chapter 12).

After luma-chroma separation takes place, the color components encoded into the chroma signal are recovered by the chroma demodulator while the luma signal proceeds directly to the DSP section on the video data path.

8.3.2 Chroma Modulation and Demodulation

The YUV color space used by the composite system is related to the RGB color space by the following mathematical relations:

$$Y = 0.299R + 0.587G + 0.114B$$
$$U = -0.147R - 0.298G + 0.436B = 0.492\,(B - Y)$$
$$V = 0.615R - 0.515G - 0.1B = 0.877\,(R - Y)$$

The chrominance or chroma signal is encoded by amplitude modulating the two orthogonal components of the sinusoidal color carrier with U and V. Orthogonality is essential since it allows U and V to remain completely independent of each other through the demodulation process. By definition the $\sin(x)$ and $\cos(x)$ functions are orthogonal; their phases are 90 degrees apart or $\sin(x) = \cos(90° - x)$. Therefore we can write

$$C = U \sin \omega t + V \cos \omega t$$

where $\omega = 2\pi f_c$ and $f_c = 3.579545$ MHz.

The composite signal is obtained by summing the luma, chroma, and timing components together.

$$V_{COMP} = Y + C + \text{Timing} = Y + U \sin \omega t + V \cos \omega t + \text{Timing}$$

On the decoder side the first step is the separation of the luma Y and chroma C components, which is done by digital comb filters or, in older equipment, by notch bandstop and bandpass filters. After separation from luma, the chroma signal needs further processing in order to recover U and V and, subsequently, R, G, and B. Multiplication of C by $2 \sin \omega t$ and $2 \cos \omega t$ yields the intermediate variables C_U and C_V.

$$C_U = C \times 2 \sin \omega t = 2(U \sin \omega t + V \cos \omega t) \sin \omega t = 2U \sin^2 \omega t + 2V \sin \omega t \cos \omega t$$
$$C_V = C \times 2 \cos \omega t = 2(U \sin \omega t + V \cos \omega t) \cos \omega t = 2U \sin \omega t \cos \omega t + 2V \cos^2 \omega t$$

If we apply the trigonometric identities

$$2 \sin^2 \omega t = 1 - \cos 2\omega t$$
$$2 \cos^2 \omega t = 1 + \cos 2\omega t$$
$$2 \sin \omega t \cos \omega t = \sin 2\omega t$$

we obtain:

$$C_U = C \times 2 \sin \omega t = U (1 - \cos 2\omega t) + V \sin 2\omega t = U - U \cos 2\omega t + V \sin 2\omega t$$
$$C_V = C \times 2 \cos \omega t = U \sin 2\omega t + V (1 + \cos 2\omega t) = V + U \sin 2\omega t + V \cos 2\omega t$$

Since the frequency of the sinusoidal components is $2f_c$ (or 7.159 MHz) and the bandwidth of U and V is at most 4.2 MHz, the U and V color components can easily be recovered by low-pass filtering C_U and C_V. The block diagram for a chroma demodulator is shown in Figure 8.20. Conversion to RGB is done by using the equations:

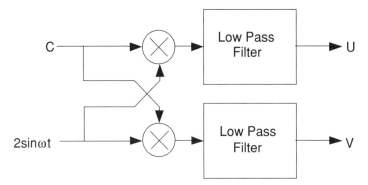

Figure 8.20 Chroma Demodulation.

$$R = Y + 1.14V$$
$$G = Y - 0.395U - 0.581V$$
$$B = Y + 2.032 \ U$$

Once in the RGB color space we can convert to YCbCr or YPbPr by using the transformations introduced in Chapter 2.

8.3.3 Contrast, Brightness, Saturation, and Hue

The DSP section of video decoder-digitizers is involved in changing the attributes of the signal. These include contrast, brightness, saturation, and hue.

For 8-bit decoders with a luminosity range from 0 to 255 (0 to 1023 for 10-bit decoders) *contrast* adjustments are made by multiplying Y with a user-programmable contrast gain. The gain values are generally between 0 and 2. If the Y range is from 16 to 235, the black level value of 16 must be subtracted first.

Adjusting the *brightness* is done by adding or subtracting from Y a brightness offset, which is also user-programmable. If earlier we subtracted the black level DC bias from the input signal, we must add it back at the end (Figure 8.21).

Image *hue* is by definition the phase of the chroma signal with respect to the color burst. To change the hue we have to add a user-programmable phase offset φ to the chroma C.

$$C_H = U \ \sin(\omega t + \varphi) + V \ \cos(\omega t + \varphi)$$

If we expand the terms

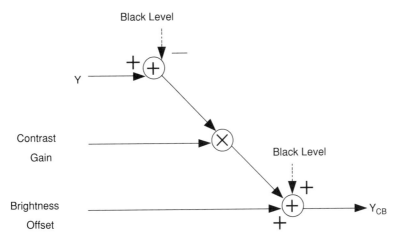

Figure 8.21   Contrast and Brightness Adjustment.

$$C_H = U \sin \omega t \cos \varphi + U \cos \omega t \sin \varphi + V \cos \omega t \cos \varphi - V \sin \omega t \sin \varphi$$

Factoring out $\sin \omega t$ and $\cos \omega t$ the expression becomes

$$C_H = (U \cos \varphi - V \sin \varphi) \sin \omega t + (U \sin \varphi + V \cos \varphi) \cos \omega t$$

Identifying the coefficients of $\sin \omega t$ and $\cos \omega t$ as the hue adjusted color components U_H and V_H, we have the expressions:

$$U_H = U \cos \varphi - V \sin \varphi$$
$$V_H = U \sin \varphi + V \cos \varphi$$

Saturation adjustment consists in multiplying the U_H and V_H color components by a common color gain factor S. (Figure 8.22.)

$$U_{HS} = U \, S \cos \varphi - V \, S \sin \varphi$$
$$V_{HS} = U \, S \sin \varphi + V \, S \cos \varphi$$

For color adjustments in the YCbCr space we must first subtract the 128 DC bias, then perform the saturation and hue adjustment, and then add 128 back. The equations, without the DC bias, are

$$Cb_{HS} = Cb \, S \cos \varphi + Cr \, S \sin\varphi$$
$$Cr_{HS} = -Cb \, S \sin \varphi + Cr \, S \cos\varphi$$

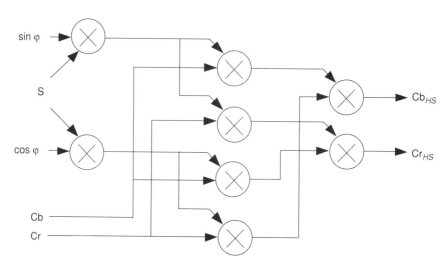

Figure 8.22   Saturation and Hue Adjustment.

The final stage in the decoder block diagram is the output formatter. Besides digitized video data this block also generates the reference signals necessary to capture, frame, and store the output image. Some of these signals are horizontal sync (HSYNC), vertical sync (VSYNC), pixel clock (PXLCLK or CLKX2), odd-even field (FIELD), active line (ACTIVE), and so on.

8.3.4 Video Decoder-Digitizer Circuit

In Figure 8.23 we have the schematic for a complete video decoder-digitizer circuit using Conexant's Bt835 IC. Like most video decoders in this class, the Bt835 is quite self-contained, with very few external components required.

Although this IC has a four-to-one analog multiplexer built in, we use only one input channel (MUX0) to keep the schematics simple. The module can accept either composite video through the J1 BNC connector or s-video through the J2 mini-DIN. When s-video is applied, the chroma input goes directly to the Cin pin, but the luma input needs software routing to the MUX0 input of the multiplexer (registers MUXS[1:0] = 00).

The module is compatible with all commonly used analog formats: NTSC, PAL, SECAM, and most of their variants. To select a particular format we set the FMT[3:0] bits of the INPUT control register to the appropriate binary value.

The analog input circuitry consists of terminator resistors (R4, R6), coupling capacitors (C3, C4), and an explicit protection network comprised of dual Schottky diodes (D1, D2) and current-limiting resistors (R3, R5).

Besides analog video, the Bt835 can also process 8-bit, 27 MHz digital parallel video streams in 4:2:2 YCbCr format. The synchronization signals are provided separately on the horizontal reset (DIG_H), vertical reset (DIG_V),

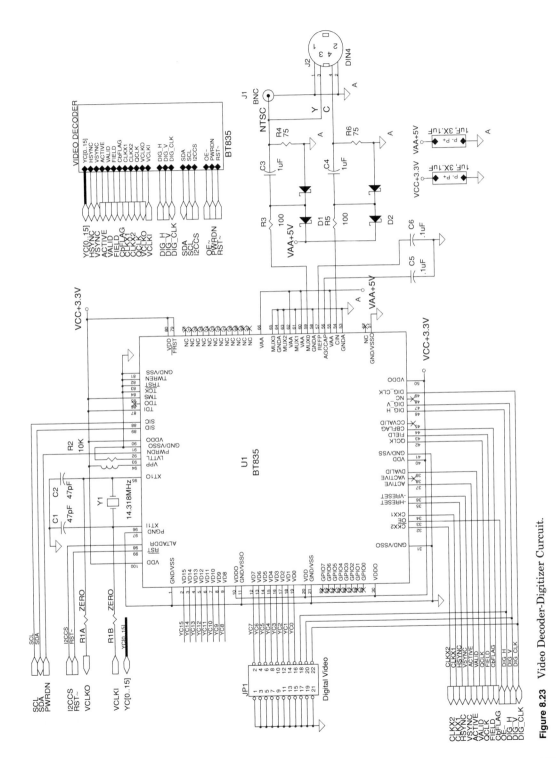

Figure 8.23 Video Decoder-Digitizer Curcuit.

187

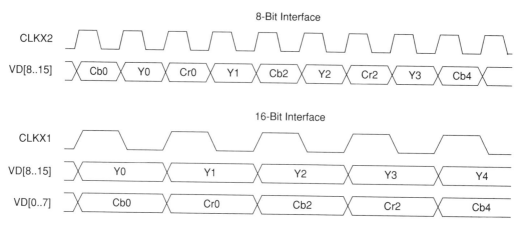

Figure 8.24 Bt835 8- and 16-Bit Output Interfaces.

and pixel clock lines (DIG_CLK). Alternately, timing data (SAV and EAV codes) can also be automatically extracted from the data streams if the streams are CCIR656 standard compliant.

The decoder can operate either in an 8-bit or 16-bit mode, with corresponding clock rates of 27 MHz and 13.5 MHz, respectively. When in 8-bit mode the digital outputs are provided on the VD[8..15] lines, with VD[0..7] either tri-stated (for analog inputs) or used as the digital input bus. In 16-bit mode the VD[0..7] bus outputs the digitized chroma, while the VD[8..15] lines provide the luma (Figure 8.24).

It should be noted that the CLKX2 and CLKX1 rates are $8f_c$ (28.63636 MHz) and $4f_c$ (14.31818 MHz), not 27 MHz and 13.5 MHz. This adds an extra margin to the rate requirements of the Sampling Theorem, and provides more time cycles for the decoder.

The PLL on board the Bt835 generates all the clocks, for all formats and standards, using a single 14.318 MHz crystal. The XT1I crystal port can also be used as an external oscillator input, while the XT1O can drive external loads, although we suggest buffering it with a Schmitt trigger first.

The module itself can be used either with an external oscillator or with a local crystal, depending on which one of the zero-ohm resistor positions (R1A or R1B) is populated. For multi decoder systems it is strongly recommended to use the oscillator from one decoder to drive all the other decoders. This minimizes the potential for clock signal interference, a major source of noise in digital video systems.

The Bt835 supports a number of different 8- and 16-bit YCbCr formats. Among them the easiest to use and the most instructive is the *synchronous pixel interface mode 1* (SPI Mode 1). The horizontal timing diagram for the 8- and 16-bit SPI Mode 1 format is shown in Figure 8.25, while the corresponding vertical timing is illustrated in Figure 8.26.

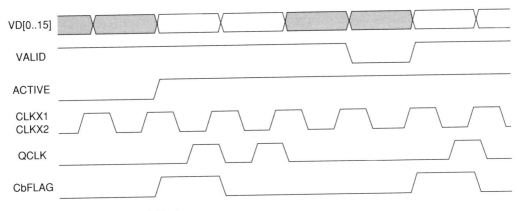

Figure 8.25   Bt835 Horizontal Timing.

The beginning of the active part of each new line is marked by a low-to-high transition on the ACTIVE pin. The timing characteristics of the ACTIVE signal are controlled by the contents of four user programmable registers inside the Bt835 chip; HDELAY, HACTIVE, VDELAY, and VACTIVE. The HDELAY and HACTIVE determine the starting point and length of the ACTIVE high signal in the horizontal direction. VDELAY and VACTIVE perform the same function in the vertical direction. Their value can be increased in half-line increments to accommodate NTSC/PAL interlaced formats.

Not all the output data generated during the active portion of the line represent valid pixels. Since the output pixel rate must be 13.5 MHz (720X480 NTSC standard) and the output data rate is 14.31818 MHz (16-bit mode YCbCr), some of the data must be discarded.

In addition, when using the built-in downscaler the number of pixels per active line shrinks even more; however, the clock rates remain unchanged. The VALID signal is the filter that tells us which output pixels are "good"—and therefore should be processed further—and which should be dropped.

CbFLAG indicates if the current C byte is Cb or Cr. It is useful for tracking chroma storage in designs with minimal or no FPGA/CPLD support. Of significant help in this case is QCLK, a CLKX1 or CLKX2 version internally gated by the ACTIVE and VALID signals. QCLK can directly drive the write clock line of FIFO type memory buffers.

The FIELD signal identifies if the current pixel belongs to an odd or even field. This may be important when we split the frame memory buffer into odd and even banks. We can also obtain a FIELD flag by registering VRESET~ on the falling edge of HRESET~, as can be inferred from Figure 8.26. As seen in the detail view for odd fields, HRESET~ precedes VRESET~ by two clock pulses.

The OE~ line is useful when we connect multiple decoders in parallel across the same output bus. Depending upon the status of two programmable bits

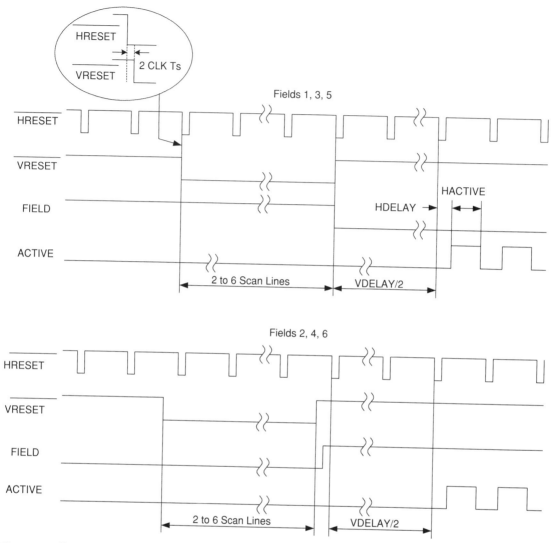

Figure 8.26   Bt835 Vertical Timing.

(OES[0..1]), QE~ can tri-state all output lines: data and timing, data and clocks, or data only.

The Bt835 interfaces to the local controller over a standard two-line I²C bus. To facilitate multi-decoder use, the on-board I²C controller can assume one of two addresses (88 or 8A), selectable by the ALTADDR line. In Mode 2 all the timing information is embedded in the output data, in what Conexant calls the ByteStream format. The mechanism consists in using the luma values between 0 and 15 and chroma values 254 and 255 for control codes. The luma-chroma pair (F06:FF0), for example, marks a VRESET to be followed by an odd field. When in VIP mode the Bt835 generates a CCIR656 type digital output, with the timing information embedded in the SAV and EAV headers (see Chapter 2).

Our selection of the Bt835 is based on its ease of use, good performance, and low cost. However, there are many other decoder-digitizers available from a variety of top manufacturers, including Philips Semiconductors, Analog Devices, and Texas Instruments.

8.4 Video Graphics Digitizers

Video graphics digitizers differ from video decoders in the number of ADCs used and in the package of functions integrated with the ADCs (Figure 8.27). Designed mainly for RGB video and VGA graphics, the digitizers have three ADCs operating at much higher speeds than the NTSC/PAL pixel rates.

The AC coupling of RGB and YUV inputs necessitates the use of clamps to bring their DC levels within the range of the ADCs. If the inputs are YCbCr the color components Cb and Cr are DC-biased at midspan (digital 128). In order to use the full resolution of the ADCs these signals have to be clamped at the ADCs' midrange voltage.

Following DC adjustments, the inputs are passed through programmable gain amplifiers (PGAs). Their function is not to compensate for line losses, but

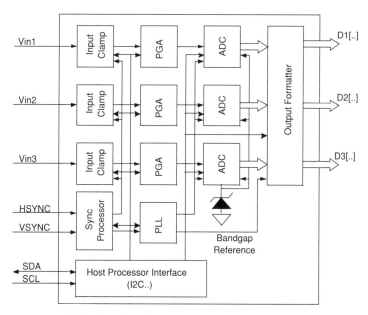

Figure 8.27 Video Graphics Digitizer Block Diagram.

to match the 0.5V to 1V input voltage levels to the full scale of the ADCs. Besides gains the PGAs also provide signal offsets. Gain and offset, when individually programmable for each RGB, YUV, YCrCb, or YPbPr channel, provide full user control over the brightness, contrast, saturation, and hue of the output image. Clock generation is done by an internal PLL locked onto the horizontal sync of the input video. HSYNC may be available either as a separate digital input or embedded into the Y or G inputs; the latter requiring the "slicing" of the sync out of the analog signal.

8.4.1 Video Graphics Digitizer Circuit

The core of the video graphics digitizer circuit is the THS8083 from Texas Instruments (Figure 8.28). Its three inputs are independently clamped, amplified, and offset before being digitized at a maximum rate of 95 Msps (mega samples per second).

Clamp timing is user programmable, and so are the minimum widths of the horizontal and vertical sync pulses; pulses shorter than the minimums are ignored. The gain of each PGA is controlled by separate course (6 bits) and fine (5 bits) registers.

Communication with the local controller is done via a two-wire I^2C bus, with the address being hexadecimal 40 and 41 for write and read operations, or alternately 42 and 43, depending on the state of the I2CA0 pin. Built-in pixel and line counters can be used by the local controller to automatically detect the format of the incoming video, while a "pixel trap" can be set to capture selected pixels and transfer them to a controller-accessible readback register.

The THS8083 formatter can be programmed to output 16 or 24 bits of data clocked at full sampling rate (Figure 8.29), or 48 bits clocked at half sampling rate (Figure 8.30). DATACLK1 is the output pixel clock and its leading edge can be used for storing the digitized video in downstream memory buffers. ADCCLK2 is driven by the ADC clock, but can optionally be inverted and/or divided by two and employed to synchronously drive external logic.

The HSYNC signal is the input horizontal sync onto which the PLL is locked. The THS8083 issues its own output sync pulse, HSD, coincident with the first valid output of the current line, and seven ADCCLK2 pulses from the rise of HSYNC. HSD can be programmed to be one sample clock long (MODE 0) or to match the length of the HSYNC input (MODE 1).

In 24-bit parallel mode, each A channel outputs the digitized versions of their matching inputs; digital CH1_OUTA[..] corresponds to analog CH1_IN, and so on. Or, looking at Figure 8.29, D1 is the digitized Red output; D2 the Green, D3 the Blue, and n the number of pixels per line. All B channels are tri-stated. This is the preferred format for RGB and SDTV/HDTV signals.

4:2:2 YUV and YCbCr inputs can only be processed in 16-bit mode, although the signals must be separately applied to all three input channels: Y to CH1_IN, U (Cb) to CH2_IN, and V (Cr) to CH3_IN. Then, CH1_OUTA[..] carries the digitized Y and CH2_OUTA[..] the digitized color, starting each

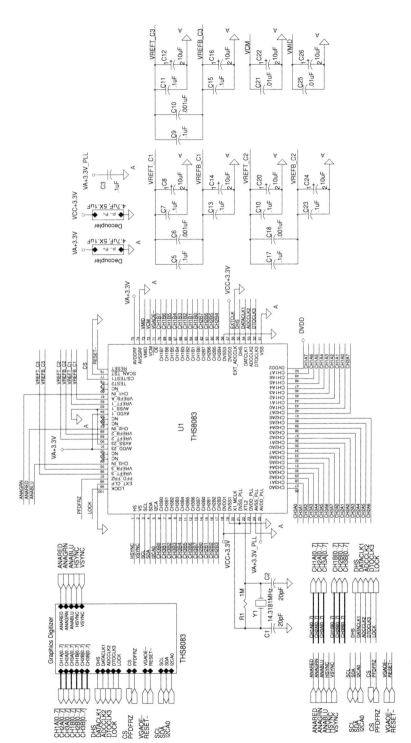

Figure 8.28 Video Graphics Digitizer Circuit.

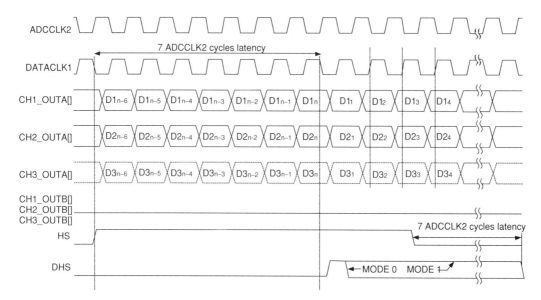

Figure 8.29   THS8083 16- and 24-Bit Modes.

line with $D2_1 = U$ (Cb) followed by $D2_2 = V$ (Cr), and so on Channels CH3_OUTA[..], CH1_OUTB[..], CH2_OUTB[..], CH3_OUTB[..] are all tri-stated.

For high-speed applications, when the THS8083 operates close to the 95 Msps limit, the 48-bit output formats are preferred. They allow the designer to separate the data into two 24-bit streams, each with a data rate of only half the sampling frequency. The access speed requirements on the downstream memory buffers can then be dropped to roughly 20 nsecs.

The THS8083 accepts RGB, YUV, YCrCb, SDTV, and HDTV input signals. If the video "content" part of all these signals is essentially the same (analog voltage), the method for providing timing information is different for each.

8.4.2 VGA, XGA, UXGA Digitizer Circuit

Computer formats and some studio RGB flavors have separate digital sync inputs. Other RGB, YUV, and SDTV formats have their timing embedded in the luma or green signals, while for HDTV we may have composite sync on all three channels, only on green, or as a separate input.

When the sync information is available as a separate HSYNC input the THS8083 PLL has a relatively easy task in locking to it, subject to the usual filter-jitter tradeoffs. We can then complete a VGA/XGA digitizer by adding a VGA connector interface to the digitizer module (Figure 8.31).

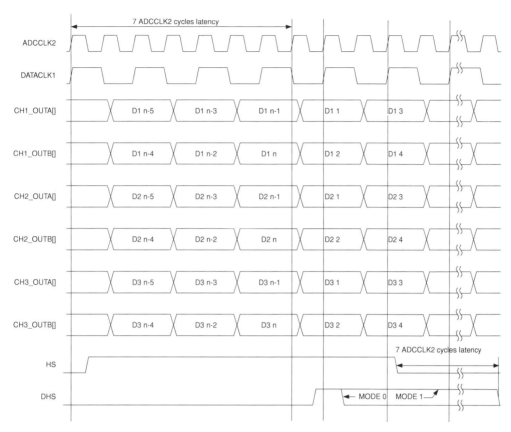

Figure 8.30   THS8083 48-Bit Mode.

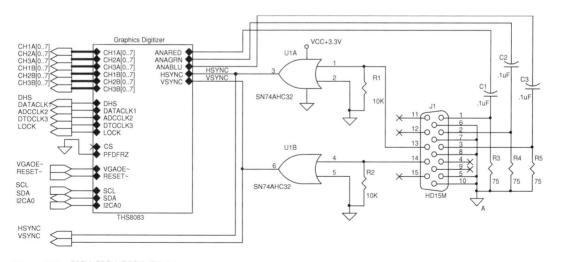

Figure 8.31   VGA/XGA/UGA Digitizer.

8.4.3 HDTV Digitizer Circuit

If no signal is present on the HSYNC pin (as reflected by internal status bit) then the IC assumes the presence of an embedded sync and slices the CH1_IN at about 150 mV. The result is a composite sync that includes HSYNC pulses and all the other VBI (vertical blanking interval) signals. However, the VBI transitions do not allow the PLL to lock. The problem can be solved by providing the composite sync CS to an external sync separator and feeding back the HSYNC and VSYNC to the THS8083 (Figure 8.32). The sync separator has to be able to handle normal and tri-level syncs, something that Elantec's EL 4511 does nicely.

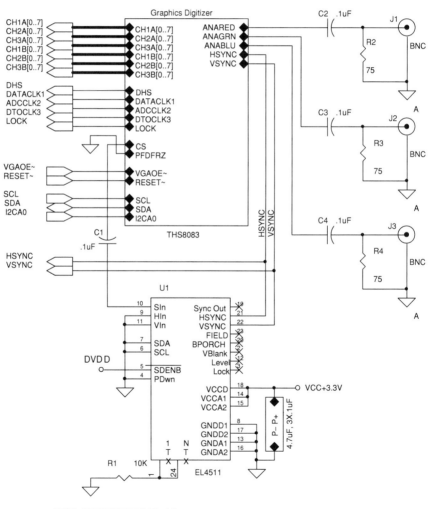

Figure 8.32   RGB, SDTV, HDTV Digitizer.

8.5 Deserializing SDI DigitalVideo

The combination of an *equalizer*, a *reclocker*, and a *deserializer* constitutes the SDI equivalent of a decoder. We covered the equalizer and reclocker circuits earlier in the book. Now we focus on the deserializer (Figure 8.33).

Once the digital input stream enters the deserializer and after translation occurs from differential to single-ended logic, it is decoded by a *SMPTE descrambler*. In accordance with the requirements of the SDI standard, the serializer encodes the signal in such a manner as to ensure a sufficient density of signal transitions for data recovery to be possible. The selected coding is NRZI, which we previously encountered in Chapter 7. Decoding it involves using two generator polynomials. The first converts it from NRZI to standard logic NRZ (non-return to zero),

$$G_2(x) = x + 1$$

and the second decodes the NRZ itself.

$$G_1(x) = x^9 + x^4 + 1$$

NRZ uses a negative voltage for 0 and a positive value for 1.

Generator polynomials are binary, with the variable x and its coefficients taking only binary values (0 or 1). As such, binary multiplication is implemented by our familiar AND gate while binary addition is nothing else but the XOR operation. Every sequence of bits can be represented by a binary poly-

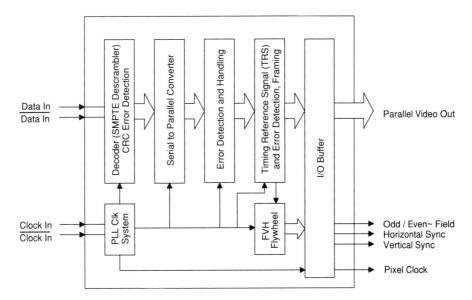

Figure 8.33   Deserializer Block Diagram.

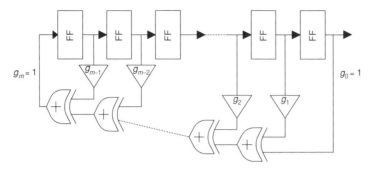

Figure 8.34 Fibonacci LFSR Polynomial Implementation.

nomial. The sequence 1000010001 for example, can be represented by the polynomial $G_1(x)$.

In general, any arbitrary binary polynomial $G(x)$ can be implemented in one of two equivalent forms; the Fibonacci (Figure 8.34) and the Galois (Figure 8.35) implementations.

$$G(x) = g_m x^m + g_{m-1} x^{m-1} + g_{m-2} x^{m-2} \cdots + g_2 x^2 + g_1 x^1 + g_0$$

Both use so called LFSR (linear feedback shift register) devices as circuit elements.

The $G_1(x)$ and $G_2(x)$ polynomials can be thought of as functions that, when applied to a serial data stream, modify it in a reversible fashion. In this case the modification translates into the NRZI to NRZ conversion.

The generator polynomials mandated by the SDI protocols lead us to the circuit in Figure 8.36.

Following NRZI descrambling, the stream may undergo a *CRC check* using a similar method. A common generator polynomial used both for CRC calculation and check is:

$$\mathrm{CRC}_{16}(x) = x^{16} + x^{12} + x^5 + 1$$

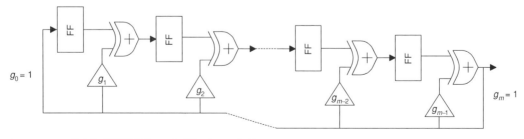

Figure 8.35 Galois LFSR Polynomial Implementation.

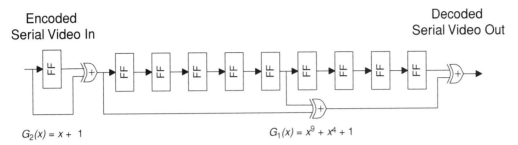

$G_2(x) = x + 1$

$G_1(x) = x^9 + x^4 + 1$

Figure 8.36   SMPTE NRZI Descrambler.

If we represent an information packet by its binary polynomial and we divide it by the CRC generator polynomial, we obtain a remainder. The binary form of the remainder is then appended to the data packet. The receiver, after separating the data and dividing it by the generator polynomial, should obtain the same remainder. In binary, division by two is a one-step shift operation. So the hardware implementation in Figure 8.37 simply performs the division of the appended packet by the CRC generator polynomial. If the transmission of the appended data packet is error-free,then value of CRC Out is zero.

Once descrambled and checked for communications errors, the serial stream is *converted to parallel*, and passed forward to *error handling, word alignment,* and *TRS extraction and processing*. The deserializer arrangement we looked at so far is not unique. The serial handling of SMPTE descrambling and CRC error check saves on circuit complexity but requires a high-speed clock. Some ICs convert the stream to parallel format first and then decode and check the data. This reduces the clock speed by a factor of ten but doubles the circuit size.

Word alignment consists in the detection of a number of consecutive TRS-ID codes and the synchronization of the local *flywheel* to the incoming stream. The flywheel is a local timing generator that once locked, produces horizontal blanking, vertical blanking, and field signals consistent with the information embedded in the input data. These are made available on the IC's output pins together with the parallel video data.

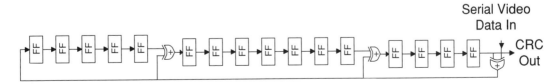

Figure 8.37   CRC-16 Decoder.

Deserializers can also go deeper into the data layers and retrieve the ancillary data transported in the blanking intervals. These may include digital audio packets, timecodes, closed captioning, and so on. Error detection and handling packets (EDH) form a different layer of error correction used in SDTV; they are also inserted in the horizontal blanking interval. Finally, the clock infrastructure that drives the deserializer is provided by a PLL block locked onto the clock signal retrieved by the reclocker.

8.5.1 Deserializer Circuit

Our deserializer module is built around Gennum's GS1561 multirate reclocking deserializer IC (Figure 8.38). This device has two separate differential inputs (DI1+, DI1– and DI2+, DI2–) that are buffered and passed through a two-to-one input multiplexer. The input circuitry also includes internal 50 Ω terminators (pins TERM1 and TERM2) that must be externally AC-grounded. The input channel selection is made by setting pin IP_SEL high for channel 1 and low for channel 2.

Each data input channel has associated with it a carrier detect input (CD1~, CD2~) that is provided either by the upstream cable equalizer that feeds into the GS1561 or by external logic. If carrier detect is high the associated channel input is considered invalid. Once a signal is qualified as valid it is applied to the internal reclocker, which recovers separate data and clock streams and realigns them to filter out jitter.

The GS1561 has two modes of operation hardware selected by the master/slave pin. If the master/slave line is high, the device automatically detects, reclocks, deserializes, and processes SD, HD, and DVB-ASI information. The locking mechanism searches the input data for TRS and DVB-ASI sync codes, alternately at 1.485 Gbps and 270 Mbps. If such data is consistently found the LOCKED output is set high, and the SMPTE_Bypass~, DVI-ASI and SD/HD~ (outputs in master mode) will reflect the input stream type. If no such data is found in four consecutive tries, the reclocker locks to the input stream without detecting TRS or DVB-ASI syncs.

In slave mode the device configuration is determined by the logic values of the SMPTE_Bypass~ and DVI-ASI input pins. To enter the bypass mode the SMPTE_bypass~ and DVB-ASI lines are both set to zero. This tells the GS1561 to expect raw "pass-through" data with no embedded timing or special coding. This mode can be used for high-speed serial data transport in non-video applications. In slave mode the SD/HD~ instructs the device to lock only on 270 Mbps SD (SD/HD~ high) or 1.485 Gbps HD (SD/HD~ low) streams.

If SMPTE_bypass is set low, but the DVB-ASI line is high the device expects DVB-ASI video data. The DVB-ASI standard originated with the European Digital Video Broadcasting (DVB) standards association as a simpler method of interconnecting MPEG-2 serial streams.

DVB-AVI is a fixed frequency interface that operates at 270 MHz and employs IBM's patented 8b/10b encoding scheme. The 8b/10b algorithm con-

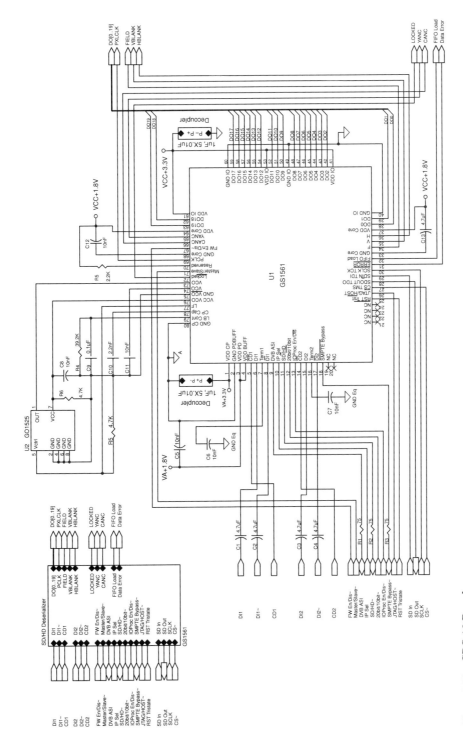

Figure 8.38 CRC-16 Decoder.

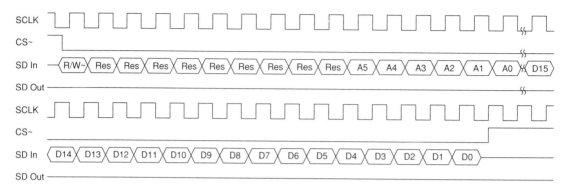

Figure 8.39 GS1561 Write Communication Protocol.

verts each byte of a data packet into a 10-bit output with the goal of equalizing the total number of 0's and 1's in the packet. 8b/10b limits the number of permitted identical consecutive bits to five and provides for the detection of single-bit errors.

Special 10-bit comma characters (K28.5 code words) are used for synchronization and padding of the stream when data is missing. The total 8b/10b overhead requires a 133 Mbps signaling rate for an effective 100 Mbps data rate. The GS1561 enters the DVB-ASI mode when it detects 32 consecutive error-free DVB-ASI words, and drops out when it tallies 32 consecutive errors. After conversion to parallel the words are 8b/10b decoded and aligned to the K28.5 sync characters. Then they are placed on the DO[10 .. 17] lines.

In SMPTE mode the words are NRZI-to-NRZ decoded, then descrambled and aligned to the TRS sync codes. The flywheel is also synchronized by the

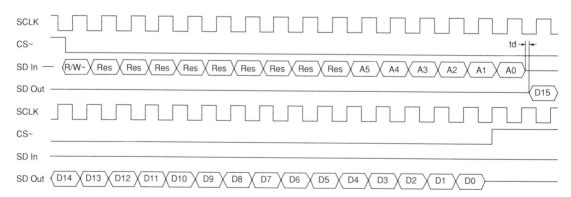

Figure 8.40 GS1561 Read Communication Protocol.

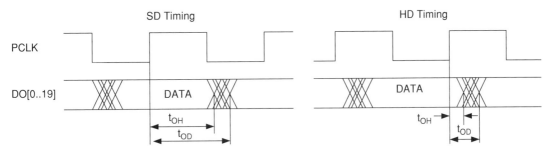

Figure 8.41   GS1561 Output Timing.

TRS codes, with the resynchronization being continuous when FW_EN/DIS~ is low and only on sync error when FW_EN/DIS~ is high.

The GS1561 can generate either 10-bit multiplexed or 20-bit nonmultiplexed output data, according to the logic level of the 20bit/10bit~ pin. Most on-board data processing can be disabled by setting the IOProc_En/Dis~ pin low. This includes EDH packet generation and insertion (SD), ancillary data preambles and checksum generation, line-based CRC (HD), TRS generation and insertion, and video format detection.

The device can communicate with the local processor through a four-wire interface consisting of an SD In input line, an SD Out output line, a clock SCLK, and a CS~ chip select input. The user interface is enabled when JTAG/Host~ input is low. The communication format consists of 16-bit command words followed by 16-bit data words. For a write cycle the first bit in the command word is 0 and is followed by 9 reserved bits and 6 register address bits (Figure 8.39).

The 16-bit data word which is to be written at the address pointed to by the command word must follow on the next 16 SCLK pulses. However, there are no maximum delay requirements between the two sets of clock pulses. Read cycle command words start with 1 bit, followed by 9 reserved bits and 6 address bits (Figure 8.40). The data emerges on the SD Out line after a 12-nsec delay. Maximum SCLK rate is 6.6 MHz. The relationship between the output lines DO[0 . . 19] and the output clock PCLK is shown in Figure 8.41. Hold time for HD is 1.5 nsec.

The VBLANK, HBLANK, and FIELD are the output vertical and horizontal blanking and field signals, as signaled by the TRS data. Y ancillary (YANC) and Cancillary (CANC) outputs indicate the presence or absence of ancillary data in the luma and chroma data streams, and Data_Error~ flags data stream errors.

Video Encoders, Graphics Digital to Analog Converters, and Serializers

9.1 Theory of Signal Reconstruction

Bringing the digitally processed video back in the analog realm is the function of video encoders and video graphics digital to analog converters (DACs). Aside from timing differences, both devices read the digital video data from memory buffers, convert it to an analog voltage, and hold it for the durations of the pixel period.

The contents of the memory buffers can be considered to be samples of a continuous video output still to be reconstructed. The encoder or the DAC are then zero-order holds for that output. If we define $x(t)$ as a signal with digital samples identical to those of the yet unknown output, then the model for the conversion to analog is that shown in Figure 9.1.

In this interpretation, the digital data read from the buffers is represented by $x_s(t)$, and the output of the encoder or DAC by $x_{ZOH}(t)$. (See Figure 9.2.)

As we saw earlier, the Fourier transform of $x_s(t)$ is related to the transform of $x(t)$ by

$$X_s(f) = \frac{1}{T} \sum_{m=-\infty}^{m=+\infty} X(f - mf_s)$$

We can describe the zero-order hold as a circuit that provides a single pulse response to an impulse input, where $q(t)$ in Figure 9.3 is that response.

If we apply the Fourier transform to $q(t)$ then:

$$F[q(t)] = Q(f) = \int_{-\infty}^{+\infty} q(t)e^{-j\omega t}dt = \int_{0}^{T} q(t)e^{-j\omega t}dt$$

Figure 9.1   Sampling and Reconstruction.

Replacing $q(t)$ and integrating between 0 and T we obtain:

$$Q(f) = \frac{1}{-j\omega}(e^{-j\omega T} - 1) = \frac{1}{-j\omega}e^{-j\omega T/2}(e^{-jwT/2} - e^{j\omega T/2})$$

Using the Euler identities:

$$Q(f) = \frac{2}{\omega}e^{-j\omega T/2}\sin\left(\frac{\omega T}{2}\right) = Te^{-j\pi fT}\operatorname{sinc}(fT)$$

Since the Fourier transform of $x_{ZOH}(t)$ is $Xs(f)\,Q(f)$, we find:

$$X_{ZOH}(f) = X_s(f)Q(f) = \sum_{m=-\infty}^{m=+\infty} e^{-j\pi fT}\operatorname{sin c}(fT)X\left(f - \frac{m}{T}\right)$$

where

$$\operatorname{sinc}(x) = \frac{\sin(\pi x)}{\pi x}$$

This formula relates the spectrum of the zero-order hold output (the output of the encoder or DAC) to the spectrum of the time domain signal we want to reconstruct.

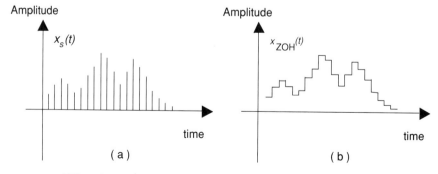

Figure 9.2   Effect of Zero-Order Hold.

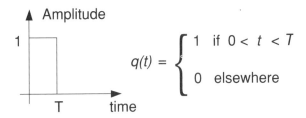

Figure 9.3 ZOH Pulse Function.

9.1.1 Frequency Spectrum of Reconstructed Signals

If f_{BW} is the highest frequency component of $X(f)$, the stepwise restoration process introduces a number of images of f_{BW} at $f_s - f_{BW}$, $f_s + f_{BW}$, $2f_s - f_{BW}$, $2f_s + f_{BW}$, and so on (Figure 9.4). As a matter of fact the whole spectrum of $x(t)$ is mirrored in the bands $f_s - f_{BW}$ to $f_s + f_{BW}$, $2f_s - f_{BW}$ to $2f_s + f_{BW}$, and so on.

We see again that for spectral images not to overlap $X(f)$, f_s must be larger than $2f_{BW}$. Restricting the bandwidth to f_{BW} is the job of output low-pass filters known as *reconstruction* filters.

Please note that the spectrum of the synthesized $x(t)$ is distorted by the envelope of the sinc(fT). This distortion cannot be filtered out by a low-pass filter, and may be quite substantial for marginal f_s values. The remedy consists in preprocessing the digital domain signal with the inverse sinc function.

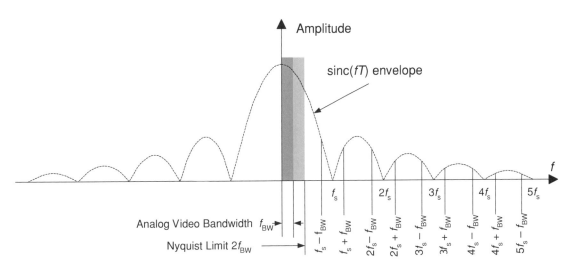

Figure 9.4 ZOH Output Spectrum.

9.2 Video Encoders

Video encoders accept a parallel digital video stream, translate it to a user-programmable color space, encode it to composite, and then convert it to a standards compatible analog format. One key process, encoding the video to composite, requires the input to be in the 4:4:4 YUV space. Since the most common digital format is 4:2:2 YCbCr, the first step must be scaling up, or upsampling, the Cb and Cr color components to 4:4:4. The block responsible for this is the 4:2:2-to-4:4:4: interpolator (Figure 9.5). Conversion to YUV is then completed by a front color space converter. The digital filter blocks that follow limit the bandwidth of the YUV and, depending on the device, may increase sharpness (provide higher gains at higher frequencies), adjust brightness (add offset), and process saturation and hue. This is also the stage where various precompensation processes take place, such as gamma correction and inverse $sin(x)/x$ compression.

Insertion of teletext, closed captioning, and copy protection codes must also be performed while in the digital domain. The data to be inserted is either built in or received from the local controller through a host interface, usually an I²C or SPI serial port.

Now the digital streams are ready for final output processing. The most common standard resolution formats are composite, s-video, and interlaced or progressive RGB. The chroma part of s-video is obtained by modulating the color subcarrier with the U and V components (see Chapter 8). To generate the composite output the chroma is mixed with the luma, timing, and color burst signals (CVBS). For s-video, the luma and timing signals are provided on one output line (Y), with the chroma and the burst on another line (C).

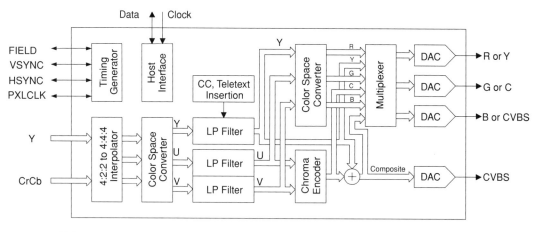

Figure 9.5 Video Encoder Block Diagram.

RGB out is the result of another space conversion process, with the RGB and YC streams making their way to the DACs through a user-controlled data multiplexer. The final stage of the encoder is the DAC bank and the associated voltage reference.

Video encoders can operate either in a master or slave mode. In master mode the standard-specific timing signals are generated internally. These signals include pixel clock (PXLCLK), horizontal sync (HSYNC), vertical sync (VSYNC), and field (FIELD). To maintain synchronicity with the video stream, the timing signals generated by the encoder must also drive the raster generator that reads the data out of the memory buffers.

When in slave mode the encoder is driven by external timing signals provided by the same raster generator that reads the memory buffers. While using the slave mode gives the designer more flexibility, the master mode requires less FPGA/CPLD resources.

HDTV encoders include further upsampling or upscaling blocks, insertion of SAV/EAV codes and ancillary data, tri-state syncs, and higher resolution/ higher speed DACs.

9.2.1 Video Encoder Circuit

For the encoder circuit we select the Bt864A YCbCr to NTSC/PAL encoder IC from Conexant, a good match to the Bt835 we used earlier. Proper operation requires for the Bt864A to be supplied a clock signal CLKX2 with a frequency twice the pixel rate. From this input the device produces the system clock, which in master mode drives the internal timing engine responsible for the generation of the NTSC/PAL reference signals. In slave mode, the system clock registers the externally generated reference signals. Selection between master and slave modes is made by strapping the SLAVE pin to the ground or power rail. Other system clock functions are to register all the input data buses and to drive all the on-board DACs, filters, and other logic circuits.

The video data inputs are sampled starting with the rising edge of the fifth CLKX2 pulse after the drop of HSYNC. The Bt864A supports two input bus modes. In the 8-bit mode the Cb and Cr bytes are interleaved with the Y bytes at twice the pixel rate (Figure 9.6).

In 16-bit mode, the Y and CbCr bytes must be available at the same time at their corresponding bus ports; they also have to be updated every pixel period. The order of the Cb and Cr bytes can be switched by setting the internal CBSWAP bit.

In master mode, the length and offset of the HSYNC pulse are user-programmable, allowing some flexibility in its use for retrieving video data out of the memory buffers. However, the starting point of the actual analog horizontal sync is fixed either at 42 or 43 CLK pulses from the pixel counter reset. HSYNC can also be defaulted to the standard 4.7 μsec width. VSYNC is active

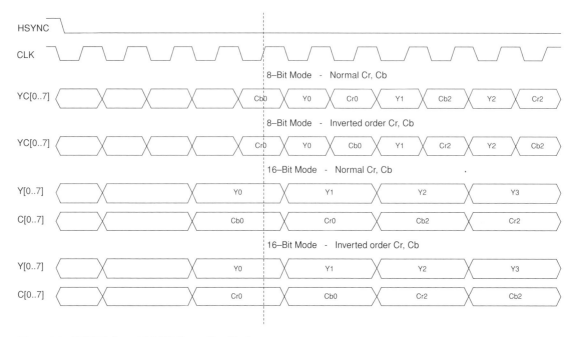

Figure 9.6 Bt964A 8- and 16-Bit Operating Modes.

for three lines in NTSC formats, and coincident falling edges for HSYNC and VSYNC mark an odd FIELD.

In slave mode, the internal pixel and line counters are driven by CLKX2 and reset by the external HSYNC and VSYNC pulses. If a falling edge of VSYNC occurs within a quarter-line of a falling edge of HSYNC, the FIELD output is cleared (odd field). The pixel rate is assumed to be 13.5 MHz, unless the number of pixels between HSYNCS matches the square pixel count. Then the square pixel 14.75 MHz clock rate is presumed.

The pipeline delay for the analog horizontal sync is 51 or 52 clocks (depending on the status of the internal SYNCDLY programmable bit).

For both master and slave modes, the delay between a new data byte or word at the inputs of Bt864A and the actual pixel voltage output is 52 CLKX2 pulses. The BLNK~ input is asserted on the rising edge of CLKX2 and complements the automatic internal (standard-specific) BLANK signal. Digital processing includes low-pass filtering, gamma correction, and $\sin(x)/x$ compensation. Data services supported by the device are closed captioning, teletext, wide screen signaling, and Macrovision anticopy protection. The I^2C address of Bt864A is 8A (write) and 8B (read) with the I2CCS (ALTADDR) pin low, and 88 (write) and 89 (read) for ALTADDR high (Figure 9.7). The default mode is 8-bit YCrCb, interlaced NTSC. After power-up the output stays in black burst mode until activated by software (EACTIVE bit set to 1).

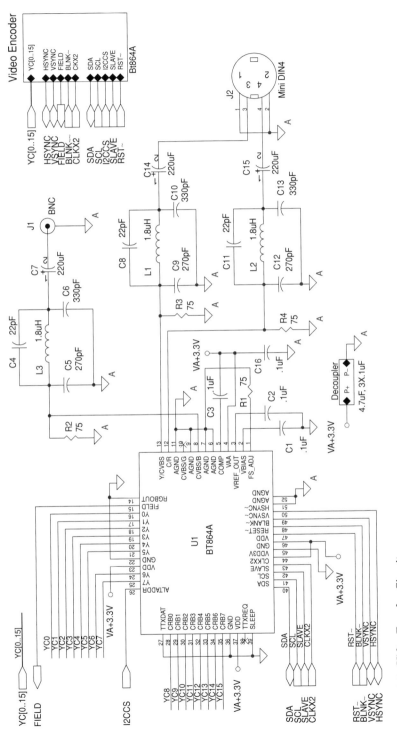

Figure 9.7 Video Encoder Circuit.

9.3 Video Graphics DAC

The block diagram for a graphics video DAC is deceptively simple (Figure 9.8). Once the user selects a particular mode of operation a *color space converter* translates the input data to the desired color format. The outputs of the registers are then stored in *holding registers* until the DACs settle. Sync and blank generation is handled differently than in video encoders. Instead of digitally inserting them into the data stream before being converted to analog, they are added in the analog domain to the outputs of the DACs. Since the DACs are current sources, the syncs and blanks are also current sources that add or subtract currents from the DAC's output. The actual video voltages are developed across the source and load termination resistors.

The main advantage to this architecture is its ability to work at higher frequencies than the all-digital approach. Current 10-bit DACs can work at frequencies up to 330 MHz.

9.3.1 Video Graphics DAC Circuit (VGA, XGA, UXGA, SDTV, HDTV)

Some of the most "designer friendly" video DACs are manufactured by Texas Instruments. The THS8134 is an 8-bit, 80 Msps, triple video DAC that is able to generate VGA/XGA as well as HDTV signals. This IC is our choice for the graphics digitizer module (Figure 9.9). Its bigger and faster version, the THS8135, has 10-bit DACs and, a maximum clock rate of 240 MHz and is almost identical in its operation to the THS8134 part.

The input buses are registered on the rising edge of the input CLK and the corresponding analog outputs emerge after a maximum delay of nine clock pulses. The output currents are programmed by means of a current bias resis-

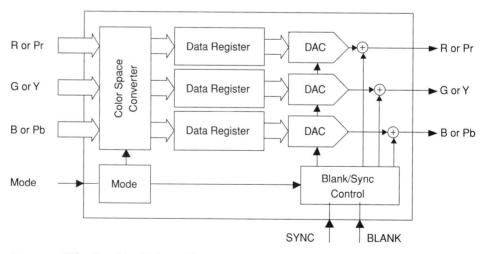

Figure 9.8 Video Graphics Digitizer Block Diagram.

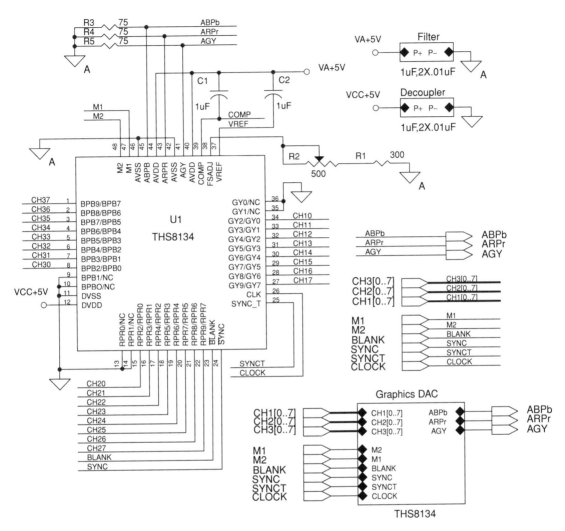

Figure 9.9   Video Graphics DAC (VGA, XGA, SDTV, HDTV).

tor connected to the FSADJ input. A nominal value of 430 Ω corresponds to a full scale output current of 26.67 mA. For normal operation, the IC expects a double 75 Ω parallel termination—one at the source and one at the load—on all three outputs.

A low on the BLANK~ input will drive all outputs to the appropriate format-dependent output blanking levels. The BLANK~ input is also registered on the raising edge of the clock. The modes of operation for the THS8134 are programmed via two input pins M1 and M2. In the bigger THS8135 the functionality of both is multiplexed, no doubt in an effort to save package pins. In THS8134 only M2 is multiplexed.

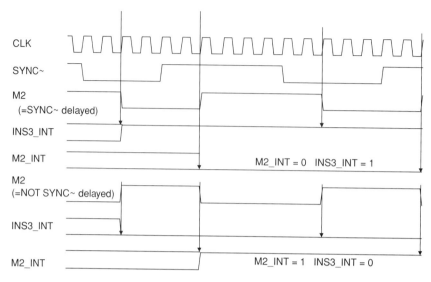

Figure 9.10   THS8134 Mode Selection.

To understand how M2 works we adopt TI's "_INT" extension which means "interpreted as." The interpretation given to M2 (that is, the value of M2_INT) depends on the last event that occurred on the SYNC~ line (Figure 9.10). If the last transition on the SYNC~ line was $0 \rightarrow 1$, then the value read at the M2 pin on the second rising edge of CLK is given to the internal variable M2_INT. Together, the asynchronous input signal M1 and M2_INT configure the device according to Table 9.1. If the last transition on SYNC~ was $1 \rightarrow 0$, the value read at the M2 pin on the second rising edge of CLK is given to a second variable INS3_INT. For INS3_INT = 0 an analog sync is inserted on the GY bus only; if INS3_INT = 1 sync is inserted in all three outputs.

TABLE 9.1   THS8134 Configurations

M1	M2_INT	Configuration	Description
0	0	4:4:4: ; RGB	Separate 8-bit R, G, B buses registered on ↑CLK. Latency 7 CLK pulses. Figure 9.11.
0	1	4:4:4: ; YPbPr	Separate 8-bit Y, Pb, Pr buses registered on ↑CLK. Latency 7 CLK pulses. Figure 9.11.
1	0	4:2:2: ; YPbPr	Two 8-bit buses used; one for Y the other for Pb/Pr registered on ↑CLK. Latency 8 CLK pulses. Sequence starts with YPb. Figure. 9.12.
1	1	4:2:2: ; YPbPr	A single 8-bit bus used for Y, Pb, and Pr registered on ↑CLK. Latency 9 CLK pulses. Sequence starts with PbYPrY. Figure 9.13.

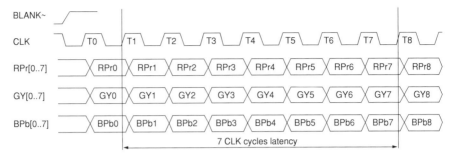

Figure 9.11   4:4:4 Output Mode.

Controlling the device with the M1 and M2 pins is easy as long as M2_INT and INS3_INT are both the same value. We simply set M2 to either 0 or 1. It is a bit more difficult to program one high and the other low. To do so we must generate different M2 values for different SYNC~ transitions. This can be done by delaying the SYNC~ for two CLK pulses before we apply it to the M2 pin, as can be seen in Figure 9.10. A straight delay generates M2_INT = 0 and INS3_INT = 1; inverting SYNC~ and then delaying it by two CLK pulses yields M2_INT = 1 and INS3_INT = 0. This can be easily achieved with a couple of embedded D flip-flops.

The type of analog sync inserted by the THDS8134 depends on the timing of SYNC~ as well as SYNC_T. Asserting SYNC~ injects a negative current into GY or all outputs (depending on INS3_INT), thus dropping the voltage across the loads.

By contrast, bringing SYNC_T high when SYNC~ is active injects a positive current into the outputs and brings them to the high tri-state level required by some HDTV formats. We can see this in Figure 9.14. When SYNC~ is inactive (high) the value of SYNC_T is irrelevant. The timing of SYNC~ and

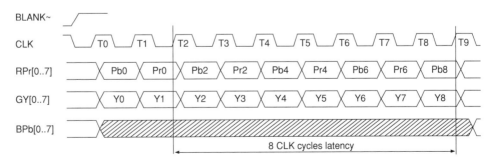

Figure 9.12   4:2:2 16-Bit Output Mode.

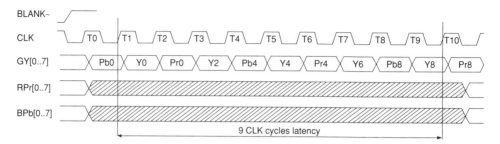

Figure 9.13   4:2:2 8-Bit Output Mode.

SYNC_T must be provided by an external timing generator synchronous to CLK and the data inputs.

9.4 Serializing SDI Video

The process of serializing digital video starts with registering the parallel input data onto the device using the pixel clock. The same clock, together with rate selection information, is used by the on-board PLL to synthesize the high-frequency clock needed to drive the serial data output (Figure 9.15). A *sync detector* block identifies the SAV and EAV headers within the incoming parallel data. These markers are used by the serializer to insert the TRS-ID codes

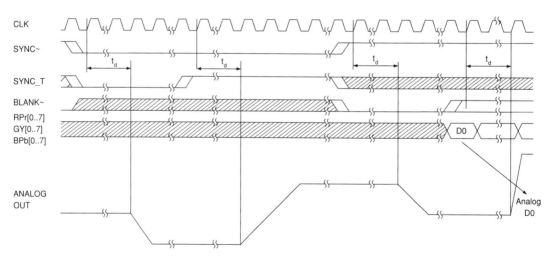

Figure 9.14   THS8134 Sync Generation.

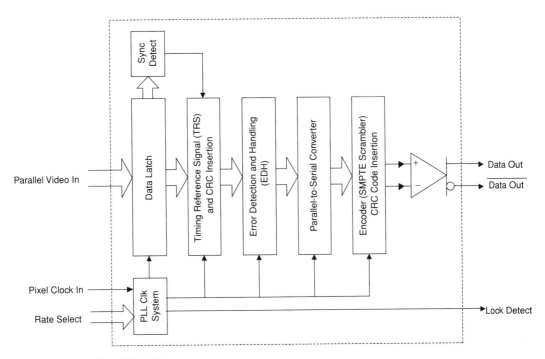

Figure 9.15 Serializer Block Diagram.

and all the ancillary data to be embedded in the stream. The error detection and handling codes (EDH) are also calculated and added.

The actual conversion to serial format employs a *parallel loading shift register* clocked at a frequency ten times higher than the input clock (assuming a 10-bit width for the input bus). CRC calculation, if done serially, uses the same shift register encoding technique we described in Chapter 8. The *SMPTE scrambler* also utilizes the same $G_1(x)$ and $G_2(x)$ generator polynomials as the deserializer, but it implements them in reverse order (Figure 9.16). Some serializers consolidate a *differential driver* output stage into the package.

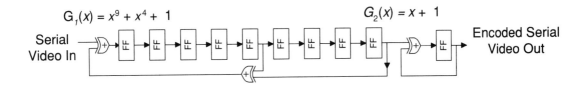

Figure 9.16 NRZI SMPTE Scrambler.

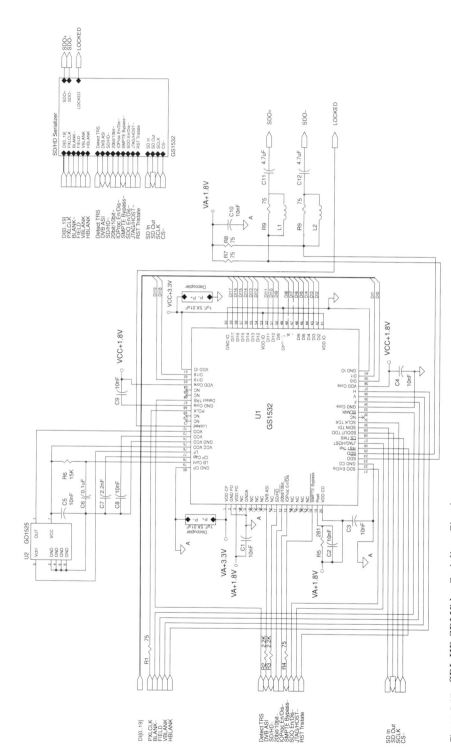

Figure 9.17 SDI, HD-SDI Video Serializer Circuit.

9.4.1 Serializer/Driver Circuit

The GS1532 multirate serializer we choose for our module supports three distinct modes of operation. In the simplest "data-through" mode the output is a serialized translation of the parallel input without any data encoding or scrambling being performed. To set the device for data-through operation both SMPTE BYPASS~ and DVB_ASI must be set to 0.

When the SMPTE BYPASS~ is low and the DVB ASI pin is high the device is compliant with the DVB–ASI standard. In this mode the TRS detection, SMPTE scrambling, and NRZI encoding are all bypassed. The input is 8b/10b encoded instead and K28.5 sync characters are automatically generated for sync and data padding (when no data is present at the input).

In SMPTE mode (SMPTE BYPASS~ and DVB ASI both 1) SMPTE scrambling and NRZI encoding are enabled. The internal flywheel synchronizes the device either to the embedded TRS-ID codes (if the Detect TRS pin is high) or the external FIELD, HBLANK, and VBLANK signals.

I/O processes available in SMPTE mode includes EDH packet generation and insertion (SD), ancillary data preambles and checksum generation, line-based CRC (HD), TRS generation and insertion, and video format detection. If pin IO_Proc_En/Dis is low, all I/O processing is suspended.

Bringing the BLANK~ input low sets the data to the standard specific blank levels, except for the TRS-ID codes. Depending on the 20bit/10bit~ pin, the GS1532 expects 20-bit nonmultiplexed luma and chroma at inputs DI[0..19], or 10-bit interleaved luma/chroma at DI[10..19] with DI[0..9] inputs ignored.

Output levels and edge timing are determined by the SD/HD~ pin, which selects between standard resolution and high resolution settings. The output swing is controlled by the value of the resistor connected at the Rset pin. A value of 281 Ω corresponds to an 800 mV swing across a 75 Ω load. The output can be put in tri-state if the SDO_En/Dis~ pin is set low.

The user interface is similar to SPI with dedicated input SD In, output SD Out, clock SCLK, and chip select CS~ ports. To activate the user interface the JTAG/HOST~ line must be set low. The GS1532, just like the GS1561 deserializer, is designed to work with Gennum's GO1525 oscillator for which it provides dedicated power and interface connections (Figure 9.17). Besides differential data output pins SDO and SDO~, the GS1532 also generates a "locked" signal that is high if the internal PLL is locked to the CLK input and the IC is in data-through mode or has detected TRS-ID or K28.5 signaling.

10.1 I/O Memory Buffers

When we look at the general structure of any digital video system we find one or more memory blocks somewhere in the data flow (Figure 10.1). If we use an external video source we need an *input processor* to convert the video signal into a parallel digital format. This block may be a decoder-digitizer, a deserializer, or a custom front-end processor. In most cases the output of the input processor is stored into an *input memory buffer*. This gives the input processor the latitude of operating in synchronicity with the timing of the incoming video without having to handle downstream timing requirements. Some systems may have more than one input source and therefore more than one input processor—memory buffer pair.

Once the inputs are stored in the input memory buffers, they may be operated on by a *local video processor*. This is the heart of any video system and is where the system creates or adds value for the user. It may consist of a logo overlay engine, an image splitter, a motion detector, or a video scaler.

No matter what the box does, its local video processor needs access to the input memory buffers. After performing its function, the local processor places final data in *output buffers*, from where the output processors (encoder, serializer, and so on) retrieve it, convert it into standard formats, and make it available to communication or display devices, recorders, and so forth.

The most common forms of memory blocks are the field and frame buffers. In the simple application in Figure 10.2 the analog composite from a video camera is digitized and stored in a memory buffer. From there the digital video data is read into the video encoder which reconstructs the composite signal and displays it on a local monitor.

The value added for this arrangement may consist in text or graphics overlay, simple format conversion (NTSC $\leftrightarrow$ PAL), or image capture. All three are related to the way the controller handles the memory buffer through the mem-

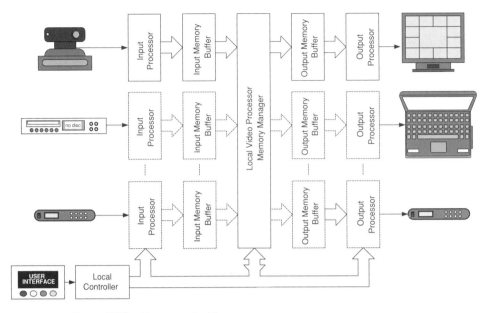

Figure 10.1   General Video Processor Architecture.

ory manager. The raster generator block may or may not be needed depending on the ability of the encoder to generate its own timing (master mode).

We can implement this design using the decoder and encoder modules we developed in Chapters 8 and 9 using both in 8-bit, 4:2:2, YCbCr modes. Can we simply connect the output of the decoder into the input of the encoder without any further buffering? Aside from the fact that all value added is gone,

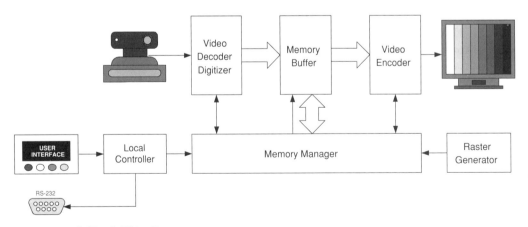

Figure 10.2   A Simple Video Processor.

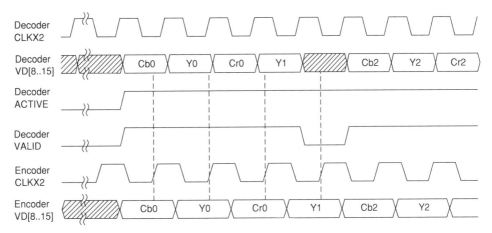

Figure 10.3 Decoder Encoder Timing Differences.

another problem becomes obvious when we look at the timing of the interface between the two (Figure 10.3). The decoder clock is 28.63636 MHz while the encoder operates at a clock speed of 27 MHz, although the average data throughput is the same for both: 27 MBps. To explain the difference we must recall that all the bytes presented to the encoder outside blanking intervals are actually "valid"—that is, they are all converted and encoded. However, in the decoder case, only the data corresponding to a high VALID signal is usable. For the same format, on average, both yield the same number of pixels.

We could try to match the encoder and decoder blocks by setting the encoder to slave mode, using a 28.63636 MHz in/27 MHz out PLL to generate the encoder clock, and feeding the encoder the HSYNC and VSYNC outputs of the decoder. Still, the different clock frequencies would result in data mis-alignment, with the encoder clock often registering invalid input data. For this arrangement the memory buffer is essential, as it is for any other case where the clock frequencies of the input and output are different.

The same situation occurs when we overlay multiple images from random sources on a single screen. Even if we use 27 MHz decoders (less common than the 28.63636 MHz variety) we could at most synchronize the output to one of the inputs, all the others requiring some sort of buffering.

10.1.1 Field Synchronization–Field Inversion

Let us start by using a field buffer between the decoder and the encoder mod-ules. Without any loss of generality we will assume that the read clock f_r is slightly faster than the write clock f_w. Because of the inherent symmetry of the design, the same conclusions would be reached if f_r is slightly slower than f_w. The address pointers specify the address where or from where the memory manager will write or read the next data word. The data to be written originates with the decoder while the data that is read is delivered to the encoder.

Each address pointer is individually reset when it reaches the end of the stored field. We start with both address pointers reset at the top of the very first field (odd) of a video stream. We then take a snapshot of the location of the read address pointer when the write address pointer returns to the top of the odd field (Figure 10.4). After the first couple of fields of invalid data (until a complete frame is written in the buffer), the read pointer races ahead of the write pointer. When the accumulated offset between the read and write pointers exceeds one field, the write pointer and the read pointer will use the same physical buffer to access different fields. While we write odd fields to the buffer we position the same data in the even field of the output raster, resulting in a field inversion. The main consequence of field inversion is a degraded image with staircased edges and stranded lines and pixels [Figure 10.5(b)].

There are two solutions to this problem, the first of which is to use a full frame buffer (Figure 10.6). With a full frame stored in memory there is always a field of the "right" type available for the encoder to read from. When the read address pointer catches up with the write pointer (or, equivalently, is a frame away from the write pointer), it reads the same frame for a second

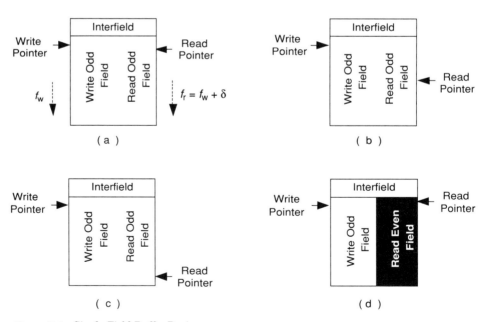

Figure 10.4 Single Field Buffer Design.

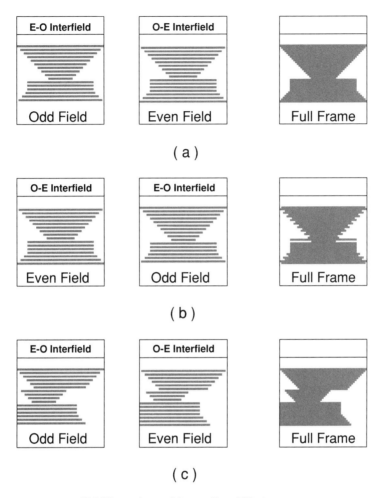

Figure 10.5   Field Inversion and Image Tear Effects.

time, before the write pointer has a chance to upgrade it. The price of this solution is that occasionally we duplicate a frame. If f_r is slower than f_w we occasionally drop a frame instead of repeating it. Either way, at 30 frames per second this is not a problem.

10.1.2 Frame Synchronization–Image Tear

This solution solves some of our problems but not all. Consider the case of a scaled-down video on top of an arbitrary background (PiP). Since the Bt835 has a built-in scaler the block diagram in Figure 10.2 is still relevant. We can fol-

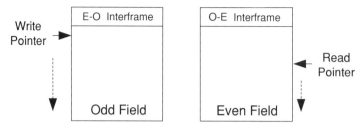

Figure 10.6   Frame Buffer.

low the process from the moment when, as it sporadically happens, the encoder and the decoder have their timing fully aligned (HSYNC, VSYNC, FIELD).

Following vertical resets both the read and write pointers start scanning their respective rasters; the encoder covers the whole screen, while the decoder only the live picture-in-picture area (Figure 10.7). Assume that when the read pointer is not within the PiP area it places a default color code on the input bus of the encoder, thus creating a background color for the PiP.

It becomes obvious that the read pointer is faster than the write pointer. After all, the decoder generates only PiP pixels while the encoder must display both PiP and background; both must take 16.67 msec per field. Since the read pointer is significantly faster it passes the write pointer quite often. As the write pointer advances, the pixels it leaves behind are part of a new PiP frame while the pixels ahead, yet to be updated, are part of the old frame. Therefore, at the top of a new PiP field the read pointer will read newly updated information, part of a new image. Once it passes the write pointer it starts reading

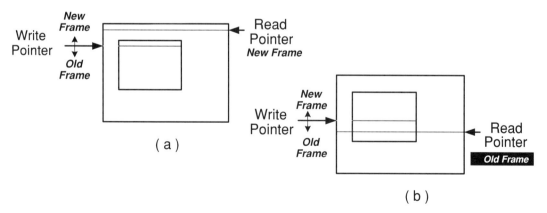

Figure 10.7   PiP Image Tear.

old data. If the video contains motion, even in moderate amounts, the image within the PiP shows a horizontal tear [Figure 10.5(c)], with the old image below the tear and the new image above it. This is a bothersome effect because of its persistence. It does not go away in the blink of an eye; it stays on for minutes at a time and is cyclic in nature.

10.1.3 Dual-Bank Full Frame Buffers

Increasing the memory to two banks, each bank frame size, eliminates this difficulty (Figure 10.8). In this memory architecture the read address pointer runs unimpeded from Odd Field–BANK A to Even Field–BANK A, to Odd Field–BANK B, to Even Field–BANK B, back to Odd Field–BANK A, and so on. However, every time the write pointer reaches the top of the Odd Field–BANK A the memory manager looks for the position of the read pointer. If it is found in Even Field–BANK A or Odd Field–BANK B then it proceeds with writing one frame starting in Odd Field–BANK A [Figure 10.8(a)].

If the read pointer is in Odd Field–BANK A or Even Field–BANK B it is deemed to be too close to the current position of the write pointer and writing is pushed to Odd Field–BANK B [Figure 10.8(b)]. In effect, this puts at least one full field between the two pointers. Memory banking is effective in all cases of pointer speed mismatch, including NTSC/PAL conversions, scan conversions, videowalls, and so forth. The situations dipicted in Figures 10.8(c) and 10.8(d) will lead to image tears.

10.1.4 Single-Bank Full Frame Buffers

When there are sufficient FPGA/CPLD resources to have full control over raster generation one may use the memory-efficient, single-bank method in Figure 10.9. Instead of judging the position of the read pointer relative to fields, each field is split in upper and lower halves.

When the write pointer is at the top of a new odd field, we start writing only if the read pointer is in the bottom half of the odd field or the top half of the even field [Figure 10.9(a)]. If not [Figure 10.9(b)], we wait for one field [Figure 10.9(c)] and then start writing. However, this will obviously introduce a field inversion for which we must compensate. If we increase the even–odd interfield spacing by one line and reduce the odd–even interfield spacing also by one line we eliminate the field inversion side effects [Figure 10.8(d)]. This method can be very effective in closed-system applications such as security boxes but may lack flexibility in open systems.

10.2 Memory Devices

Depending on the type of application we have in mind and our level of experience, we can classify memory buffer devices according to three criteria: flex-

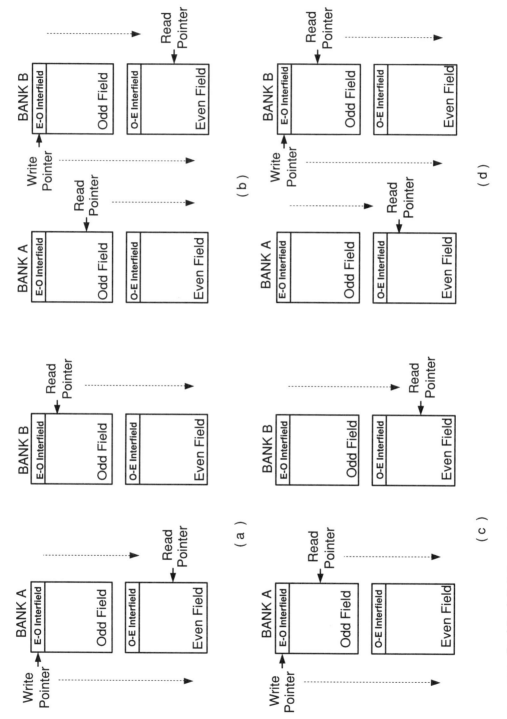

Figure 10.8 Dual Bank Full Frame Memory Buffers.

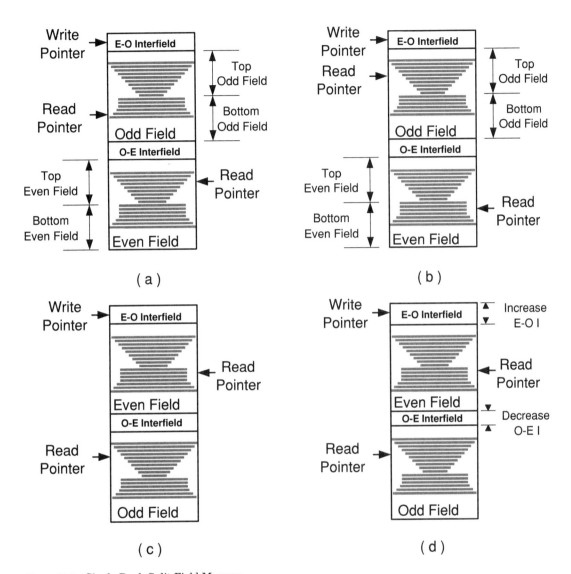

Figure 10.9 Single-Bank Split Field Memory.

ibility, cost, and ease of use. At the extremes the most flexible memory technology is the fast static RAM; the least expensive per stored byte is the synchronous dynamic random access memory (SDRAM) and the easiest to use is the FIFO DRAM.

10.2.1 Static RAM

The internal structure of a static random access memory (SRAM) chip is shown in Figure 10.10. The actual information storage is done in a two-dimensional array of memory cells, each cell consisting of six MOS transistors (6T cell).

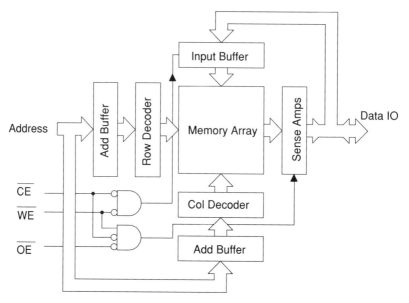

Figure 10.10   SRAM Block Diagram.

The site of every cell is uniquely defined by its row (word) and column (bit) position in the array. To write to a particular location, the corresponding row and column addresses are applied and internally decoded. Once the decoding is stable, the gated chip enable (CE~) and write enable (WE~) signals open the input buffer and the addressed memory cell registers the input line [Figure 10.11(a)]. Registration takes place on the rising edge of the output of the AND gate (CE~, WE~). This allows the alternate write timing in Figure 10.11(b) where the WE~ is held low and the CE~ is used to enable the input buffer.

To read a memory cell we first set up the row and column addresses, and then enable the output sense amplifiers by asserting the OE~ (output enable) line. The sense amplifiers have two roles; they bring the rather small amplitude bit line voltages to logic levels and they limit the output currents.

10.2.1.1 Multi-Image Vertical Display with RAM Buffers   What gives SRAMs an advantage in some video applications is their ability to access with equal ease any memory cell at any time. This is not true about SDRAMs or FIFO DRAMs.

Consider the application in Figure 10.12, where we combine three video streams into one image and display it on a vertical 16:9 plasma screen. Each image is defined by the coordinates of two opposing corners (x_{LL}^i, y_{LL}^i) and (x_{UR}^i, y_{UR}^i). For convenience we will select our inputs to be 480p (progressive, 59.94 Hz, 720 pixels by 480 lines) and our output 720p (progressive, 59.94 Hz,

1280 pixels by 720 lines). Since we only have 1280 pixels per output line we have to crop each input image to 426 lines. This way, after we rotate each input image by 90 degrees and tile them together, we cover 1278 pixels per line (3×426). This is a good match given that some additional cropping will be needed anyway if borders are added.

The correspondence between the input image pixels and output image pixels can easily be established with the help of the image coordinate systems in Figure 10.13. As the 720p scan progresses, for each line, the memory

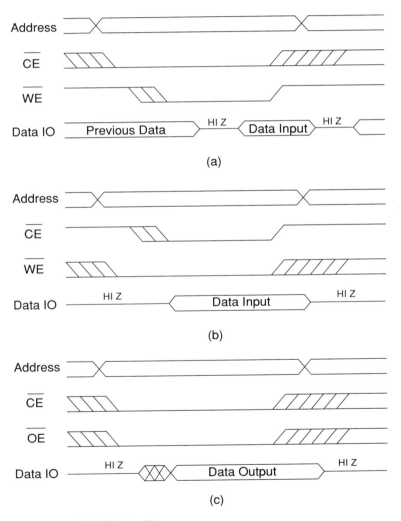

Figure 10.11   RAM Timing Diagram.

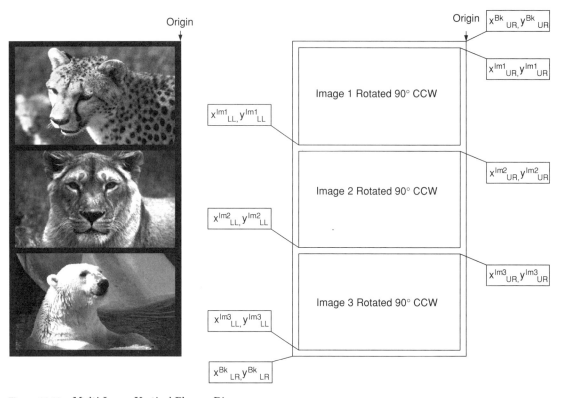

Figure 10.12   Multi-Image Vertical Plasma Dispay.

manager takes the data to be displayed from the memory buffer of the first image when output pixel count xo ≤ 426, then from the memory buffer of the second image (as long as xo ≤ 852), and finally from the memory buffer of the third image.

The equations relating the 720p output pixel and line coordinates (xo, yo) to the (xi_1, yi_1), (xi_2, yi_2), and (xi_3, yi_3) coordinates of the 480p input images are given by:

$$
\begin{cases}
1 \le xo \le 426 & \Rightarrow \begin{cases} xo = yi_1 \\ yo = 721 - xi_1 \end{cases} \begin{cases} xi_1 = 721 - yo \\ yi_1 = xo \end{cases} \\[2em]
427 \le xo \le 852 & \Rightarrow \begin{cases} xo = 426 + yi_2 \\ yo = 721 - xi_2 \end{cases} \begin{cases} xi_2 = 721 - yo \\ yi_2 = xo - 426 \end{cases} \\[2em]
853 \le xo \le 1278 & \Rightarrow \begin{cases} xo = 852 + yi_3 \\ yo = 721 - xi_3 \end{cases} \begin{cases} xi_3 = 721 - yo \\ yi_3 = xo - 852 \end{cases}
\end{cases}
$$

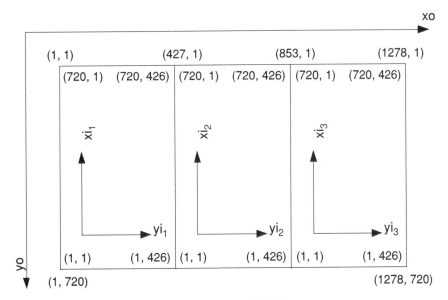

Figure 10.13   Scanning Order for Vertical Plasma Display.

To implement this design we can use two RAM banks for every input, each having a 512KX16 RAM IC. As the input decoder writes into one bank the output encoder is supplied data from the other.

Let's look closer at the RAM access process for one of the memory banks of the first video input. As we can see in Figure 10.14, the YC information for each row is written to consecutive RAM locations arrived at by incrementing the A[0..9] address lines. When the end of a line is reached, A[0..9] is cleared and A[10..A18] is incremented by one. A[0..9] can be regarded as the pixel address pointer while A[10..18] is the line address pointer. When the line address is 479 and the pixel address reaches 719 both address ranges are cleared, and the memory manager gets ready for a possible bank swap (see section 10.1 in this chapter).

On the output side, information will only be read out when the current 720p pixel overlaps the first video area, which is one-third of the time. As can be seen in Figure 10.15 each output line consists of three input columns. To read the data out in the proper order, the A[0..9] is set to 719 and the A[10..18] is cleared. Then the column that corresponds to the output ⅓-line is read pseudo-sequentially by keeping A[0..9] constant and incrementing A[10..18]. When the end of the column is reached, A[0..9] is decremented by one and A[10..18] is cleared. The process continues until the end of the column for row 720 is reached. Then A[0..9] is set to 719, A[10..18] is cleared and everything starts over again. This type of address hopping can only be done with RAMs.

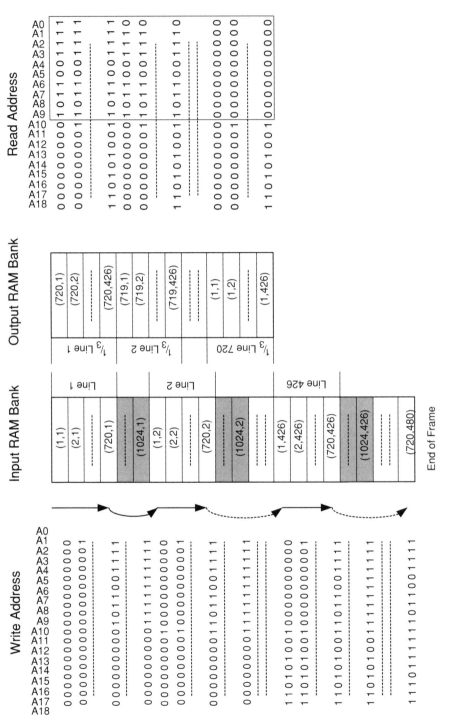

Figure 10.14 RAM Memory Addressing for Vertical Plasma Display.

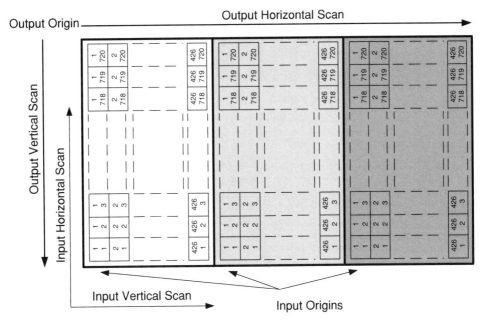

Figure 10.15   Vertical Plasma Display Pixel Counts.

10.2.2 DRAM

When dynamic random access memories (DRAMs) first became available they represented a huge leap in memory density when compared to SRAMs. This is due in large part to the structure of the DRAM memory cell. Instead of the six transistors found in a SRAM cell, DRAMs employ one transistor and one leaky storage capacitor. The silicon area occupied by such a simplified arrangement is much smaller, and therefore the DRAM cell density is a lot higher than that of the SRAM. However, there is a price to pay. DRAM speeds are lower, and there is significant overhead required to maintain the charge on the storage capacitor.

DRAM maintenance is accomplished by a built-in "refresh" controller that periodically scans and rewrites all memory cells. When cells are refreshed they are not available to external devices, a de facto delay not present in RAMs. Further delays are incurred because of address line multiplexing, which is used in DRAMs as a way of reducing pin numbers and therefore packaging costs (Figure 10.16).

The way DRAMs operate can be seen in Figures 10.17 and 10.18. If we organize our video raster to match the internal organization of the DRAM memory array, then columns correspond to pixels and rows to video lines. The input address is latched first on the row address register by the falling edge of RAS~ (row address strobe), and then on the column address register on the falling edge of CAS~ (column address strobe).

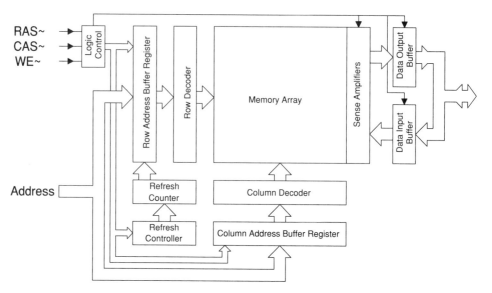

Figure 10.16 DRAM Block Diagram.

For a read operation, the first data can be read from the output buffer of the device after an access time of about 100 nsec from the falling edge of RAS~ (we used Oki semiconductor DRAM data as an example). In fast page mode, consecutive data words can be read just by toggling the CAS~ line (Figure 10.17); the corresponding read times are as low as 35 nsec.

Fast page write operation is similar to fast page read (Figure 10.18). After initial access and first data word write operation, consecutive writes are performed by toggling the CAS~ line. Refresh cycles need to be performed every 16 to 128 msec, and take about 100 nsec per row. They could be done during video blanking intervals. Although random address read and write operations require the full access time of 100 nsec, fast page access is down to pixel time levels.

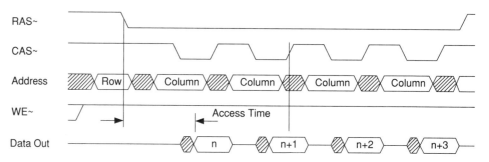

Figure 10.17 DRAM Read Cycle—Fast Page Mode.

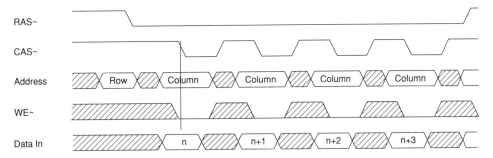

Figure 10.18 DRAM Write Cycle—Fast Page Mode.

It is clear that, by carefully managing the needs of the video raster and the operational idiosyncrasies of DRAMs, they can be successfully used in video applications, especially where large memory buffers are required. Still, the local controller is burdened with asynchronously placing addresses on the DRAM pins and holding them there for prescribed periods of time. It must also strobe the RAS~ and CAS~ lines in proper sequences for read, write, and refresh operations. While the DRAM precharges the lines, decodes the address, senses the data, and routes it to the output buffers, the controller must simply wait.

The need for closer synchronization between the DRAM and the memory controller led to the development of SDRAMs.

10.2.3 SDRAM

The architecture of synchronous DRAMs or SDRAMs includes two separate memory arrays (banks), synchronous input and output interfaces, and a com-

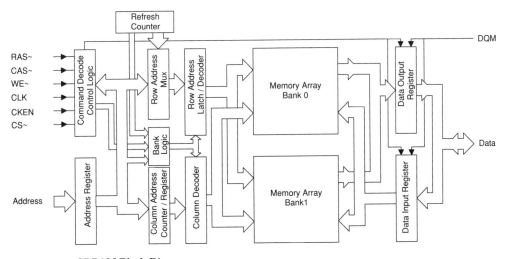

Figure 10.19 SDRAM Block Diagram.

TABLE 10.1   Examples of SDRAM Commands

Current State	CS~	RAS~	CAS~	WE~	Command
Any	0	1	1	1	No Operation
Idle	0	0	1	1	Select and Activate Row
Row Active	0	1	0	0	Select Column and Start Write Burst
Write	0	1	1	0	Terminate Write Burst

mand decoder (Figure 10.19). In an SDRAM all the I/O lines, (data, address, and control) are synchronized to the system clock (CLK) by means of I/O latches. In this new environment the old RAS~, CAS~, CS~, and WE~ signals lose their traditional DRAM meanings and, when grouped together and latched, become SDRAM commands.

Table 10.1 exemplifies some commands used in Micron's 64Mb SDRAMs. Once a command and the associated address are registered by the device, the controller is free to perform other tasks and may return to the SDRAM after the command has been executed.

The full page read and write burst modes are the modes employed most often in video applications. After a page (row) is opened and after issuing a WRITE or READ command, writing information on the page or reading it out is done automatically on consecutive clock pulses (Figures 10.20 and 10.21). In these modes, SDRAM-based designs can be very fast, with access time as low as 6 nsec.

The two separate memory arrays add even more flexibility to the SDRAM. They become a natural venue for implementing the dual-bank memory

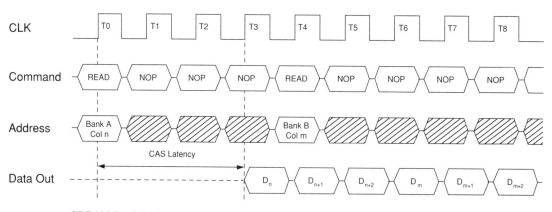

Figure 10.20   SDRAM Read Cycles with Bank Switching.

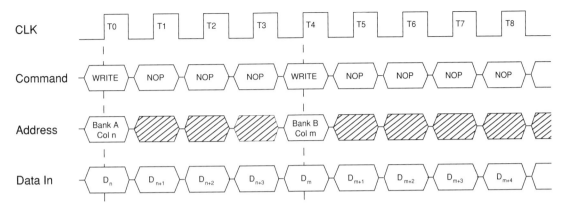

Figure 10.21   SDRAM Write Cycles with Bank Switching.

designs mentioned earlier; alternately, they can be used in interleaved mode to further boost the operational speed of the device. These kinds of capabilities have made SDRAM the preferred memory technology for video applications. However, if what you need is fast random access to any location in memory, SDRAMs may fall short in some applications, such as high-resolution live vertical displays.

As can be seen in Figure 10.22, a random access cycle in SDRAMs is broken down into a number of separate commands, executed in series. This brings the access time in the 30 to 40 nsec range, much slower than the access time for fast RAMs. The command structure required by random access is also more complicated, making the FPGA/CPLD memory manager more complex.

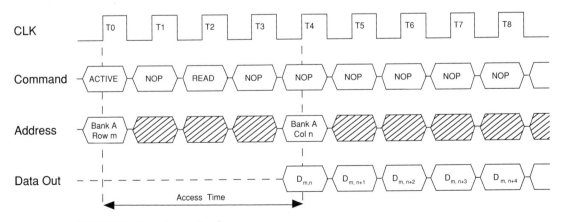

Figure 10.22   SDRAM Random Access Read.

10.2.4 FIFO DRAM

If you do not need random access and you just want the simplest and quickest memory buffer solution available, the FIFO DRAM is the answer. FIFO DRAMs generate their own internal read and write addresses, in a FIFO type arrangement. The address counters can only be externally incremented or reset, which is not a problem if your system does not need random access.

Internal data processing starts with the write clock pulses (WCLK) storing the input data in a pair of serial-in parallel-out shift registers. When the shift registers are full (one complete line), the contents are passed on to a write line buffer and from there to the memory array, at the line location pointed to by the read/write/refresh controller (Figure 10.23). Partially filled lines that do not make it to the array can be read from separate partial line shift registers. After each shift register transfer the write line address pointer is incremented. You can reset this pointer at any time by asserting the WRST write reset line (Figure 10.24).

The read operation starts with applying a read reset signal to the RRST line. This clears the read address counter and retrieves the first line of data

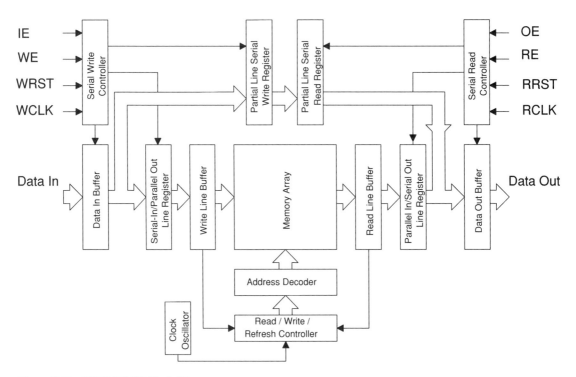

Figure 10.23   FIFO DRAM Block Diagram.

IE, WE Logic High

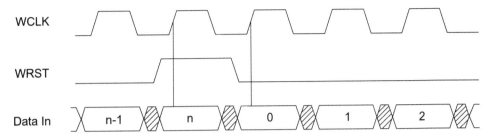

Figure 10.24   FIFO DRAM Write Reset and Write Cycles.

from the memory array. The line is then stored in a parallel-in serial-out shift register, from where it is retrieved by consecutive RCLK pulses and placed in a data out buffer. With continuing read pulses the shift register is emptied, and a new line is read out of the array.

As the timing diagrams in Figures 10.24 and 10.25 show, the only control signals needed to operate a FIFO-based memory buffer are WCLK, RCLK, WRST, and RRST. The first two correspond directly to write and read pixel clocks while the last two can be generated by the corresponding vertical interval signals. There are no address lines and no memory refreshing (all done internally) to manage. There are some minor limitations, but any of the tear prevention techniques we covered, if implemented, also satisfy the FIFO restrictions.

10.2.4.1 FIFO DRAM Circuit The FIFO DRAM circuit module in Figure 10.26 uses two OKI MS81V10160 640K×16 devices. Each IC fully implements a 4:2:2 YCrCb memory buffer, both for NTSC and PAL standards, and both for 4:3 and 16:9 screen layouts. With read and write access times of 12 nsec the module can handle interlaced (with consecutive field storage) as well as pro-

OE, RE Logic High

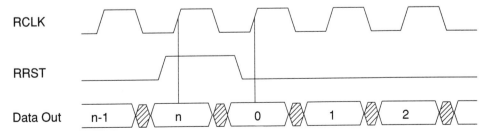

Figure 10.25   FIFO DRAM Read Reset and Read Cycles.

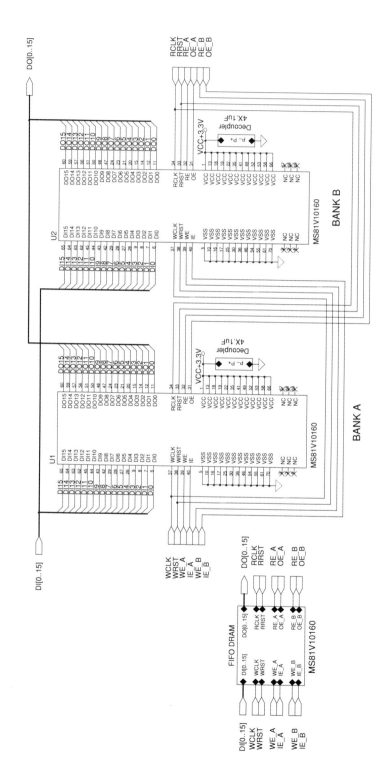

Figure 10.26 FIFO DRAM Memory Circuit.

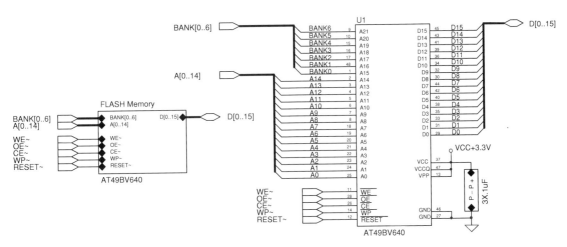

Figure 10.27   FLASH Memory Circuit.

gressive formats. The ICs can also be cascaded for larger capacity storage, which makes them usable through 1080i, 74.25 MHz data rates.

10.2.5 FLASH Memory

In a different class entirely, but with significant applications in video design, we find high-capacity FLASH memories. A single 4MX16 IC, such as the AT49BV640 from Atmel, can store almost twelve full uncompressed NTSC frames (Figure 10.27). FLASH memory operation has been simplified to a point where, once initialized, reading from it is identical with reading a RAM (Figure 10.28). The read access time for the AT49BV640 is about 70 nsec.

Internally, the device is segmented into memory blocks or sectors that can be individually erased and programmed. Memory locations in a given sector

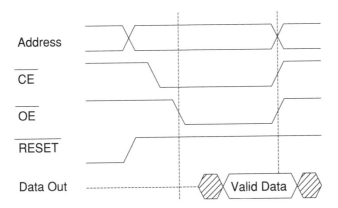

Figure 10.28   FLASH Read Cycle.

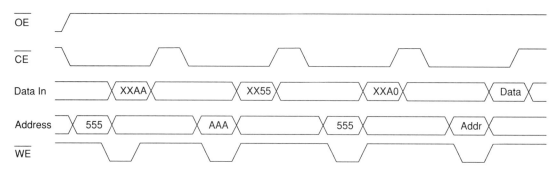

Figure 10.29   Example of FLASH Memory Write Sequence.

cannot be programmed unless the sector is erased first. Writing to the FLASH involves sending it a series of commands, followed by a wait period (Figure 10.29). The local controller can verify if a write cycle is completed by polling the internal registers of the FLASH.

The word programming time for the Atmel part is 22 μsec. The erase sequence for the AT49BV640 is shown in Figure 10.30 and the typical sector erase time is 100 to 500 msec.

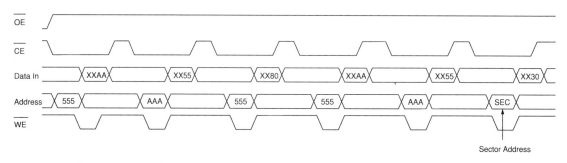

Figure 10.30   Example of FLASH Memory Erase Sequence.

Combining Video Images

11.1 Overlay Processor

Image tiling is one of the most common applications of digital video. The process of combining multiple images on a single screen starts with digitizing the images and scaling them to the desired proportions. An overlay template is then created to tell the system where each image is to appear on the screen and what its dimensions should be.

Image shaping, cropping, and scaling—all part of image tiling—are independent operations. We may zoom into a scene and simultaneously shrink the overall image by cropping it in the horizontal and vertical directions. If the pictures overlap we need to communicate the priority of each image: which one is on top, which is next, and so on. Each image may have a distinctive frame of width and color (or pattern) to be programmed in by the user. We can then add special effects. Shadows, graphics, text, and chroma-keyed images can all be overlayed on top of live video streams.

The functional block that performs the actual image mixing is the *overlay processor*. The internal block diagram of an overlay processor is shown in Figure 11.1, in the context of a four-channel image combiner. The data on which the overlay processor operates consists of the digitized and scaled versions of the four input videos; these are stored in four dual-ported frame buffers.

According to an *overlay template* provided by the user through the local controller, the overlay processor reads pixel data from the memory buffers and routes it, through a bus multiplexer, to the output video encoder or graphics DAC. The overlay template is read according to the current pixel and line addresses, which are provided by line and pixel counters inside the *raster generator*. In addition to video data the encoder also requires standards-specific reference signals such as pixel clock, vertical and horizontal syncs, blanking and field information. These are synthesized by the raster generator as well.

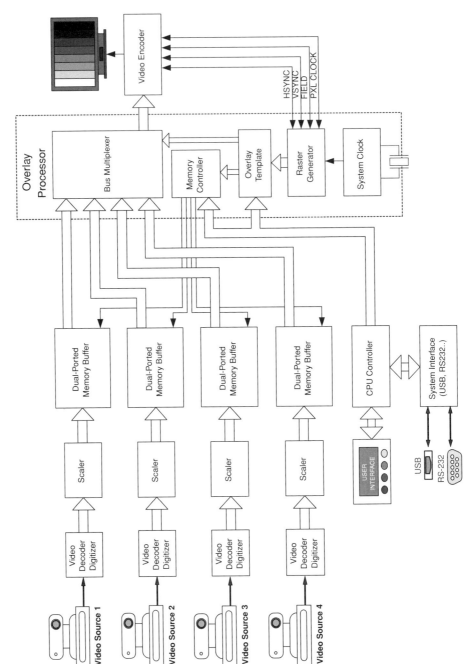

Figure 11.1 Four-Channel Video Processor.

Encoders that support a so-called "master mode" generate their own video timing signals that can be fed back to the overlay processor, thus simplifying the FPGA or CPLD firmware. All the digital circuitry is synchronous and is driven by a system clock with a frequency multiple that of the pixel clock.

11.2 Raster Generation for Progressive Video

As discussed in Chapter 1, an image can be scanned either in a progressive or an interlaced pattern. The screen organization for a progressive scan 525 line, 60 Hz, 720 × 480 pixels video format is shown in Figure 11.2. Its simpler layout makes progressive type images easier to combine than its interlaced counterparts, although it requires significantly higher processing speeds.

A raster generator (Figure 11.3) has two major functions. When driven by a clock with a frequency equal to (or a multiple of) the desired pixel rate, it generates the screen address for the current pixel. The address consists of the current line number (given by *line count[..]*) and the current pixel rank (*pixel count[..]*). This address is generally used to fetch the values (RGB, YCbCr...) of the current pixel from a video buffer and deliver them to a video encoder or another IC device. The raster generator also produces all the reference signals required to properly position the current pixel. They are horizontal sync (HSYNC), vertical sync (VSYNC), pixel clock (PXLCLK), and blanking intervals (BLANK). On occasion the components of BLANK, horizontal blanking (HBLANK) and vertical blanking (VBLANK) are also made available. The reference signals are generated by comparing the values of the line

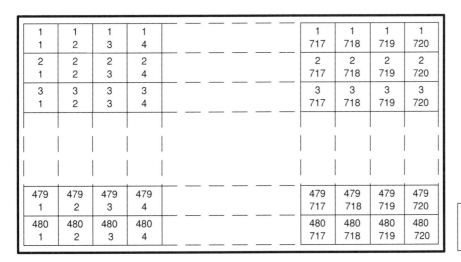
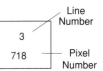

Figure 11.2   Progressive 720 × 480 NTSC Pixel Count.

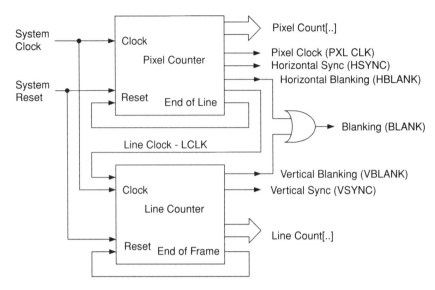

Figure 11.3 Progressive Raster Generator.

and pixel counters to landmarks on the timing diagram of the video format we are synthesizing.

Figure 11.4 shows a timing diagram of a typical progressive format. Even if theoretically any edge of any signal can be used as a starting point for the counters, most designs use either the leading edge of HBLANK or the falling edge of HSYNC as origin (pixel count[..] = 0, line count[..] = 0). We will use the first one.

At the start of a new line, HBLANK, HSYNC, and LCLK are all high. LCLK, or line clock, is the auxiliary clock driving the line counter. When the value of the pixel counter reaches the desired length of the line clock pulse x_{LW}, LCLK is set low. The pixel counter then arrives at x_{HFD}, the horizontal front delay value, where it drops the HSYNC to zero starting the horizontal sync pulse. At $x_{HFD} + x_{HSW}$, after the horizontal sync pulse is completed, the HSYNC is restored high. When the elapsed time equals the desired horizontal front blanking time x_{FB} HBLANK is brought low, and the first active pixel is processed.

The count continues until the *end of line* value is reached. At this point the counter is reset and the cycle is repeated. The visible part of the line is between the falling edge of HBLANK and the *end of line* marker. A similar process is carried out along the vertical axis, this time driven by the line clock LCLK. With each new pixel line, the line counter is advanced and decoded.

On the first line VBLANK and VSYNC are set high. Then, after the vertical front delay y_{VFD}, VSYNC is brought low for the duration of y_{VSW} ; VBLANK pulse is completed at y_{FB} and the *end of frame* is reached at y_{EFR}. For the line

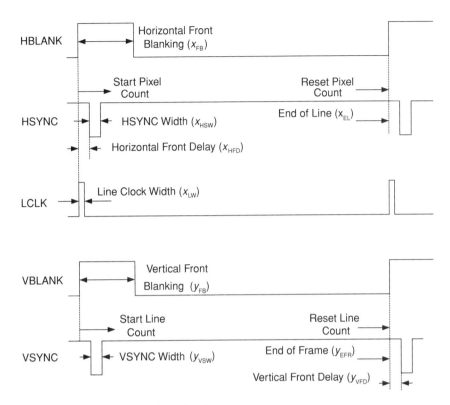

Figure 11.4 Progressive Vertical Signals.

when VSYNC is asserted, VSYNC leads the corresponding HSYNC pulse by the horizontal front delay x_{HFD}. This phase relationship may not always hold and may need adjustments for some formats. The way the various landmarks frame a video image can be seen in Figure 11.5. The origin of this raster is in the top left corner. The image itself starts at (x_{UL}, x_{UL}—upper left) and is scanned left to right and top to bottom until the *end of frame* at (x_{LR}, y_{LR}—lower right).

11.3 Raster Generation for Interlaced Video

Generating the pixel addresses and reference signals for interlaced formats is more complicated. The format itself is more complex. Instead of a single frame with a single scanning pattern, we have two fields, each half-frame, shifted in time by half a line (vertically by one line and horizontally by minus half a line). The pixel structure itself is the same as for the progressive format (after all, a screen is a screen); the order of reading and writing the pixels is different (Figure 11.6).

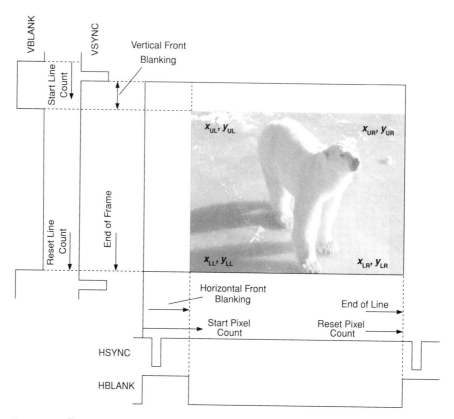

Figure 11.5 Progressive Image Frame.

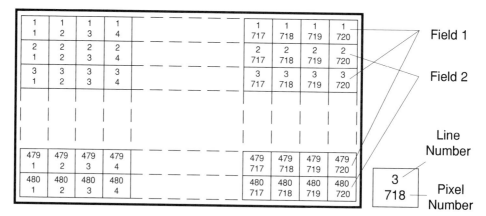

Figure 11.6 Interlaced 720X480 NTSC Pixel Count.

We first write line 1, then line 3, continuing until line 479. Then we write line 2, followed by line 4 and, at the end, 480. The advantage of interlaced video is its lower refresh rate. So where are the half-lines and timing headaches? The answer is found in the block diagram of the interlaced raster generator (Figure 11.7) and its associated timing diagram (Figure 11.8). The first change we see when comparing against the progressive raster generator is the replacement of the line counter with a half-line counter. This is needed to produce the two half-lines in the raster: one at the end of the first field and one at the beginning of the second. To drive the half-line counter we need a half-line clock (1/2LCLK); it is created by the pixel counter by adding a second pulse in the middle of the pixel line. Since active pixels start at the end of the horizontal front blanking interval x_{FB}, the second half-line clock pulse is not in the center of the visible screen, but shifted to the left.

There are significant changes also on the vertical. The vertical blanking signal now has two separate segments: one at the top of the screen (vertical front blanking) and one separating the two fields (vertical retrace blanking). There are also two VSYNC pulses: the first close to the beginning of the frame, the other at the beginning of the second field. The second VSYNC pulse coincides with the vertical retrace action in old CRT tubes.

If we look at the outputs of our raster generator block diagram we notice the presence of two more signals. The first is F and identifies the current field. Its

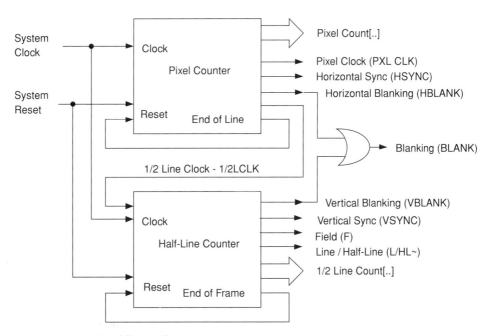

Figure 11.7 Interlaced Raster Generator.

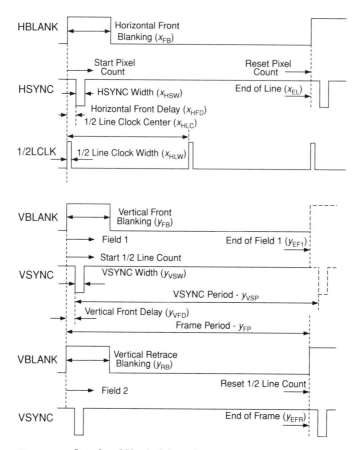

Figure 11.8 Interlaced Vertical Signals.

standard values are 0 for odd fields, and 1 for even fields. Since inadvertently switching the fields results in a jittery, "staircased" image, it is very important to associate the correct field flag F to the pixel data.

The second new signal is the line/half-line flag (L/HL~). To understand its role we can fast-forward a bit to image overlays. Assume that we want to display a rectangular image somewhere in the center of the screen. Identifying the range of lines and pixels the image occupies on a progressive raster is straightforward: pixels 340 to 500 on lines 200 to 320, for example.

For an interlaced raster the image takes the same horizontal range. Vertically, however, the image is not only split into fields—the pixels on each line are split among two half-line counts. The pixels up to 430 (for 858 total pixels per line) belong to the "front" half-line while the pixels after 430 belong to the "back" half-line. What the L/HL~ does is tell us which half-line the current pixel is on.

Why did we choose 430 as our midline instead of 429 (858/2)? If we use a 4:2:2 or 4:1:1 format we must form groups of two or four pixels in order to keep

$Y^{Bk}_{i,j}$	$Cb^{Bk}_{i,j}$	$Y^{Bk}_{i,j+1}$	$Cr^{Bk}_{i,j}$	$Y^{Bk}_{i,j+2}$	$Cb^{Bk}_{i,j+2}$	$Y^{Bk}_{i,j+3}$	$Cr^{Bk}_{i,j+2}$	$Y^{Bk}_{i,j+4}$	$Cb^{Bk}_{i,j+4}$	$Y^{Bk}_{i,j+5}$	$Cr^{Bk}_{i,j+4}$
$Y^{Bk}_{i+1,j}$	$Cb^{Bk}_{i+1,j}$	$Y^{Bk}_{i+1,j+1}$	$Cr^{Bk}_{i+1,j}$	$Y^{Bk}_{i+1,j+2}$	$Cb^{Bk}_{i+1,j+2}$	$Y^{Bk}_{i+1,j+3}$	$Cr^{Bk}_{i+1,j+2}$	$Y^{Bk}_{i+1,j+4}$	$Cb^{Bk}_{i+1,j+4}$	$Y^{Bk}_{i+1,j+5}$	$Cr^{Bk}_{i+1,j+4}$
$Y^{Bk}_{i+2,j}$	$Cb^{Bk}_{i+2,j}$	$Y^{Bk}_{i+2,j+1}$	$Cr^{Bk}_{i+2,j}$	$Y^{Bo}_{1,1}$	$Cb^{Bo}_{1,1}$	$Y^{Bo}_{1,2}$	$Cr^{Bo}_{1,1}$	$Y^{Bo}_{1,3}$	$Cb^{Bo}_{1,3}$	$Y^{Bo}_{1,4}$	$Cr^{Bo}_{1,3}$
$Y^{Bk}_{i+3,j}$	$Cb^{Bk}_{i+3,j}$	$Y^{Bk}_{i+3,j+1}$	$Cr^{Bk}_{i+3,j}$	$Y^{Bo}_{2,1}$	$Cb^{Bo}_{2,1}$	$Y^{Bo}_{2,2}$	$Cr^{Bo}_{2,1}$	$Y^{Bo}_{2,3}$	$Cb^{Bo}_{2,3}$	$Y^{Bo}_{2,4}$	$Cr^{Bo}_{2,3}$
$Y^{Bk}_{i+4,j}$	$Cb^{Bk}_{i+4,j}$	$Y^{Bk}_{i+4,j+1}$	$Cr^{Bk}_{i+4,j}$	$Y^{Bo}_{3,1}$	$Cb^{Bo}_{3,1}$	$Y^{Bo}_{3,2}$	$Cr^{Bo}_{3,1}$	$Y^{In}_{1,1}$	$Cb^{In}_{1,1}$	$Y^{In}_{1,2}$	$Cr^{In}_{1,1}$
$Y^{Bk}_{i+5,j}$	$Cb^{Bk}_{i+5,j}$	$Y^{Bk}_{i+5,j+1}$	$Cr^{Bk}_{i+5,j}$	$Y^{Bo}_{4,1}$	$Cb^{Bo}_{4,1}$	$Y^{Bo}_{4,2}$	$Cr^{Bo}_{4,1}$	$Y^{In}_{2,1}$	$Cb^{In}_{2,1}$	$Y^{In}_{2,2}$	$Cr^{In}_{2,1}$

Figure 11.9   4:2:2 Image Border Detail.

the luma and chroma information together (Figure 11.9). Therefore the number of pixels from the first active pixel to the midline point has to be divisible by two or by four unless, of course, we work with 4:4:4 video.

For 720 active pixels per line the first active pixel is 139. The number of pixels to midline is $429 - 138 = 291$. The closest number divisible by four is 292; when added back to 138, it yields 430. The proper position, then, for the beginning of the center half-line clock is pixel 430, one pixel off from the exact 429 value. Here we have to be careful. We can only move the center half-line clock pulse at most a few pixels before falling into the "not quite NTSC or PAL" format; that may result in signal locking problems for the monitor display.

The framing of an interlaced display is illustrated in Figure 11.10. We do not see any half-lines because only the lines 23 to 262 of field 1 and 286 to 525 of field 2 are displayed; the others, including the half-lines, are blanked out in digital formats or used for synchronization, equalization, and so on in analog formats.

11.4 Raster Generator Design for Interlaced Video

The module diagram and the flowchart for an interlaced raster generator are shown in Figures 11.11 and 11.12.

The operation starts with a power-up reset which sets all counters to zero. After a first evaluation of reference parameters (HSYNC, VSYNC...), the counters are checked against external resets and *end-of-pixel/end-of-frame* conditions. We need to keep the option of externally resetting our counters for applications that require resynchronization to other video sources (genlocking, vertical interval switching, and so on). If no *end-of-pixel/end-of-frame* is detected the counters are incremented and the *pixel[..]* and *half_line[..]* values are updated.

When returning to the top of the flowchart (next clock cycle) a new set of reference parameters are calculated based on the new counter values. If external resets or *end of pixel/end of frame* conditions exist, the counter(s) are reset and the cycle is restarted.

The auxiliary signal half-line flag is used to narrow the half-line clock to a single clock pulse. It is set to 1 on the first system clock which finds the half-

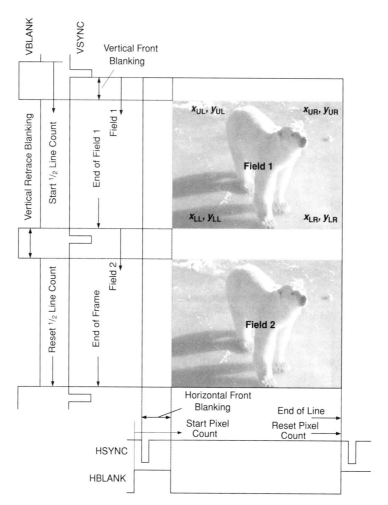

Figure 11.10 Interlaced Image Frame.

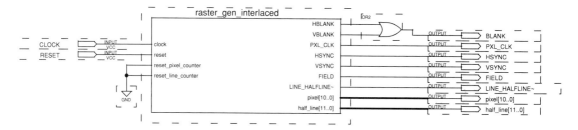

Figure 11.11   Interlaced Raster Generator Block (Altera MAX+plus II).

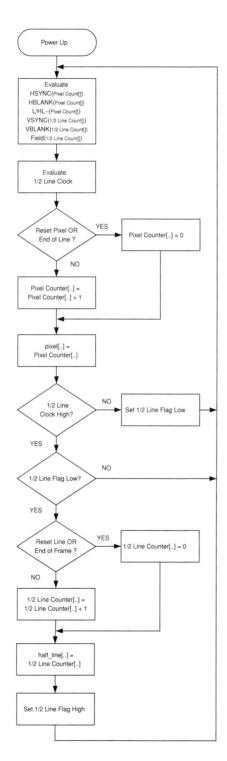

Figure 11.12 Interlaced Raster Generator Flowchart.

line clock high. If the half-line clock stays high for multiple system clocks the flag stops the half-line counter from being incremented more than once.

In a high-speed synchronous implementation the half-line flag is not needed. In slower or multiple-chip implementations the cumulative delays between the pixel and half-line counters may be sufficient for the half-line counters to miss a clock pulse. If we stretch the half-line clock on the pixel counter side and filter at the line counter location, the delay is accounted for. This artifice is more educational than imperative for this design.

11.4.1 Interlaced Raster Generation Design Example

An example of actual raster generator code is given below. We used for it a subset of commands from Altera's AHDL language that is close enough to pseudo-code to be understandable to non-FPGA/CPLD users. This code actually runs on Altera's freeware MAX+plus II Baseline compiler.

Simply store the code segment in a text file called *raster_generator_interlaced.tdf*, assign the device you want to implement it on and its pin configuration (Assign Menu), compile it using the MAX+plus II Baseline compiler, and you are ready to program it on a chip. Of course, you can use the CPLD EPM7512 schematics in Chapter 15 for your hardware.

To complete the code you only need to add the appropriate standard-specific (NTSC, PAL...) parameters to the CONSTANT section. The language is easy to understand. After declaring what signals you want to use as inputs and outputs, you need to specify what kind of hardware blocks you are using.

We selected the D flip-flop (DFF) as our elementary storage element. The required signal connections for a DFF are clock (*variable_name*.clk), reset (*variable_name*.clrn), input (*variable_name*.d), and output (*variable_name*.q). The AHDL logic and comparator operators are:

```
Priority  Operator/Comparator

1            -       (negative)
1            !       (NOT)
2            +       (addition)
2            -       (subtraction)
3            ==      (equal to)
3            !=      (not equal to)
3            <       (less than)
3            <=      (less than or equal to)
3            >       (greater than)
3            >=      (greater than or equal to)
4            &       (AND)
4            !&      (NAND)
5            $       (XOR)
5            !$      (XNOR)
6            #       (OR)
6            !#      (NOR)
```

If you are familiar with the operation of the IF__ELSIF__ELSE__END IF statement from C or BASIC you are ready to translate the flowchart of Figure 11.12 into AHDL code.

```
% Raster Generator Interlaced Format %
CONSTANT     x_fb              =           ;%horizontal front blanking%
CONSTANT     x_hsw             =           ;%HSYNC width%
CONSTANT     x_el              =           ;%end of line%
CONSTANT     x_hfd             =           ;%horizontal front delay%
CONSTANT     x_hlw             =           ;%half line clock pulse width%
CONSTANT     x_hlc             =           ;%half line clock center pulse position%
CONSTANT     y_fb              =           ;%vertical front blanking%
CONSTANT     y_vsw             =           ;%VSYNC width%
CONSTANT     y_ef1             =           ;%end of field 1%
CONSTANT     y_vfd             =           ;%vertical front delay%
CONSTANT     y_vsp             =           ;%VSYNC period%
CONSTANT     y_rb              =           ;%vertical retrace blanking%
CONSTANT     y_efr             =           ;%end of frame%


SUBDESIGN raster_gen_interlaced
(
     clock,  reset                          :INPUT;
     reset_pixel_counter, reset_line_counter :INPUT;
     HBLANK                                 :OUTPUT;
     VBLANK                                 :OUTPUT;
     PXL_CLK                                :OUTPUT;
     HSYNC                                  :OUTPUT;
     VSYNC                                  :OUTPUT;
     FIELD                                  :OUTPUT;
     LINE_HALFLINE~                         :OUTPUT;
     pixel[10..0]                           :OUTPUT;
     half_line[11..0]                       :OUTPUT;

)
VARIABLE
     pixel_count[10..0]                     :DFF;
     1/2_line_clock                         :DFF;
     1/2_line_count[11..0]                  :DFF;
     1/2_line_flag                          :DFF;
     HSYNC                                  :DFF;
     VSYNC                                  :DFF;
     VBLANK                                 :DFF;
     FIELD                                  :DFF;
     LINE_HALFLINE~                         :DFF;
```

```
BEGIN
    pixel_count[].clk              =    clock;
    pixel_count[].clrn             =    !reset_pixel_counter & !reset;
    1/2_line_clock.clk             =    clock;
    1/2_line_clock.clrn            =    !reset;
    1/2_line_count[].clk           =    clock;
    1/2_line_count[].clrn          =    !reset;
    HSYNC.clk                      =    clock;
    HSYNC.clrn                     =    !reset;
    HBLANK.clk                     =    clock;
    HBLANK.clrn                    =    !reset;
    VSYNC.clk                      =    clock;
    VSYNC.clrn                     =    !reset;
    LINE_HALFLINE~,clk             =    clock;
    LINE_HALFLINE~,clrn            =    !reset;
    1/2_line_flag.clk              =    clock;
    1/2_line_flag.clrn                  !reset;


        PXL_CLK                    =    clock;

        1/2_line_clock.d           =    (((pixel_count[].q <= x_hlw))#
                                        ((pixel_count[].q > x_hlc)&
                                        (pixel_count[].q <= x_hlc + x_hlw)));

        HSYNC.d                    =    !((pixel_count[].q > x_hfd)&
                                        (pixel_count[].q <= x_hfd + x_hsw));

        HBLANK.d                   =    (pixel_count[].q <= x_fb);

        LINE_HALFLINE~.d           =    (pixel_count[].q <= x_hlc);

        VSYNC.d                    =    !(((1/2_line_count[].q >y_vfd)&
                                        (1/2_line_count[].q <= y_vfd + y_vsw))#
                                        ((1/2_line_count[].q > y_vfd + y_vsp)&
                                        (1/2_line_count[].q <= y_vfd + y_vsp
                                            + y_vsw)));

        VBLANK.d                   =    ((1/2_line_count[].q <= y_fb)#
                                        ((1/2_line_count[].q > y_ef1)&
                                        (1/2_line_count[].q <= y_ef1 + y_rb)));

        FIELD.d                    =    !(1/2_line_count[].q <= y_ef1);

        IF    reset_pixel_counter THEN
              pixel_count[].d           =      GND;
```

```
    ELSIF  (pixel_count[].q == x_el) THEN
           pixel_count[].d        =        GND;
    ELSE
           pixel_count[].d        =        pixel_count[].q + 1;
    END IF;
           pixel[]                =        pixel_count[];


    IF    1/2_line_clock.q THEN
          IF       !1/2_line_flag.q THEN
                   IF       reset_line_counter THEN
                            1/2_line_count[].d      =           GND;
                   ELSIF    (1/2_line_count[].q == y_efr) THEN
                            1/2_line_count[].d      =           GND;
                   ELSE
                            1/2_line_count[].d      = 1/2_line_count[].q + 1;
                   END IF;
                   1/2_line_flag.d            =      VCC;
          ELSE
                   1/2_line_count[].d         =      1/2_line_count[].q;
                   1/2_line_flag.d            =      1/2_line_flag.q;
          END IF;
        ELSE
            1/2_line_count[].d      =               1/2_line_count[].q;
            1/2_line_flag.d         =               GND;
    END IF;
    half_line[]              =      1/2_line_count[].q;
END;
```

Designing the flowchart and the corresponding code for a progressive format is considerably simpler, and we leave it as an optional exercise for the reader. For a complete NTSC timing generator you may use Altera's NTSC component from the mega_lpm library (Figure 11.13).

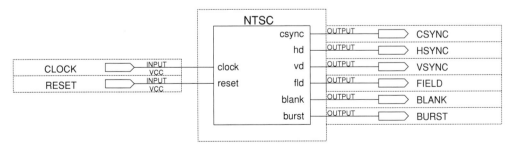

Figure 11.13 NTSC Raster Generator Component (Altera).

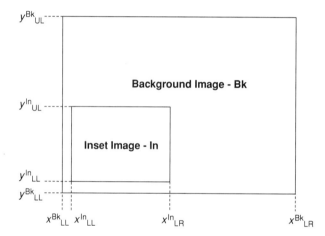

Figure 11.14   PiP Frame.

11.5 The Picture-in-Picture Overlay

The simplest overlay format is the ubiquitous PIP (picture-in-picture). It has become so common that video IC manufacturers offer this capability in a single chip package (Philips, Micronas...). However, the image quality is rather poor and their PIP formats are limited to only a few small sizes. For our analysis we choose the simpler progressive format to work with. As we see in Figure 11.14 the background image is a rectangle fully specified by the $(x^{Bk}_{UL}, y^{Bk}_{UL})$ and the $(x^{Bk}_{LR}\ y^{Bk}_{LR})$ corners. The inset image is similarly described by its $(x^{In}_{UL}, y^{In}_{UL})$ and the $(x^{In}_{LR}\ y^{In}_{LR})$ corners.

The memory arrangement for this example is shown in Figure 11.15. Images are retrieved from two memory buffers, one for the background and one for the inset. An address pointer (RE or read enable) locates the next pixel in the order of the raster scan. The pixel pointed to is displayed if the corresponding

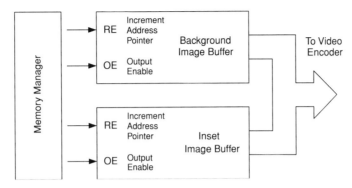

Figure 11.15 PiP Block Diagram.

output enable (OE) is active. If OE is inactive the output of the buffer is tri-stated. The outputs of the memory buffers are connected together and sent to a video encoder where they are converted to analog video.

When the pixel and line counts are within the area of the inset image then

$$(x^{\text{In}}_{\text{UL}} < \text{pixel}[..] \leq x^{\text{In}}_{\text{LR}}) \text{ and } (y^{\text{In}}_{\text{UL}} < \text{line}[..] \leq y^{\text{In}}_{\text{LR}})$$

and the inset image is displayed by asserting, for one pixel clock, the output enable of the inset memory buffer. A pixel clock consists of one pulse for 16- and 24-bit buses, and two pulses for 8-bit data buses. The address pointer of the inset memory buffer is then incremented to the next pixel location. The address pointer of the background memory buffer is also incremented even if its output enable is inactive during this period (the background is not visible at this location). This is because the background memory buffer contains the whole raster of the background image, including the pixels screened by the inset. These pixels must be scanned when we display the inset. Otherwise, when we emerge from the boundaries of the inset and start to display the background again, the background pixels will be out of order.

When the pixel and line counts are outside the inset image but within the limits of the background image then

$$(x^{\text{Bk}}_{\text{UL}} < \text{pixel}[..] \leq x^{\text{Bk}}_{\text{LR}}) \text{ and } (y^{\text{Bk}}_{\text{UL}} < \text{line}[..] \leq y^{\text{Bk}}_{\text{LR}})$$

The output enable of the background buffer is now active for one pixel period and, subsequently, the background address pointer is incremented. The address pointer of the inset is not incremented because there are no inset pix-els to go over; none of the inset pixels is masked. Also, since we are outside the inset area, the inset output enable is inactive.

If neither of the above expressions is true, we are outside the viewable area and the image is blanked.

Translating this algorithm in pseudo-code

```
IF ((xIn_UL < pixel[..] <= xIn_LR) & (yIn_UL < line[..] <= yIn_LR)) THEN
    display inset, advance inset and background address pointers
ELSIF ((xBk_UL < pixel[..] <= xBk_LR) & (yBk_UL < line[..] <= yBk_LR)) THEN
    display background, advance background address pointer
ELSE
    blank the image
END IF;
```

11.5.1 Adding Borders

Adding a border to the PIP is a straightforward extension of the algorithm (Figures 11.16 and 11.17). The counters are checked first against the corners of the inset, then against the corners of the border, and finally against the background limits.

```
IF ((xIn_UL < pixel[..] <= xIn_LR) & (yIn_UL < line[..] <= yIn_LR)) THEN
    display inset, advance inset and background address pointers
ELSIF ((xBo_UL < pixel[..] <= xBo_LR) & (yBo_UL < line[..] <= yBo_LR)) THEN
    display border, advance background address pointer
ELSIF ((xBk_UL < pixel[..] <= xBk_LR) & (yBk_UL < line[..] <= yBk_LR)) THEN
    display background, advance background address pointer
ELSE
    blank the image
END IF;
```

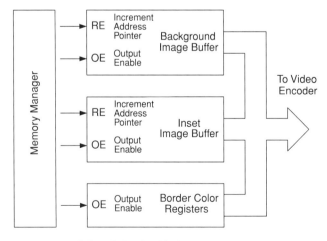

Figure 11.16   PiP with Border Block Diagram.

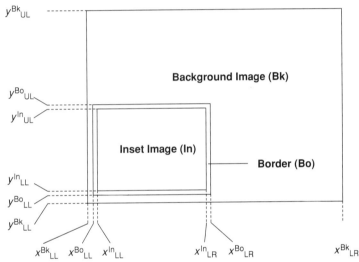

Figure 11.17 PiP with Border Frame.

The border color is placed in a border color register by the local controller. It consists of a limited number of YCbCr or RGB combinations which are recalled from a prestored color palette.

11.5.2 Adding Shadows

One of the more interesting effects we can add to a PiP is to overlay the shadow of a framed inset picture onto the background image (Figure 11.18). To first approximation the shadow can be generated by lowering the luminosity of the visible part of the shadow frame by a fixed amount (Figure 11.19).

This works very well as long as the shadow is not very deep and the luminosity does not approach black level. For shadow synthesis we submit each background pixel both to a subtraction and limiting block, and a pipeline register.

If the counters are in the shadow region we enable the output of the front register of the subtraction block. For background pixels not in the shadow we enable the output of the pipeline register associated with the background image buffer. Both the inset and background pipeline registers are used to compensate for the latency of the subtraction block. The border color is fixed and therefore no compensation is needed.

The pseudo-code is again self-explanatory.

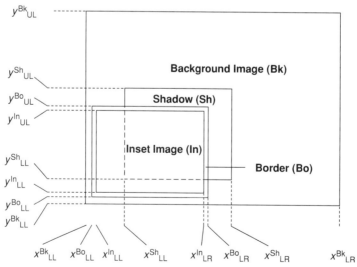

Figure 11.18 PiP with Border and Shadow Frame.

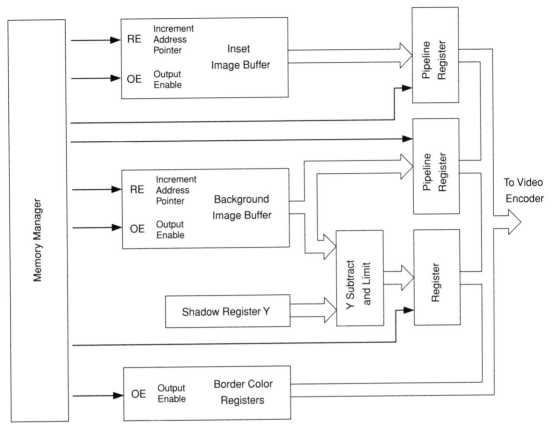

Figure 11.19 PiP with Border and Shadow Block Diagram.

```
IF     ((xInUL < pixel[..] <= xInLR) & (yInUL < line[..] <= yInLR)) THEN
           display inset, advance inset and background address pointers
ELSIF  ((xBoUL < pixel[..] <= xBoLR) & (yBoUL < line[..] <= yBoLR)) THEN
           display border, advance background address pointer
ELSIF  ((xShUL < pixel[..] <= xShLR) & (yShUL < line[..] <= yShLR)) THEN
           display background with reduced luma, advance background pointer
ELSIF  ((xBkUL < pixel[..] <= xBkLR) & (yBkUL < line[..] <= yBkLR)) THEN
           display background, advance background address pointer
ELSE
       blank the image
END IF;
```

For studio quality work this approach to shadow generation may not yield the desired results. The problem is that real environmental shadows are the result of less *incident light* reaching the reflective surface. If the reflectivity of a pixel is R_{pxl} and the intensity of the incident light is I_{light}, the intensity of the reflected light reaching our eyes is $R_{pxl} \times I_{light}$. In the shadow the incident light decreases by a fixed amount ΔI_{light}, and therefore the reflected intensity becomes $R_{pxl} \times (I_{light} - \Delta I_{light})$. The net loss of intensity is then $R_{pxl} \times \Delta I_{light}$.

As we can see, the decrease in intensity is not constant but proportional to the local reflectivity or, in our case, the background pixel value. Therefore, in order to obtain a perfect shadow, we have to read the intensity of a background pixel, multiply it by a shadow parameter equivalent to ΔI_{light}, then subtract the result from the original pixel luma value. This method obviously requires more processing power which can only be afforded by high-end special effects applications.

11.6 Graphics Key Overlay

Text, logos, graphics, and even animation can also be overlaid on a live video background (Figure 11.20). The process starts with creating a graphics file in the same color space as the background video and storing that file in a graphics buffer. For animations, a series of such graphics files are created and stored, together with a script file describing when and where the individual files are to be displayed.

Each graphics frame consists of a bitmapped foreground image on top of a keyed color background; for simplicity in Figure 11.20 we chose "0" as our key color code. No part of the foreground image can contain the key color. Consequently, before transferring it to the graphics buffer the foreground image is filtered out. If any key codes are found, they are slightly modified so the corresponding pixels are electronically distinguishable from the key color but not visibly different.

The overall shape of the image in the graphics buffer is a rectangle that, as we saw in the preceding sections, can be uniquely described by the coordinates of two opposing corners.

The method of overlaying this rectangle on a background image is similar to what we had before; the pixel and line counter outputs are compared against the borders of the graphics image. If not within the range of the graphics, the background image is displayed.

Inside the graphics box the processing is different, however. The color of each pixel in the graphics range is compared with the key color value. If they are the same, we display the background pixel; if not we display the foreground graphics pixel.

Pipeline registers delay the arrival of the pixels at the output multiplexer, so they reach it in the same time as the proper key color detect output. To eliminate the pipeline registers, some designs use a "graphics template" file containing 0 bits for background image pixels and 1 bits for graphics image pixels. The template is scanned one pixel period ahead of the graphics buffer. Its single bit output is used to steer the 2-to-1 multiplexer directly, eliminating the key color detection block and the associated pipeline registers.

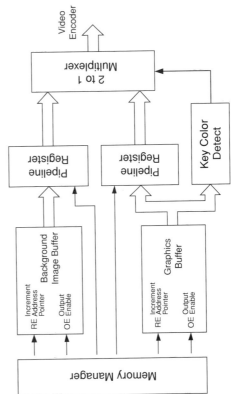

Graphics Memory Buffer

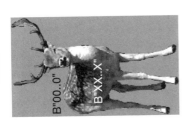

Graphics Template

B"00..0"

B"XX..X"

B"00..0"

B"11..1"

Video Encoder

2 to 1 Multiplexer

Pipeline Register

Pipeline Register

Key Color Detect

RE Increment Address Pointer

Background Image Buffer

OE Output Enable

RE Increment Address Pointer

Graphics Buffer

OE Output Enable

Memory Manager

Figure 11.20   Graphics Keying Process.

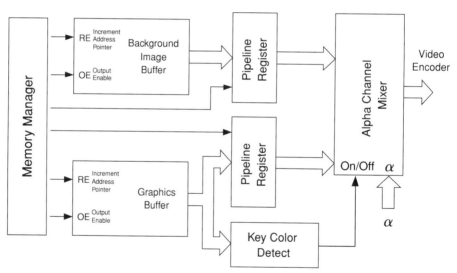

Figure 11.21   Transparent Graphics Keying.

11.6.1 Adding Transparency

We can take the graphics overlay process a step further by making the overlay transparent (Figure 11.21). To achieve it we replace the hard switching of the 2-to-1 multiplexer with an alpha mixer. Its transfer function is given by the equations

$$Y = \alpha\,(Y^{In} - 16) + (1 - \alpha)\,(Y^{Bk} - 16) + 16$$
$$Cb = \alpha\,(Cb^{In} - 128) + (1 - \alpha)\,(Cb^{Bk} - 128) + 128$$
$$Cr = \alpha\,(Cr^{In} - 128) + (1 - \alpha)\,(Cr^{Bk} - 128) + 128$$

where α is the transparency of the graphics image and is valued between 0 and 1.

In a simplified design only a few α values are allowed, and the eight most significant bits of the products $\alpha \times (\ldots)$ and $(1 - \alpha) \times (\ldots)$ are stored in a multiplication look-up table. The alpha channel mixer can then be implemented as a multiple adder. One caution: Colors do not mix as well as intensities do, and the result of adding two chromas of comparable values ($0.4 < \alpha < 0.6$) may lead to an unpleasantly harsh image. We can reduce this effect by using an α value for intensity mixing, and a different α value, away from the center range, for color.

11.7 Multiple Video Overlays

The easiest way to obtain a multiwindow display is to simply cascade a number of PiP processors. However, the image quality of the lower priority windows, the ones at the bottom of the image stack, will be degraded by the multiple video encoding and decoding steps.

11.7.1 Design Definition

This problem is resolved if we process all images at the same time. Let's look at the display arrangement in Figure 11.22. We have a total of four overlapping images, each with a different size and color border. In this case the overlay processor must recognize eight individual rectangular objects within a user-defined priority structure. As seen earlier, we can identify each of these objects by the (x,y) coordinates of an opposing pair of corners.

We can then proceed to compare the outputs of the pixel and line counters with the pixel and line ranges of the top image and, if within their limits, display the top image; if not, we compare with the ranges of the border for the top image, and so on.

It all works if we know which one is the top image, the top border, the next image, the next border, and so forth.

Since image priorities are user programmable, our design must accommodate all such possible image stacks. If code space is not a concern, we can just duplicate the comparison tree for all image permutations. However, this code runs at pixel rates on FPGA or CPLD ICs, where code space is at a high premium. We have to look for a more efficient solution.

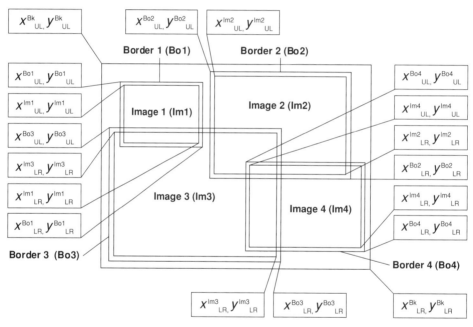

Figure 11.22   Multiple Video Frame.

11.7.2 Overlay Processor

We start by changing our perspective on the output screen. Instead of looking at it as eight overlapping rectangles, we treat it as real estate split into 289 parcels. To each parcel we assign a unique set of memory parameter values. To be specific, the parameters of most interest are the read enables (RE) and the output enables (OE) of the image buffers and border registers. The block diagram for this implementation is illustrated in Figure 11.23. When the user

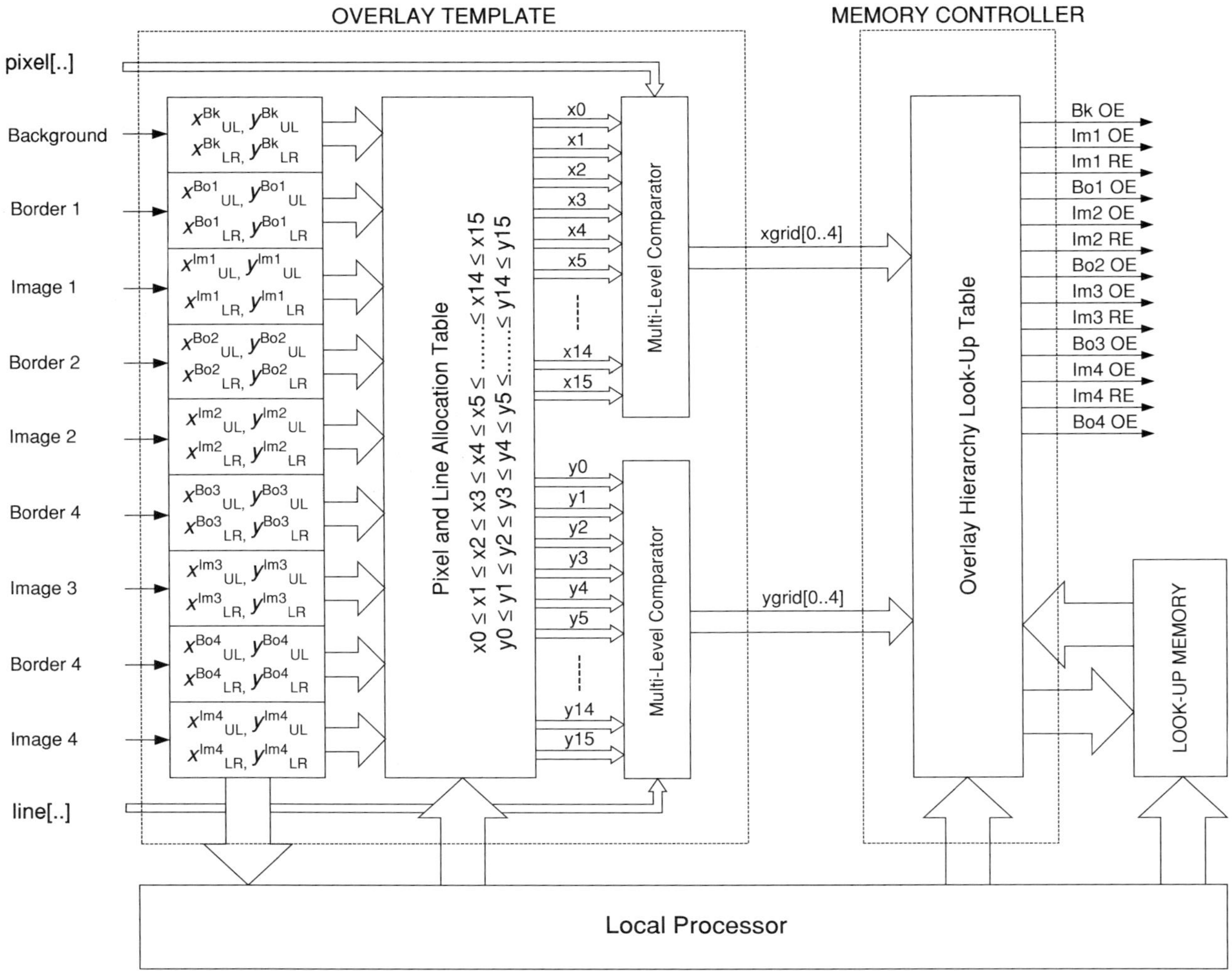

Figure 11.23 Overlay Template and Memory Processor Block Diagram.

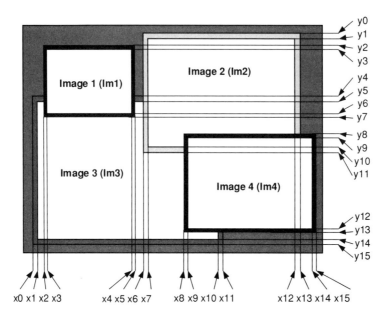

Figure 11.24 Reordering of Coordinates.

enters a new template command, a new set of corner coordinates is generated and stored in the local processor memory.

The local processor then organizes the x coordinates and, separately, the y coordinates of the corners in increasing or decreasing order (Figure 11.24) and stores them in a *pixel and line allocation table*. At the same time it recalculates and stores in the *overlay hierarchy look-up table* the values for the RE and the OE, for each image buffer, for each of the 289 image segments (parcels). Since the size of the look-up table is $289 \times 13 = 3757$ bits, we can reduce the cost of the design by using an external fast static RAM for the actual data storage.

As the pixel and line counts progress, their values are compared with the 16 x- and 16 y-ordered corner values. Based on the results, the corresponding multi-level comparators generate the address (*xgrid[..]*, *ygrid[..]*) of the screen segment the current pixel is in. For example, if (*xgrid[..]*, *ygrid[..]*) = (3, 3) we are inside the border of image 1, while if (*xgrid[..]*, *ygrid[..]*) = (8, 4) we display image 2 (Figure 11.23). Together, *xgrid[..]* and *ygrid[..]* form the address to the overlay hierarchy look-up table. The outputs of the look-up table directly control the image memory buffers and the border registers.

11.7.3 Multiple Video Overlays—Design Example

Figure 11.25 shows a practical implementation of this algorithm. The ordered corner coordinates are stored by the local controller in the *template registers*. The pixel and line addresses produced by the *raster_gen_non_interlaced* block are then compared with the stored template by the *xy_grids* processor. As a

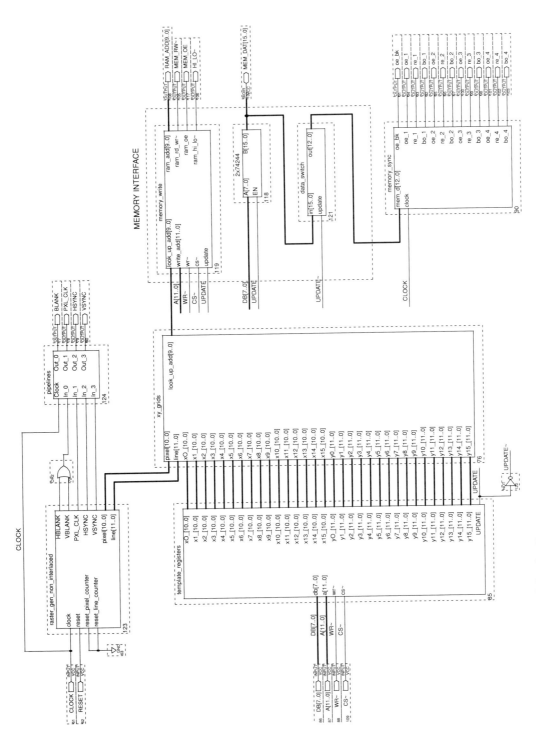

Figure 11.25 CPLD-Based Overlay Processor.

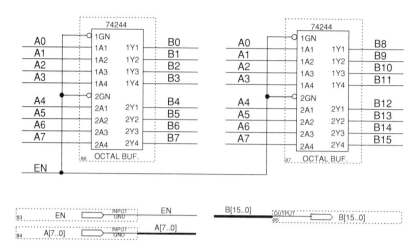

Figure 11.26 Data Steering Circuit 2×74244

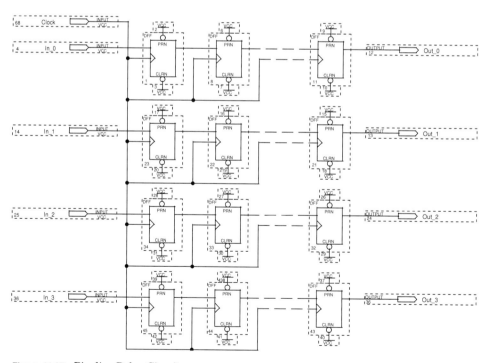

Figure 11.27 Pipeline Delay Circuit.

result of the comparison the *xy_grids* block produces the *look_up_add[9..0]* address for the external 16 bit look-up RAM.

To change the contents of the external RAM the local controller brings the *UPDATE* line high. This configures the *memory interface* for direct RAM access, so the controller can update the RAM. When the UPDATE line is brought low the control of the RAM is given back to the *xy_grids* block.

For each new pixel position the system identifies to which image segment the pixel belongs; it generates the address of the parameters characterizing that segment, reads the parameters from the external look-up RAM, and delivers them as control signals RE and OE to the image memory buffers. The memory sync block registers the outputs to the pixel clock. Examples of the codes and circuitry contained by each block are presented on the following pages and in Figures 11.26 and 11.27.

```
        XY GRID GENERATOR BLOCK


SUBDESIGN xy_grids
(
        pixel[10..0]            : INPUT;
        line[11..0]             : INPUT;

        x0_[10..0]              : INPUT;
        x1_[10..0]              : INPUT;
        x2_[10..0]              : INPUT;
        x3_[10..0]              : INPUT;
        x4_[10..0]              : INPUT;
        x5_[10..0]              : INPUT;
        x6_[10..0]              : INPUT;
        x7_[10..0]              : INPUT;
        x8_[10..0]              : INPUT;
        x9_[10..0]              : INPUT;
        x10_[10..0]             : INPUT;
        x11_[10..0]             : INPUT;
        x12_[10..0]             : INPUT;
        x13_[10..0]             : INPUT;
        x14_[10..0]             : INPUT;
        x15_[10..0]             : INPUT;

        y0_[11..0]              : INPUT;
        y1_[11..0]              : INPUT;
        y2_[11..0]              : INPUT;
        y3_[11..0]              : INPUT;
        y4_[11..0]              : INPUT;
        y5_[11..0]              : INPUT;
        y6_[11..0]              : INPUT;
```

```
        y7_[11..0]              :INPUT;
        y8_[11..0]              :INPUT;
        y9_[11..0]              :INPUT;
        y10_[11..0]             :INPUT;
        y11_[11..0]             :INPUT;
        y12_[11..0]             :INPUT;
        y13_[11..0]             :INPUT;
        y14_[11..0]             :INPUT;
        y15_[11..0]             :INPUT;
        look_up_add[9..0]       :OUTPUT;
)
BEGIN
    IF      (pixel[] < x0_[]) THEN
                look_up_add[4..0] = 0;
    ELSIF (pixel[] < x1_[]) THEN
                look_up_add[4..0] = 1;
    ELSIF (pixel[] < x2_[]) THEN
                look_up_add[4..0] = 2;
    ELSIF (pixel[] < x3_[]) THEN
                look_up_add[4..0] = 3;
    ELSIF (pixel[] < x4_[]) THEN
                look_up_add[4..0] = 4;
    ELSIF (pixel[] < x5_[]) THEN
                look_up_add[4..0] = 5;
    ELSIF (pixel[] < x6_[]) THEN
                look_up_add[4..0] = 6;
    ELSIF (pixel[] < x7_[]) THEN
                look_up_add[4..0] = 7;
    ELSIF (pixel[] < x8_[]) THEN
                look_up_add[4..0] = 8;
    ELSIF (pixel[] < x9_[]) THEN
                look_up_add[4..0] = 9;
    ELSIF (pixel[] < x10_[]) THEN
                look_up_add[4..0] = 10;
    ELSIF (pixel[] < x11_[]) THEN
                look_up_add[4..0] = 11;
    ELSIF (pixel[] < x12_[]) THEN
                look_up_add[4..0] = 12;
    ELSIF (pixel[] < x13_[]) THEN
                look_up_add[4..0] = 13;
    ELSIF (pixel[] < x14_[]) THEN
                look_up_add[4..0] = 14;
    ELSIF (pixel[] < x15_[]) THEN
                look_up_add[4..0] = 15;
```

```
ELSE
        look_up_add[4..0] = 16;
END IF;

IF      (line[] < y0_[]) THEN
        look_up_add[9..5] = 0;
ELSIF (line[] < y1_[]) THEN
        look_up_add[9..5] = 1;
ELSIF (line[] < y2_[]) THEN
        look_up_add[9..5] = 2;
ELSIF (line[] < y3_[]) THEN
        look_up_add[9..5] = 3;
ELSIF (line[] < y4_[]) THEN
        look_up_add[9..5] = 4;
ELSIF (line[] < y5_[]) THEN
        look_up_add[9..5] = 5;
ELSIF (line[] < y6_[]) THEN
        look_up_add[9..5] = 6;
ELSIF (line[] < y7_[]) THEN
        look_up_add[9..5] = 7;
ELSIF (line[] < y8_[]) THEN
        look_up_add[9..5] = 8;
ELSIF (line[] < y9_[]) THEN
        look_up_add[9..5] = 9;
ELSIF (line[] < y10_[]) THEN
        look_up_add[9..5] = 10;
ELSIF (line[] < y11_[]) THEN
        look_up_add[9..5] = 11;
ELSIF (line[] < y12_[]) THEN
        look_up_add[9..5] = 12;
ELSIF (line[] < y13_[]) THEN
        look_up_add[9..5] = 13;
ELSIF (line[] < y14_[]) THEN
        look_up_add[9..5] = 14;
ELSIF (line[] < y15_[]) THEN
        look_up_add[9..5] = 15;
ELSE
        look_up_add[9..5] = 16;
END IF;

END;
```

```
                         TEMPLATE STORAGE REGISTERS

CONSTANT                              Kx      = ;% pixel offset%
CONSTANT                              Ky      = ;%line offset%

SUBDESIGN template_registers
(
        db[7..0]                  :INPUT;
        a[11..0]                  :INPUT;
        wr~, cs~                  :INPUT;
        x0_[10..0]                :OUTPUT;
        x1_[10..0]                :OUTPUT;
        x2_[10..0]                :OUTPUT;
        x3_[10..0]                :OUTPUT;
        x4_[10..0]                :OUTPUT;
        x5_[10..0]                :OUTPUT;
        x6_[10..0]                :OUTPUT;
        x7_[10..0]                :OUTPUT;
        x8_[10..0]                :OUTPUT;
        x9_[10..0]                :OUTPUT;
        x10_[10..0]               :OUTPUT;
        x11_[10..0]               :OUTPUT;
        x12_[10..0]               :OUTPUT;
        x13_[10..0]               :OUTPUT;
        x14_[10..0]               :OUTPUT;
        x15_[10..0]               :OUTPUT;

        y0_[11..0]                :OUTPUT;
        y1_[11..0]                :OUTPUT;
        y2_[11..0]                :OUTPUT;
        y3_[11..0]                :OUTPUT;
        y4_[11..0]                :OUTPUT;
        y5_[11..0]                :OUTPUT;
        y6_[11..0]                :OUTPUT;
        y7_[11..0]                :OUTPUT;
        y8_[11..0]                :OUTPUT;
        y9_[11..0]                :OUTPUT;
        y10_[11..0]               :OUTPUT;
        y11_[11..0]               :OUTPUT;
        y12_[11..0]               :OUTPUT;
        y13_[11..0]               :OUTPUT;
        y14_[11..0]               :OUTPUT;
        y15_[11..0]               :OUTPUT;
        UPDATE                    :OUTPUT;
)
```

```
VARIABLE
        sx0_[7..0], sx1_[7..0], sx2_[7..0], sx3_[7..0]          :DFF;
        sx4_[7..0], sx5_[7..0], sx6_[7..0], sx7_[7..0]          :DFF;
        sx8_[7..0], sx9_[7..0], sx10_[7..0], sx11_[7..0]        :DFF;
        sx12_[7..0], sx13_[7..0], sx14_[7..0], sx15_[7..0]      :DFF;
        sy0_[7..0], sy1_[7..0], sy2_[7..0], sy3_[7..0]          :DFF;
        sy4_[7..0], sy5_[7..0], sy6_[7..0], sy7_[7..0]          :DFF;
        sy8_[7..0], sy9_[7..0], sy10_[7..0], sy11_[7..0]        :DFF;
        sy12_[7..0], sy13_[7..0], sy14_[7..0], sy15_[7..0]      :DFF;
        UPDATE                                                  :DFF;


BEGIN
        sx0_[].clk          = !cs~&!a11&!a5&!a4&!a3&!a2&!a1&!a0&!wr~;
        sx1_[].clk          = !cs~&!a11&!a5&!a4&!a3&!a2&!a1&a0&!wr~;
        sx2_[].clk          = !cs~&!a11&!a5&!a4&!a3&!a2&a1&!a0&!wr~;
        sx3_[].clk          = !cs~&!a11&!a5&!a4&!a3&!a2&a1&a0&!wr~;
        sx4_[].clk          = !cs~&!a11&!a5&!a4&!a3&a2&!a1&!a0&!wr~;
        sx5_[].clk          = !cs~&!a11&!a5&!a4&!a3&a2&!a1&a0&!wr~;
        sx6_[].clk          = !cs~&!a11&!a5&!a4&!a3&a2&a1&!a0&!wr~;
        sx7_[].clk          = !cs~&!a11&!a5&!a4&!a3&a2&a1&a0&!wr~;
        sx8_[].clk          = !cs~&!a11&!a5&!a4&a3&!a2&!a1&!a0&!wr~;
        sx9_[].clk          = !cs~&!a11&!a5&!a4&a3&!a2&!a1&a0&!wr~;
        sx10_[].clk         = !cs~&!a11&!a5&!a4&a3&!a2&a1&!a0&!wr~;
        sx11_[].clk         = !cs~&!a11&!a5&!a4&a3&!a2&a1&a0&!wr~;
        sx12_[].clk         = !cs~&!a11&!a5&!a4&a3&a2&!a1&!a0&!wr~;
        sx13_[].clk         = !cs~&!a11&!a5&!a4&a3&a2&!a1&a0&!wr~;
        sx14_[].clk         = !cs~&!a11&!a5&!a4&a3&a2&a1&!a0&!wr~;
        sx15_[].clk         = !cs~&!a11&!a5&!a4&a3&a2&a1&a0&!wr~;
        sy0_[].clk          = !cs~&!a11&!a5&a4&!a3&!a2&!a1&!a0&!wr~;
        sy1_[].clk          = !cs~&!a11&!a5&a4&!a3&!a2&!a1&a0&!wr~;
        sy2_[].clk          = !cs~&!a11&!a5&a4&!a3&!a2&a1&!a0&!wr~;
        sy3_[].clk          = !cs~&!a11&!a5&a4&!a3&!a2&a1&a0&!wr~;
        sy4_[].clk          = !cs~&!a11&!a5&a4&!a3&a2&!a1&!a0&!wr~;
        sy5_[].clk          = !cs~&!a11&!a5&a4&!a3&a2&!a1&a0&!wr~;
        sy6_[].clk          = !cs~&!a11&!a5&a4&!a3&a2&a1&!a0&!wr~;
        sy7_[].clk          = !cs~&!a11&!a5&a4&!a3&a2&a1&a0&!wr~;
        sy8_[].clk          = !cs~&!a11&!a5&a4&a3&!a2&!a1&!a0&!wr~;
        sy9_[].clk          = !cs~&!a11&!a5&a4&a3&!a2&!a1&a0&!wr~;
        sy10_[].clk         = !cs~&!a11&!a5&a4&a3&!a2&a1&!a0&!wr~;
        sy11_[].clk         = !cs~&!a11&!a5&a4&a3&!a2&a1&a0&!wr~;
        sy12_[].clk         = !cs~&!a11&!a5&a4&a3&a2&!a1&!a0&!wr~;
        sy13_[].clk         = !cs~&!a11&!a5&a4&a3&a2&!a1&a0&!wr~;
        sy14_[].clk         = !cs~&!a11&!a5&a4&a3&a2&a1&!a0&!wr~;
        sy15_[].clk         = !cs~&!a11&!a5&a4&a3&a2&a1&a0&!wr~;
        UPDATE.clk          = !cs~&!a11&a5&!a4&!a3&!a2&!a1&!a0&!wr~;
```

```
sx0_[].clrn              =         vcc;
sx1_[].clrn              =         vcc;
sx2_[].clrn              =         vcc;
sx3_[].clrn              =         vcc;
sx4_[].clrn              =         vcc;
sx5_[].clrn              =         vcc;
sx6_[].clrn              =         vcc;
sx7_[].clrn              =         vcc;
sx8_[].clrn              =         vcc;
sx9_[].clrn              =         vcc;
sx10_[].clrn             =         vcc;
sx11_[].clrn             =         vcc;
sx12_[].clrn             =         vcc;
sx13_[].clrn             =         vcc;
sx14_[].clrn             =         vcc;
sx15_[].clrn             =         vcc;
sy0_[].clrn              =         vcc;
sy1_[].clrn              =         vcc;
sy2_[].clrn              =         vcc;
sy3_[].clrn              =         vcc;
sy4_[].clrn              =         vcc;
sy5_[].clrn              =         vcc;
sy6_[].clrn              =         vcc;
sy7_[].clrn              =         vcc;
sy8_[].clrn              =         vcc;
sy9_[].clrn              =         vcc;
sy10_[].clrn             =         vcc;
sy11_[].clrn             =         vcc;
sy12_[].clrn             =         vcc;
sy13_[].clrn             =         vcc;
sy14_[].clrn             =         vcc;
sy15_[].clrn             =         vcc;
UPDATE.clrn              =         vcc;

sx0_[].d                 =         db[];
sx1_[].d                 =         db[];
sx2_[].d                 =         db[];
sx3_[].d                 =         db[];
sx4_[].d                 =         db[];
sx5_[].d                 =         db[];
sx6_[].d                 =         db[];
sx7_[].d                 =         db[];
sx8_[].d                 =         db[];
sx9_[].d                 =         db[];
```

```
sx10_[].d          =          db[];
sx11_[].d          =          db[];
sx12_[].d          =          db[];
sx13_[].d          =          db[];
sx14_[].d          =          db[];
sx15_[].d          =          db[];
sy0_[].d           =          db[];
sy1_[].d           =          db[];
sy2_[].d           =          db[];
sy3_[].d           =          db[];
sy4_[].d           =          db[];
sy5_[].d           =          db[];
sy6_[].d           =          db[];
sy7_[].d           =          db[];
sy8_[].d           =          db[];
sy9_[].d           =          db[];
sy10_[].d          =          db[];
sy11_[].d          =          db[];
sy12_[].d          =          db[];
sy13_[].d          =          db[];
sy14_[].d          =          db[];
sy15_[].d          =          db[];
UPDATE.d           =          db[0];

x0_[10..3]         =          sx0_[7..0].q;
x1_[10..3]         =          sx1_[7..0].q;
x2_[10..3]         =          sx2_[7..0].q;
x3_[10..3]         =          sx3_[7..0].q;
x4_[10..3]         =          sx4_[7..0].q;
x5_[10..3]         =          sx5_[7..0].q;
x6_[10..3]         =          sx6_[7..0].q;
x7_[10..3]         =          sx7_[7..0].q;
x8_[10..3]         =          sx8_[7..0].q;
x9_[10..3]         =          sx9_[7..0].q;
x10_[10..3]        =          sx10_[7..0].q;
x11_[10..3]        =          sx11_[7..0].q;
x12_[10..3]        =          sx12_[7..0].q;
x13_[10..3]        =          sx13_[7..0].q;
x14_[10..3]        =          sx14_[7..0].q;
x15_[10..3]        =          sx15_[7..0].q;
y0_[11..4]         =          sy0_[7..0].q;
y1_[11..4]         =          sy1_[7..0].q;
y2_[11..4]         =          sy2_[7..0].q;
y3_[11..4]         =          sy3_[7..0].q;
y4_[11..4]         =          sy4_[7..0].q;
```

```
y5_[11..4]              =       sy5_[7..0].q;
y6_[11..4]              =       sy6_[7..0].q;
y7_[11..4]              =       sy7_[7..0].q;
y8_[11..4]              =       sy8_[7..0].q;
y9_[11..4]              =       sy9_[7..0].q;
y10_[11..4]             =       sy10_[7..0].q;
y11_[11..4]             =       sy11_[7..0].q;
y12_[11..4]             =       sy12_[7..0].q;
y13_[11..4]             =       sy13_[7..0].q;
y14_[11..4]             =       sy14_[7..0].q;
y15_[11..4]             =       sy15_[7..0].q;

x0_[2..0]               =       Kx;
x1_[2..0]               =       Kx;
x2_[2..0]               =       Kx;
x3_[2..0]               =       Kx;
x4_[2..0]               =       Kx;
x5_[2..0]               =       Kx;
x6_[2..0]               =       Kx;
x7_[2..0]               =       Kx;
x8_[2..0]               =       Kx;
x9_[2..0]               =       Kx;
x10_[2..0]              =       Kx;
x11_[2..0]              =       Kx;
x12_[2..0]              =       Kx;
x13_[2..0]              =       Kx;
x14_[2..0]              =       Kx;
x15_[2..0]              =       Kx;
y0_[3..0]               =       Ky;
y1_[3..0]               =       Ky;
y2_[3..0]               =       Ky;
y3_[3..0]               =       Ky;
y4_[3..0]               =       Ky;
y5_[3..0]               =       Ky;
y6_[3..0]               =       Ky;
y7_[3..0]               =       Ky;
y8_[3..0]               =       Ky;
y9_[3..0]               =       Ky;
y10_[3..0]              =       Ky;
y11_[3..0]              =       Ky;
y12_[3..0]              =       Ky;
y13_[3..0]              =       Ky;
y14_[3..0]              =       Ky;
y15_[3..0]              =       Ky;


END;
```

NON-INTERLACED RASTER GENERATOR BLOCK

```
% Raster Generator Non-Interlaced Format %

CONSTANT    x_fb          =                  ;%horizontal front blanking%
CONSTANT    x_hsw         =                  ;%HSYNC width%
CONSTANT    x_el          =                  ;%end of line%
CONSTANT    x_hfd         =                  ;%horizontal front delay%
CONSTANT    x_lw          =                  ;%half line clock pulse width%
CONSTANT    x_hlc         =                  ;%half line clock center pulse position%


CONSTANT    y_fb          =                  ;%vertical front blanking%
CONSTANT    y_vsw         =                  ;%VSYNC width%
CONSTANT    y_vfd         =                  ;%vertical front delay%
CONSTANT    y_vsp         =                  ;%VSYNC period%
CONSTANT    y_rb          =                  ;%vertical retrace blanking%
CONSTANT    y_efr         =                  ;%end of frame%


SUBDESIGN raster_gen_non_interlaced
(
  clock, reset                       :INPUT;
  reset_pixel_counter, reset_line_counter :INPUT;
  HBLANK                             :OUTPUT;
  VBLANK                             :OUTPUT;
  PXL_CLK                            :OUTPUT;
  HSYNC                              :OUTPUT;
  VSYNC                              :OUTPUT;
  pixel[10..0]                       :OUTPUT;
  line[11..0]                        :OUTPUT;


  )
VARIABLE
  pixel_count[10..0]                 :DFF;
  line_clock                         :DFF;
  line_count[11..0]                  :DFF;
  HSYNC                              :DFF;
  HBLANK                             :DFF;
  VSYNC                              :DFF;
  VBLANK                             :DFF;
  line_flag                          :DFF;

BEGIN
  pixel_count[].clk        =      clock;
  pixel_count[].clrn       =      !reset_pixel_counter & !reset;
  line_clock.clk           =      clock;
  line_clock.clrn          =       !reset;
```

```
line_count[].clk                 =        clock;
line_count[].clrn                =        !reset;
line_flag.clk                    =        clock;
line_flag.clrn                   =        !reset;


HSYNC.clk                        =        clock;
HSYNC.clrn                       =        !reset;
HBLANK.clk                       =        clock;
HBLANK.clrn                      =        !reset;
VSYNC.clk                        =        clock;
VSYNC.clrn                       =        !reset;
VBLANK.clk                       =        clock;
VBLANK.clrn                      =        !reset;

  PXL_CLK                        =        clock;

  line_clock.d                   =        !(pixel_count[].q <= x_lw);

  HSYNC.d                        =        !((pixel_count[].q > x_hfd)&
                                          (pixel_count[].q <= x_hfd + x_hsw));

  HBLANK.d                       =        (pixel_count[].q <= x_fb);

  VSYNC.d                        =        !((line_count[].q > y_vfd)&
                                          (line_count[].q <= y_vfd + y_vsw));

  VBLANK.d                       =        (line_count[].q <= y_fb);

IF      reset_pixel_counter THEN
        pixel_count[] d          =        GND;
ELSIF (pixel_count[].q == x_el) THEN
        pixel_count[].d          =        GND;
ELSE
        pixel_count[].d          =        pixel_count[].q + 1;
END IF;
pixel[]                          =        pixel_count[];

IF      line_clock.q THEN
        IF      line_flag.q THEN
                IF      reset_line_counter THEN
                        line_count[].d        =        GND;
                ELSIF   (line_count[].q == y_efr) THEN
                        line_count[].d              =        GND;
                ELSE
                        line_count[].d                   =        line_count[].q + 1;
                END IF;
```

```
                          line_flag.d                          =         VCC;
            ELSE
                                     line_count[].d            =         line_count[].q;
                                     line_flag.d               =         line_flag.q;
            END IF;
            ELSE
            line_count[].d                  =         line_count[].q;
            line_flag.d                     =         GND;
            END IF;
            line[]           =           line_count[];

            END;
```

MEMORY WRITE MANAGER

```
SUBDESIGN memory_write
(
    look_up_add[9..0]      :INPUT;
    write_add[11..0]       :INPUT;
    wr~, cs~               :INPUT;
    update                 :INPUT;
    ram_add[9..0]          :OUTPUT;
    ram_rd_wr~             :OUTPUT;
    ram_oe                 :OUTPUT;
    ram_hi_lo~             :OUTPUT;
)
BEGIN

    ram_add[9..0]          =     update & write_add[9..0] # !update & look_up_add[9..0];
    ram_rd_wr~             =     !update # wr~ # cs~ # !write_add[11];
    ram_oe                 =     !update;
    ram_hi_lo~             =     write_add[10];

END;
```

MEMORY DATA SYNCHRONIZATION BLOCK

```
SUBDESIGN memory_sync
(
    mem_d[12..0]           :INPUT;
    clock                  :INPUT;
    oe_bk                  :OUTPUT;
    oe_1                   :OUTPUT;
    re_1                   :OUTPUT;
    bo_1                   :OUTPUT;
    oe_2                   :OUTPUT;
```

```
    re_2                    :OUTPUT;
    bo_2                    :OUTPUT;
    oe_3                    :OUTPUT;
    re_3                    :OUTPUT;
    bo_3                    :OUTPUT;
    oe_4                    :OUTPUT;
    re_4                    :OUTPUT;
    bo_4                    :OUTPUT;
)
VARIABLE
    smem[12..0]             :DFF;
            BEGIN
                smem[].clk          =           clock;
                smem[].clrn         =           vcc;
                smem[].d            =           mem_d[];
                oe_bk               =           smem0.q;
                oe_1                =           smem1.q;
                re_1                =           smem2.q;
                bo_1                =           smem3.q;
                oe_2                =           smem4.q;
                re_2                =           smem5.q;
                bo_2                =           smem6.q;
                oe_3                =           smem7.q;
                re_3                =           smem8.q;
                bo_3                =           smem9.q;
                oe_4                =           smem10.q;
                re_4                =           smem11.q;
                bo_4                =           smem12.q;
            END;


                    DATA SWITCH BLOCK


            SUBDESIGN data_switch
            (
                in[15..0]                   :INPUT;
                update                      :INPUT;
                out[12..0]                  :OUTPUT;
            )
            BEGIN
                IF update then
                        out[]               =           GND;
                ELSE
                        out[12..0]          =           in[12..0];
                END IF;
            END;
```

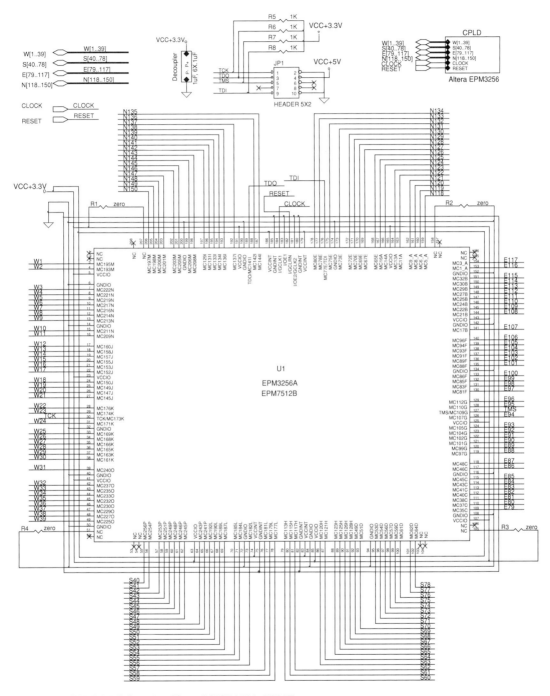

Figure 11.28   CPLD Module using Altera MAX3000A–7000B.

11.8 CPLD Circuit

The video processing unit of any video system is the FPGA or CPLD section. That is where the raster generators, memory managers, color registers, and all other processors are implemented. The principal reason is that FPGAs and CPLDs are fast enough for these tasks while microprocessors and even most DSPs are not. Figure 11.28 illustrates a CPLD module based on Altera's general purpose MAX3000A and MAX7000B series integrated circuits. This module can be populated either with a MAX EPM7512BQC208, a MAX EPM3256AQC208, or a MAX EPM3512AQC208 IC. When using the 7000 part the R1, R2, R3, and R4 jumpers must be installed, and pins 6, 40, 84, and 108 are not available for use (they are grounds for the 3000 parts).

In-circuit programming is done through the JP1 header using Altera's ByteBlaster MV parallel port cable. The software necessary for design and development as well as device programming is available, free of charge, from Altera (Quartus II Web Edition and MAX+PLUS II BASELINE). The CPLD module described in Chapter 15 includes the ByteBlaster circuit, and therefore requires only a parallel cable and connection to the PC printer port to program.

Image Scalers

12.1 Quick Theory of Digital Filters

Image scalers are special purpose digital filters designed to render a two-dimensional input image at a higher or lower spatial resolution without compromising the integrity of the image. Practical filter design relies on the connection between the Z-transform (which we will introduce shortly) and controlled circuit delay elements such as flip-flops. But first let us review the the math background.

12.1.1 The Laplace and Z-Transforms

If we expand the Fourier transform we introduced in Chapter 8 to the complex domain we arrive at the even more powerful *Laplace transform:*

$$L[x(t)] = X(s) = \int\limits_{-\infty}^{+\infty} x(t)e^{-st}dt$$

where

$$s = \sigma + j\omega$$

The *Inverse Laplace Transform* is given by

$$x(t) = \frac{1}{2\pi j}\oint X(s)e^{st}ds$$

The integral is taken around a contour that contains all the singularities of *X(s)*.

Laplace transforms are mainly used to solve high-order linear differential equations. By using them such equations are reduced to algebraic identities with readily available solutions. The properties of the Laplace transform are similar to those of the Fourier transform.

- $L[x(t - \tau)\, u(t - \tau)] = e^{-s\tau}\, X(s)$

- $L\left[\dfrac{d^n}{dt^n}\, x(t)\right] = s^n X(s) - \displaystyle\sum_{i=0}^{i=n-1} s^{n-i-1} \dfrac{d^i}{dt^i}\, x(t)\,\Big|_{t\,=\,0}$

- $L\left[\displaystyle\int_{-\infty}^{t} x(\tau)d\tau\right] = \dfrac{X(s)}{s^n} + \dfrac{1}{s}\displaystyle\int_{-\infty}^{0} x(\tau)d\tau$

The Laplace transform is ideal for the analysis and synthesis of continuous systems; however, it is not suitable for discrete systems.

Let $x_s(t)$ be the function we obtained when we multiply a continuous function $x(t)$ by the pulse series $P(t)$. We can see that $x_s(t)$ is the "sampled" version of $x(t)$, with $P(t)$ being the sampling function. Then

$$x_s(t) = \sum_{n=-\infty}^{n=+\infty} x(nT)\delta(t - nT)$$

If we apply the Laplace transform to $x_s(t)$ we obtain

$$X_s(s) = \sum_{n=-\infty}^{n=+\infty} x(nT)e^{-nTs}$$

Since the range of the sum is from minus infinity to infinity, and taking the mathematical liberty of assuming that T is an integer, we can make a change of variable and replace nT by n. If we then substitute z for e^{sT} we obtain

$$X_s(z) = \sum_{n=-\infty}^{n=+\infty} x(n)z^{-n}$$

This formula represents the *Z-transform* of *x(n)*. The Z-transform is our primary tool when working with discrete (digitized) systems. Its general definition is

$$Z[x(n)] = X(z) = \sum_{n=-\infty}^{n=+\infty} x(n)z^{-n}$$

The power of the Z-transform rests in its ability to reduce a *difference* equation to an algebraic equation, much in the same way the Laplace transform reduced differential equations to algebraic ones. The relevant property in this case is

$$Z[x(n - m)] = z^{-m}X(z)$$

where m represents signal delay measured in number of samples.

12.1.2 Digital Filters

12.1.2.1 Transfer Function
Filters are two-port systems (one input and one output) with a frequency-dependent transfer function. The transfer function of a system is the unique mathematical statement that relates the output of the system to the input.

In Figure 12.1 we have a system characterized by a transfer function $H[..]$. Its output $y(t)$ can then be described in terms of its input $x(t)$ as

$$y(t) = H[x(t)]$$

A system that is both homogeneous and additive is by definition a *linear system*. For the $H[..]$ transfer function, this condition translates to

$$H[ax_1(t) + bx_2(t)] = aH[x_1(t)] + bH[x_2(t)]$$

A *time-invariant* system is one that remains constant over time or

$$y(t) = H[x(t)] \rightarrow y(t - \tau) = H[x(t - \tau)]$$

A *causal system* has the additional property of having all its outputs dependent only on its present and past inputs.

If the response of a linear system with a transfer function $H[..]$ to a $\delta(t)$ impulse input is $h(t - \tau)$, then its response to an arbitrary input $x(t)$ is given by the convolution integral

$$y(t) = \int_{-\infty}^{+\infty} x(\tau)h(t - \tau)d\tau = \int_{-\infty}^{+\infty} x(t - \lambda)d\lambda$$

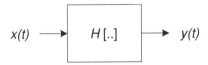

Figure 12.1 Time Domain Filter.

The function $h(t)$ is known as the system's impulse response.

For a linear, time-invariant, causal, discrete system the current output $y(n)$ depends only on the present input $x(n)$ and its past inputs $x(n-i)$ and outputs $y(n-j)$, where n is the rank of the current sample. We can write this dependency in the following form:

$$y(n) + b_1y(n-1) + b_2y(n-2) + \cdots b_ky(n-k) = a_0x(n) + a_1x(n-1) + a_2x(n-2) + \cdots + a_lx(n-1)$$

If we apply the Z-transform to both sides we obtain

$$(1 + b_1z^{-1} + b_2z^{-2} + \cdots + b_kz^{-k})\, Y(z) = (a_0 + a_1z^{-1} + a_2z^{-2} + \cdots + a_lz^{-l})\, X(z)$$

The transfer function of the system in the z domain can now be easily calculated

$$H(z) = \frac{a_0 + a_1z^{-1} + a_2z^{-2} + \cdots + a_lz^{-l}}{1 + b_1z^{-1} + b_2z^{-2} + \cdots + b_k z^{-k}} = \frac{Y(z)}{X(z)}$$

To implement this system in hardware we must first isolate the present output

$$y(n) = -b_1y(n-1) - b_2y(n-2) - \cdots b_ky(n-k) + a_0x(n) + a_1x(n-1) + a_2x(n-2) + \cdots + a_lx(n-1)$$

Recalling the properties of the z-transform

$$Z[x(n-1)] = z^{-1}X(z) = z^{-1}Z[x(n)]$$

As we can see, z^{-1} is the z domain transfer function of a system block that delays an input sequence $x(n)$ by one sample (Figure 12.2).

In practice, this exotic-sounding block is little more than a register with a flip-flop for each $x(n)$ bit.

12.1.2.2 Direct Recursive Realization

Using the z^{-1} delay blocks, we can easily translate the difference equation of $y(n)$ into a circuit capable of evaluating it in real time (Figure 12.3). This circuit is called the *direct recursive realization*

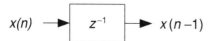

Figure 12.2 Elementary Delay Block.

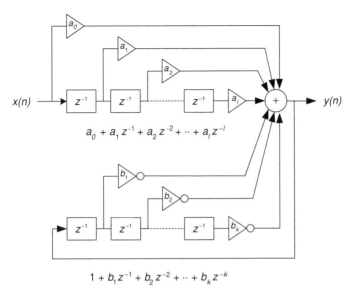

$$a_0 + a_1 z^{-1} + a_2 z^{-2} + \cdots + a_l z^{-l}$$

$$1 + b_1 z^{-1} + b_2 z^{-2} + \cdots + b_k z^{-k}$$

Figure 12.3   Direct Recursive Filter Realization.

of the $H(z)$ system. The a_i and b_i blocks are constant noninverting and inverting gains, respectively.

12.1.2.3 Canonic Recursive Realization   The somewhat more economical rearrangement in Figure 12.4 uses fewer delay blocks but requires two summing circuits. This is known as a direct *canonic recursive realization* of $y(n)$.

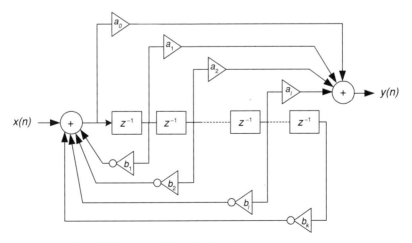

Figure 12.4   Canonic Recursive Filter Realization.

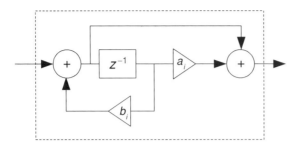

Figure 12.5   Filter Product Term.

12.1.2.4 Cascade Realization   There are other indirect ways of encoding a transfer function into hardware. If we break $H(z)$ into a product of smaller partial transfer functions $H_i(z)$, then

$$H(z) = aH_1(z)\ H_2(z)\cdots H_1(z)$$

The individual $H_i(z)$ functions are found by factoring out the numerator and denominator of $H(z)$. Their general form is

$$H_i(z) = \frac{1 + a_i z^{-1}}{1 + b_i z^{-1}}$$

and can be implemented using a single z^{-1} delay block (Figure 12.5).

The complete transfer function is then reduced to hardware by connecting the $H_i(z)$ blocks in series (Figure 12.6).

This is the *cascade or series canonic recursive realization* of $H(z)$.

12.1.2.5 Parallel Realization   If instead of factoring $H(z)$ we expand it into first-order partial fractions $H_i(z)$, then $H(z)$ can be written as

$$H(z) = a + H_1(z) + H_2(z) + \cdots + H_1(z)$$

with $H_i(z)$ given by

$$H_i(z) = \frac{a}{1 + b_i z^{-1}}$$

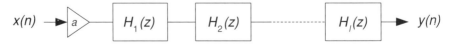

Figure 12.6   Cascade Canonic Recursive Realization.

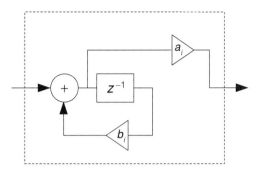

Figure 12.7   Filter Sum Term.

The circuit that realizes $H_i(z)$ is shown in Figure 12.7.

The *parallel canonic recursive realization* of $H(z)$ is illustrated in Figure 12.8.

12.1.3 The Digital Comb Filter

If $H(z)$ is the Z-transform of a single-notch bandstop or bandpass filter, a comb version of this filter is obtained by replacing z with z^L. If z^{-1} corresponds to a single sample delay, z^{-L} represents *an L samples delay*. The difference equation for a comb filter can be written in general form as

$$y(n) = x(n) + ax(n - M) - by(n - N)$$

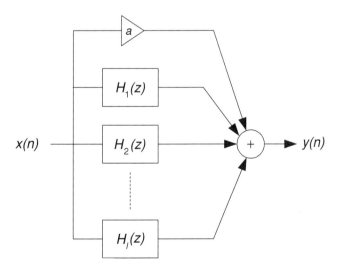

Figure 12.8   Parallel Canonic Recursive Realization.

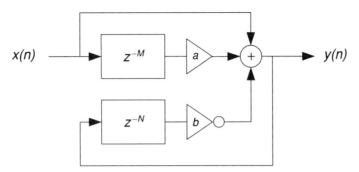

Figure 12.9

where n is the rank of the current sample, $n - i$ denotes a delay of i samples with respect to the current sample, and a and b are feedforward and feedback coefficients, respectively.

The direct recursive realization of this comb filter is shown in Figure 12.9. Reducing this diagram to practice involves the use of registers (one for each z^{-1}), two scalar multipliers, and one adder.

12.2 Up-Sampling and Interpolation (Up-Scaling)

To scale up an image by an integer factor n we must insert $n - 1$ new lines between every pair of lines of the original image, and then insert $n - 1$ new pixels between every single pixel pair on each line. The new pixels have to be given values that "average well" within their immediate two-dimensional neighborhood. This up-sampling and interpolation procedure is, in essence, a digital filtering process.

Let us introduce the up-sampler block $\uparrow U$, also known as an oversampler (Figure 12.10).

It has a transfer function defined by

$$x_u(n) = \begin{cases} x\left(\frac{n}{U}\right) & \text{for } n = 0, \pm U, \pm 2U \ldots \\ 0 & \text{for other } n \text{ values} \end{cases}$$

It is apparent that $\uparrow U$ introduces $U - 1$ zero samples between any two values of the discrete input function $x(n)$. To do so, the sampling rate of the up-

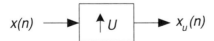

Figure 12.10 Up-Sampler Block.

sampler f_{u_s} has to be U times higher than the sampling rate of $x(n)$. Applying the Z-transform we get

$$X_u(z) = \sum_{n=-\infty}^{n=+\infty} x_u(n)z^{-n} = \sum_{\text{Int}(n/U)} x\left(\frac{n}{U}\right)z^{-n}$$

Since all the sum terms that are not a multiple of U yield zeros we can limit the sum to $\text{Int}(n/U)$. If we change our variable to m, where $n = Um$, then

$$X_u(z) = \sum_{m=-\infty}^{m=+\infty} x[m]z^{-Um} = X(z^U)$$

If we set $z = e^{j\omega}$ then

$$X_u(\omega) = X(e^{jU w}) = X(U\omega)$$

The frequency spectrum of an up-sampled version $x_u(n)$ of a discrete function $x(n)$ is a horizontally scaled (compressed) version of the frequency spectrum of $x(n)$. The scaling factor is the up-sampling rate U. For a numerical example, let's pick $U = 4$. Then

$$x_u(n) = \begin{cases} x\left(\frac{n}{4}\right) & \text{for } n = 0, \pm 4, \pm 8 \cdots \\ 0 & \text{for other } n \text{ values} \end{cases}$$

Repeating our previous analysis we find

$$X_u(z) = \sum_{n=-\infty}^{n=+\infty} x_u(n)z^{-n} = \sum_{\text{Int}(n/4)} x\left(\frac{n}{4}\right)z^{-n}$$

After an $n = 4m$ change in variable,

$$X_u(z) = \sum_{m=-\infty}^{m=+\infty} x[m]z^{-4m} = X(z^4)$$

Replacing $z = e^{jw}$ brings us to

$$X_u(\omega) = X(4\omega)$$

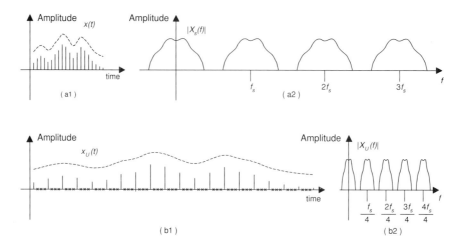

Figure 12.11   Spectrum of Up-Sampled Function $x_u(t)$.

12.2.1 Spectrum of Up-Sampled Signals

If the discrete function $x(n)$ is obtained by sampling a continuous function $x(t)$ then, as we saw in Chapter 8, its frequency spectrum is $X(f)$ plus a sequence of its spectral images [Figures 12.11(a1) and (a2)]. For convenience we replaced the angular frequency ω by the more intuitive time frequency f_s; $\omega = 2\pi f_s$.

Furthermore, if we up-sample $x(n)$ with an up-sample factor $U = 4$, the function is horizontally stretched by a factor of 4 and its frequency spectrum is compressed by a factor of 4 [Figures 12.11(b1) and (b2)].

12.2.2 Anti-Imaging Filters

The scaled-up version of $x(t)$ can then be isolated by filtering out all the high spectral images. This gives the inserted zeros their final interpolated values.

The ideal anti-imaging filter must have a rectangular transfer function (in the frequency domain) that passes through only the center band of the up-sampled spectrum.

$$H_u(\omega) = \begin{cases} U & \text{if } |\omega| < \frac{\pi}{U} \\ 0 & \text{if } \frac{\pi}{U} \leq |\omega| \leq \pi \end{cases}$$

Figure 12.12   Anti-Imaging Filter.

The impulse response of such filter is found to be

$$h_u(n) = \frac{\sin\left(\dfrac{n}{U}\right)}{\left(\dfrac{n}{u}\right)} = \text{sinc}\left(\frac{n}{u}\right)$$

12.3 Down-Sampling and Decimation (Down-Scaling)

Scaling down an image by an integer factor D involves retaining only every Dth line and every Dth pixel on each line; we eliminate the rest. The value of the pixels we keep has to be changed to reflect the "weight" of the discarded neighborhood pixels.

A subsampler or $\downarrow D$ down-sampler is a block that decimates an input sequence by a factor of D (Figure 12.13).

Its transfer function is simply

$$x_d(n) = x(nD)$$

It is clear that the sampling rate of a down sampler f_{d_s} is D times smaller than the f_s sampling rate of $x(n)$. If we apply the Z-transform to the transfer function we obtain

$$X_d(z) = \sum_{n=-\infty}^{n=+\infty} x[nD]z^{-n}$$

This expression fails to connect directly $X_d(z)$ to $X(z)$. However, if we introduce an auxiliary function $x_{\text{aux}}(n)$ defined by

$$x_{\text{aux}}(n) = c(n)\, x(n)$$

where

$$c(n) = \begin{cases} 1 & \text{for } n = 0, \pm D, \pm 2D \cdots \\ 0 & \text{for other } n \text{ values} \end{cases}$$

and therefore

$$x_{aux}(n) = \begin{cases} x(n) & \text{for } n = 0, \pm D, \pm 2D \cdots \\ 0 & \text{for other } n \text{ values} \end{cases}$$

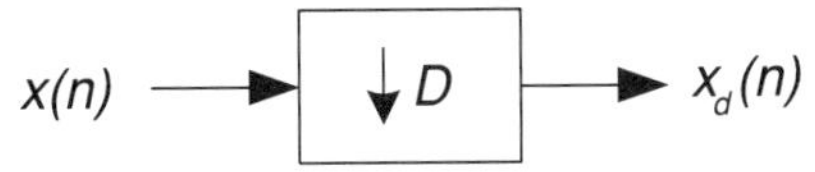

Figure 12.13 Down-Sampler Block.

Then we can rewrite

$$X_d(z) = \sum_{n=-\infty}^{n=+\infty} x(nD)z^{-n} = \sum_{n=-\infty}^{n=+\infty} x_{aux}(nD)z^{-n}$$

A change of variable

$$k = nD$$

brings us to

$$X_d(z) = \sum_{k=-\infty}^{k=+\infty} x_{aux}(k)z^{-\frac{k}{D}} = X_{aux}\left(z^{\frac{1}{D}}\right)$$

This equation relates the $X_d(z)$ to the auxiliary Z-transform $X_{aux}(z)$. If we rewrite $c(n)$ as

$$c(n) = \frac{1}{D} \sum_{k=0}^{k=D-1} W_D^{kn} \text{ where } W_D = e^{-\frac{j2\pi}{D}}$$

then

$$X_{aux}(z) = \sum_{n=-\infty}^{n=+\infty} c(n)x(n)z^{-n} = \frac{1}{D} \sum_{n=-\infty}^{n=+\infty} \left(\sum_{k=0}^{k=D-1} W_D^{kn} \right) x(n)z^{-n}$$

After rearranging the summation terms, we obtain

$$X_{aux}(z) = \frac{1}{D} \sum_{k=0}^{k=D-1} \left(\sum_{n=-\infty}^{n=+\infty} x(n)W_D^{kn}z^{-n} \right) = \frac{1}{D} \sum_{k=0}^{k=D-1} X(zW_M^{-k}) = \frac{1}{D} \sum_{k=0}^{k=D-1} X\left(ze^{-\frac{j2\pi k}{D}}\right)$$

And we finally arrive at the connection between $X_d(z)$ and $X(z)$

$$X_d(z) = X_{aux}\left(z^{\frac{1}{D}}\right) = \frac{1}{D} \sum_{k=0}^{k=D-1} X\left(z^{\frac{1}{D}}e^{-\frac{j2\pi k}{D}}\right)$$

12.3.1 Spectrum of Down-Sampled Signals

The frequency spectrum of $x_d(n)$ is found by reducing z to $e^{j\omega}$.

$$X_d(\omega) = \frac{1}{D} \sum_{k=0}^{k=D-1} X\left(\frac{\omega - 2\pi k}{D}\right)$$

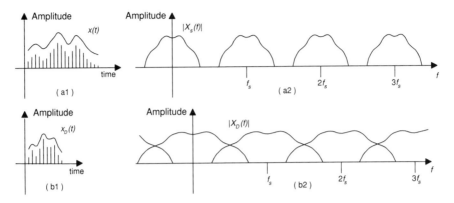

Figure 12.14   Spectrum of Up-Sampled Function $x_D(t)$.

When we decimate a discrete function $x(n)$, the spectrum of the new function $x_d(n)$ is a stretched-out version of the original spectrum plus a number of its spectral images. The expansion factor is equal to the down-sampling coefficient D. As an example let us pick $D = 2$. Then the spectrum of $x_d(n)$ is given by

$$X_d(\omega) = \frac{1}{2}\left[X\left(\frac{\omega}{2}\right) + X\left(\frac{\omega}{2} - \pi\right)\right]$$

The spectrum is given by the superposition of two horizontally expanded (by a factor of 2) versions of the original spectrum, properly shifted, and scaled. The expansion factor is in this case 2. However, since the original signal $x(n)$ was obtained by sampling it with the pulse function $P(t)$, $X(\omega)$ already contains a theoretically unlimited number of spectral images that are also duplicated by the down-sampler (Figure 12.14).

If the different components of the $X_d(\omega)$ spectrum overlap, then recovering a unique $x_d(t)$ function becomes impossible. This problem is known as *signal aliasing* and can affect any sampled system unless special precautions are taken. The solution to aliasing is to preprocess the sequence $x(n)$ so when it passes through a down-sampler its frequency spectrum remains free of overlaps.

12.3.2 Anti-Aliasing Filters

The circuit that performs this function is the anti-aliasing filter (Figure 12.15). The transfer function of an ideal anti-aliasing filter is given by

$$H_D(\omega) = \begin{cases} 1 & \text{if } |\omega| < \frac{\pi}{D} \\ 0 & \text{if } \frac{\pi}{D} \leq |\omega| \leq \pi \end{cases}$$

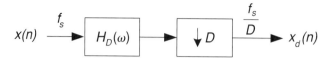

Figure 12.15   Anti-Aliasing Filter.

12.4 Rational Number Sampling (Scaling)

What do we do when the desired image scale factor is not an integer but a rational number? We cascade a $\uparrow U$ up-sampler and its associated anti-imaging filter with a $\downarrow D$ down-sampler and its anti-aliasing preprocessor (Figure 12.16).

The scaling factor is given by the ratio of the U and D parameters. The transfer function for the ideal rational number sampler is

$$H_{UD}(\omega) = H_U(\omega)H_D(\omega) = \begin{cases} U & \text{if } |\omega| < \min\left(\frac{\pi}{U}, \frac{\pi}{D}\right) \\ 0 & \text{if } \min\left(\frac{\pi}{U}, \frac{\pi}{D}\right) \le |\omega| \le \max\left(\frac{\pi}{U}, \frac{\pi}{D}\right) \end{cases}$$

12.5 Multirate Processing

One major problem with the scaling algorithms we have looked at thus far is that many of them require very high clock rates. Scaling up an NTSC image (27 MHz pixel clock) by a factor of 9 (3 horizontal × 3 vertical) would require circuitry that operates at 2.43 GHz!

To address this problem let us go back to the general equation for the Z-transform $H(z)$ of a transfer function $h(n)$, and rearrange the summation terms.

$$H(z) = \sum_{n=-\infty}^{n=+\infty} h(n)z^{-n} = \sum_{m=-\infty}^{m=+\infty} h(2m)z^{-2m} + \sum_{m=-\infty}^{m=+\infty} h(2m+1)z^{-(2m+1)}$$

If we factor out z^{-1} from the second sum we obtain

$$H(z) = \sum_{n=-\infty}^{n=+\infty} h(2m)z^{-2m} + z^{-1}\sum_{m=-\infty}^{m=+\infty} h(2m+1)z^{-2m} = H_0(z^2) + z^{-1}H_1(z^2)$$

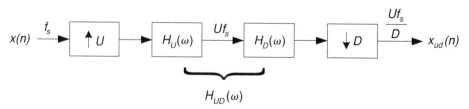

Figure 12.16   Rational Number Sampling.

where the expressions for $H_0(z)$ and $H_1(z)$ are given by

$$H_0(z) = \sum_{m=-\infty}^{m=+\infty} h(2m)z^{-m}$$

$$H_1(z) = \sum_{m=-\infty}^{m=+\infty} h(2m+1)z^{-m}$$

This is known as a two-phase decomposition of $H(z)$. It is obvious that a function $H(z)$ can be decomposed in a number of different ways.

12.5.1 Polyphase Decomposition

The general format for a *polyphase decomposition* of $H(z)$ is given by

$$H(z) = \sum_{k=0}^{k=M-1} z^{-k} H_k(Z^M)$$

where $H_k(z)$ are by definition the *polyphase components* of $H(z)$.

$$H_k(z) = \sum_{m=-\infty}^{m=+\infty} h(Mm+k)z^{-Mm}$$

To understand what polyphase decomposition achieves, let us look at a transfer function with a finite number of terms—say, ten—and decompose it into two phases.

$$H(z) = \sum_{n=0}^{n=9} h(z)z^{-n} = H_0(z^2) + z^{-1}H_1(z^2)$$

The corresponding polyphase components are

$$H_0(z) = \sum_{m=0}^{m=4} h(2m)z^{-m} = h(0) + h(2)z^{-1} + h(4)z^{-2} + h(6)z^{-3} + h(8)z^{-4}$$

and

$$H_1(z) = \sum_{m=0}^{m=4} h(2m+1)z^{-m} = h(1) + h(3)z^{-1} + h(5)z^{-2} + h(7)z^{-3} + h(9)z^{-4}$$

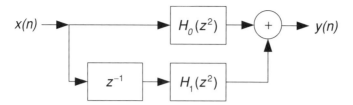

Figure 12.17   Two-Phase Filter.

Figure 12.17   illustrates a possible hardware realization of this filter.

12.5.2 Polyphase Decimator

To use it as an anti-aliasing filter for a down-sampler we must place it in front of the $\downarrow D$ block (Figure 12.18).

For simplicity of presentation we will set $D = 2$ (Figure 12.19).

If we replace the $H(z)$ block with its two-phase filter realization (Figure 12.17) we obtain Figure 12.20.

At this point we need to introduce two important mathematical equivalencies known as the *Noble Identities 1 and 2*. Their proof is beyond the scope of our book.

Noble Identity 1

$$x(n) \to \boxed{\downarrow D} \to \boxed{H(z)} \to y(n) \Leftrightarrow x(n) \to \boxed{H(z^M)} \to \boxed{\downarrow D} \to y(n)$$

Noble Identity 2

$$x(n) \to \boxed{H(z)} \to \boxed{\uparrow U} \to y(n) \Leftrightarrow x(n) \to \boxed{\uparrow U} \to \boxed{H(z^M)} \to y(n)$$

Using the first Noble identity, we now switch the positions of the $\downarrow 2$ down-sampler and the $H_0(z)$ and $H_1(z)$ components (Figure 12.21). If we follow the input signal path we note that each $\downarrow 2$ block only processes every other input sample of $x(n)$. After all, this is what a $\downarrow 2$ down-sampler does. Since the z^{-1} is

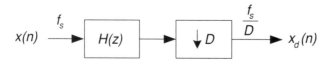

Figure 12.18

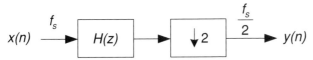

Figure 12.19

essentially a one-sample delay, the bottom $\downarrow 2$ block processes the input samples skipped by the top $\downarrow 2$ block. When combined they take care of all the $x(n)$ input samples. In turn, each $H_i(z)$ component sees a down-sampled signal at its input and therefore it operates only at half the sampling rate.

The operation of the z^{-1} block and the two $\downarrow 2$ blocks is clearly equivalent to that of a much simpler front commutator that switches its position with every sampling pulse (Figure 12.22). This filter is known as a *decimator*.

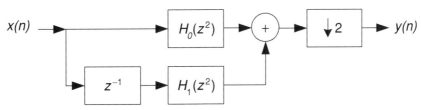

Figure 12.20

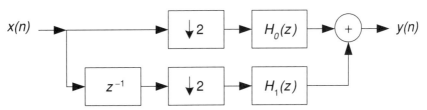

Figure 12.21

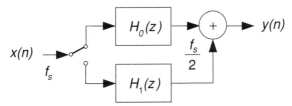

Figure 12.22   Decimator Filter

Figure 12.23

12.5.3 Polyphase Interpolator

Moving on to up-samplers we observe that a ↑2 block would normally operate at $2f_s$, where f_s is the input sampling frequency. By decomposing the transfer function of its anti-imaging filter into two phases, we get the circuit realization in Figure 12.24. After distributing the ↑2 function the realization changes to Figure 12.25.

With the help of the second Noble identity we can swap forward the ↑2 blocks (Figure 12.26). The ↑2 blocks and the summing block can then be replaced by a simpler commutator circuit (Figure 12.27). Although the commutator still operates at $2f_s$ most of the complex circuitry of the filter is driven at the lower f_s rate. This final filter configuration is called an *interpolator*.

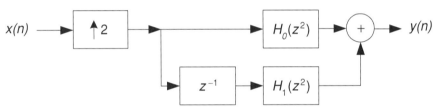

Figure 12.24

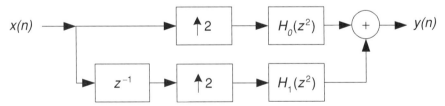

Figure 12.25

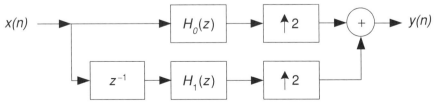

Figure 12.26

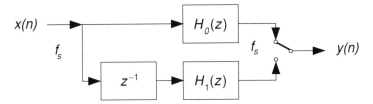

Figure 12.27   Interpolator Filters

12.6 Image Scalers

Most common image downsizing engines consist of a horizontal scaler followed by a vertical scaler. This ordering minimizes the line buffer requirements for image reducing operations (Figure 12.28). For zoom-up processors, such as videowall controllers, the order is reversed.

Each scaler is a rational number interpolator/decimator combination with the appropriate filters built in. However, the horizontal and vertical scalers do not operate fully independent of each other. Since many samples are thrown out in the decimation process, the interpolator needs to calculate only those samples that the decimator will keep. Therefore, logic circuits that control the interpolator's output switch on the basis of the desired scaling ratio are also part of each scaler section.

The number of exact scaling ratios attainable by a given scaler is related to the complexity of its filter realization. For an exact scaling factor of 5/7 for example the interpolator will have to create seven different phases (by inserting six zeros between every pair of input samples), from which the decimator will pick every fifth sample (Figure 12.29).

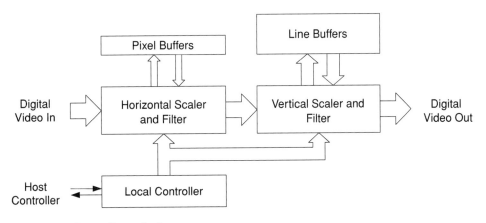

Figure 12.28   Image Down-Scaler.

Figure 12.29 Sampling Scheme for 5/7 Scaling Ratio.

Dynamically allocating filter phases for any conceivable scaling factor is not a practical solution. Instead ICs provide a fixed number of phases and "closest-fit" or "best-fit" algorithms for selecting the proper phase for each sample. The "closest-fit" approach is illustrated in Figure 12.30.

As we saw earlier, the interpolator and decimator circuits are essentially digital filters. A simplified architecture for such a filter is shown in Figure 12.31. The input data is synchronously sampled by a chain of cascaded registers, each register as wide as the data. Each register executes a one-sample delay and thus represents a z^{-1} block. The current and past input data samples are then multiplied by filter coefficients calculated (or retrieved from look-up tables) by a local controller. Their values correspond to the desired image scale factor and to the necessary degree of filtering. An output adder completes the "sum-of-products" structure of the filter. Each multiplier junction corresponds to a *"tap"* along the chain of z^{-1} delay registers.

High-speed polyphase filters can be designed in a similar manner. However, when the speed of the IC is significantly higher than data speed, a single sum-of-products tree can be used to implement multiple phases. This practice is common in video scalers where the pixel clock is 13.5 MHz while the filters operate at close to 100 MHz. To take advantage of this speed differential, the tap coefficient registers are replaced by tap register banks. Each bank contains the coefficients associated with that tap for all filter phases (Figure 12.32).

When a new sample is clocked in the register chain the *phase controller* resets all register banks to *phase 0*. The phase 0 coefficients are multiplied by the data and the products are sent to the adder where they are picked up by the output commutator switch. Then the phase controller advances all banks to their *phase 1* position. Now the phase 1 coefficients are multiplied by the same data and the

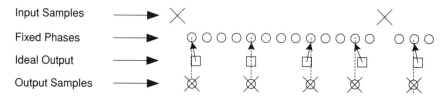

Figure 12.30 Classes Fit Approximations for 5/7 Scaling Ratio.

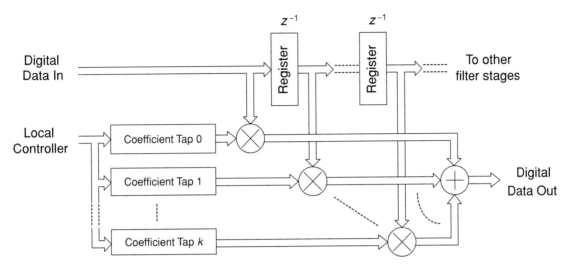

Figure 12.31   Digital Filter Hardware Architecture.

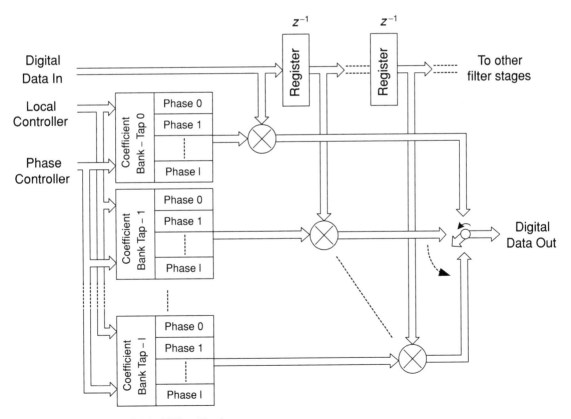

Figure 12.32   Polyphase Digital Filter Hardware.

results make their way to the adder and output switch. This cycle repeats until all phases are exhausted and a new input sample is clocked in.

12.7 Commercial Image Scaler ICs

Although you can obtain a crude form of scaling just by letting your FPGA/CPLD decimate fields and pixels (see Chapter 3), applications for such low-quality scalors rarely go beyond security and traffic monitoring. The first level of respectable quality scaling is provided as an auxiliary function by some decoder-digitizers, such as Conexant's Bt835 and the SAA7118 from Philips. In the Bt835 (Figure 12.33) the digitized luma signal is first band limited by a 3 MHz anti-aliasing filter and then is horizontally scaled by a 6-tap, 32-phase down-sampler. The output of the horizontal scaler is stored in a multiline buffer from where it can be accessed by a 2-tap vertical scaler. Finally, spectral images are removed by an output 2- to 5-tap programmable digital filter.

As we noted earlier, the chroma scaling process is simpler, requiring only two 2-tap filters, one for horizontal scaling and the other for vertical scaling. The image quality produced by the Bt835 built-in scaler is satisfactory to good (especially for certain "sweet spot" scaling factors). Programming the device is relatively easy. In the horizontal direction we start with the desired horizontal image pixel size, H_{desired} and translate it into an HSCALE parameter according to the formula

$$\text{HSCALE} = 4096\ [(910/H_{\text{desired}}) - 1]$$

which is then stored in the chip's HSCALE register.

In the vertical direction we start with a desired scaling ratio S_{desired}, then calculate and store in the register VSCALE the expression

$$\text{VSCALE} = [0 \times 10000 - 512\ (S_{\text{desired}} - 1)]\ \&\ 0 \times 1\text{FFF}$$

Image cropping is implemented independent of the resizing function by using the HDELAY, HACTIVE, VDELAY, and VACTIVE resisters mentioned in Chapter 8. The Bt835 is ideally suited for security, videoconference, and dis-

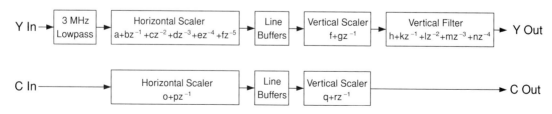

Figure 12.33   Bt835 Down-Scaler.

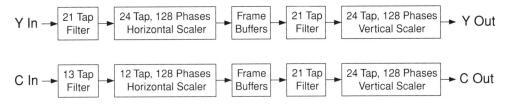

Figure 12.34 GF9320 Down-Scaler.

tance learning markets. For studio-grade applications a scaler IC needs some serious additional resources, especially in the areas of processing power and image storage. A solution for high-end image resizing is provided by Gennum's GF9320 Scaling Processor.

Figure 12.34 illustrates the GF9320 down-scaler chain, although the IC is also capable of up-scaling all the way to an output resolution of 2048 × 2048 (the order of the horizontal and vertical scalers is reversed in this case). The lower number of taps along the chroma horizontal scaling path reflects the splitting of resources needed to accommodate the 4:2:2 format the scaler is designed to handle.

In terms of memory requirements, a fully expanded GF9320 requires four memory banks, each comprised of 5 SDRAM ICs, as can be seen in Figures 12.35 through 12.37. All the memory management functions are handled internally by the IC. Communication with the local or host controller is done via a serial four-wire interface (DATA IN, DATA OUT, S CLK, S RESET~). The GF9320 expects a 20- or 10-bit (YIN[0..9], CIN[0..9]) digital stream, at input pixel or double pixel rate (CK IN). The input must contain TRS codes for internal image synchronization. If this chip is used with a source that only provides video data and HBLANK, VBLANK, FIELD, and PXLCLK reference signals, an FPGA or CPLD should buffer the inputs and insert the SAV, EAV, and TRS codes prior to submission to the scaler.

The outputs of the IC can be programmed as RGB or YCbCr in both 4:4:4 and 4:2:2 formats. Standard signals HBLANK, VBLANK, FIELD, and OUT PXLCLK as well as inserted TRS codes are available to provide output reference information to downstream buffers and processors. OUT PXLCLK is derived from the input clock CK OUT, according to the programmed scaling parameters. The vertical clock CK V is to be supplied by the higher rate of CK IN and CK OUT input signals.

The FILM FR (input film sequence reset) is used in support of film rate conversion functions and the OF RESET (output frame reset) as an external synchronization signal for the output data.

Output image size is determined by ZOOM parameters provided by the host or local controller. To calculate the ZOOM parameters we start with the desired horizontal and vertical START and STOP pixel counts for the given

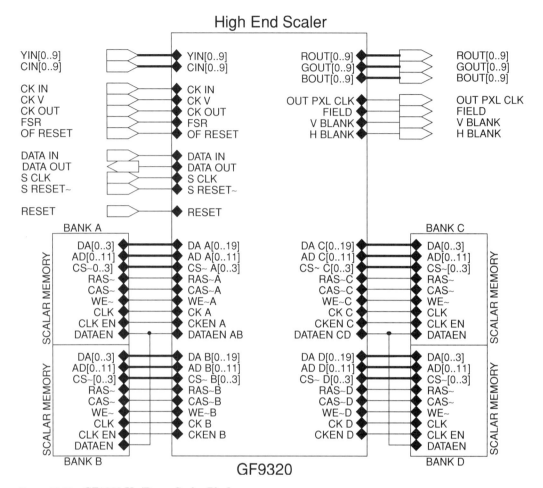

Figure 12.35   GF9320 Up/Down Scaler Block.

input and desired output image windows. Then we derive the length of the live line and frame using the formulas

$$HLIVE = HSTOP - HSTART + 1$$
$$VLIVE = VSTOP - VSTART + 1$$

where the notation is obvious. Zoom parameters are then defined by

$$H\_ZOOM\_RATIO = 524288 \; (IN\_HLIVE/OUT\_HLIVE)$$
$$V\_ZOOM\_RATIO = 524288 \; (IN\_VLIVE/OUT\_VLIVE)$$

Again, the notation is self-explanatory.

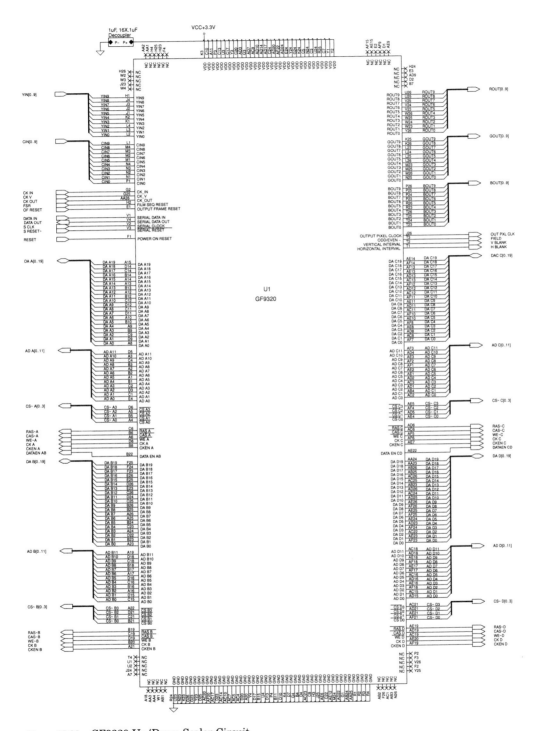

Figure 12.36 GF9320 Up/Down Scaler Circuit.

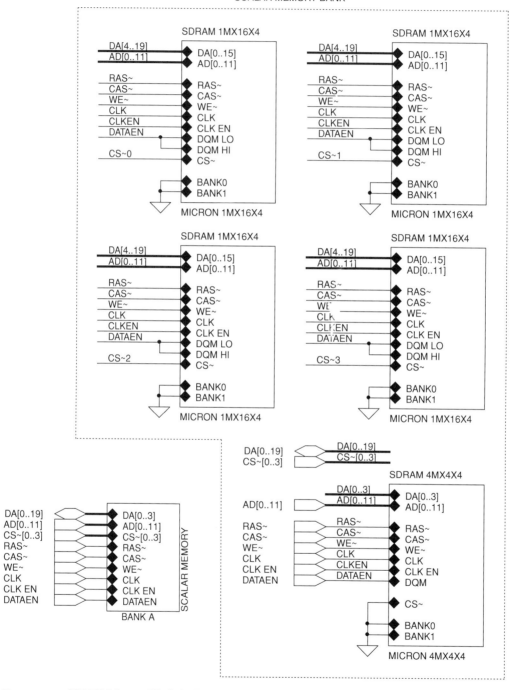

Figure 12.37   GF9320 Memory Block Architecture.

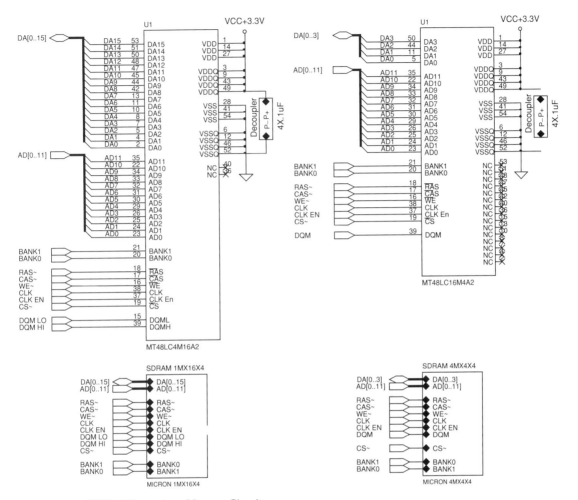

Figure 12.38   GF9320 Elementary Memory Circuits.

This is just a very brief presentation of what the GF9320 can do and how it works. The true capabilities of this IC can only be appreciated after programming the myriad of control options it offers and after evaluating the resulting image on a good high-resolution monitor. One final note: When the clock rates approach the upper limit of roughly 90 MHz, with distances between the GF9320 and the 20 memory chips associated with it inevitably in the range of a few inches, most of the data lines need series termination resistors.

A possible hardwear implementation for a GF9320 based scalor module is shown in Figures 12.35 through 12.38.

Image Compression and Decompression

The best known video compression methods to date are specified by the JPEG standard, for still-image applications, and by the MPEG standard, for video and audio applications. While JPEG is the leading technology in the digital photography marketplace, the MPEG standard has gained almost universal acceptance in studio and HDTV, as well as streaming video applications. MP3, an MPEG subset, controls the streaming audio and personal audio segments. Since its inception the MPEG standard has gone through a number of revisions; MPEG-2 is now the dominant version. A newer arrival, MPEG-4, is a low-frequency adaptation of MPEG-2 and is designed the meet the needs of the bandwidth-strapped wireless market (Cell phones, Bluetooth and Wi-Fi).

13.1 JPEG Compression and Decompression

13.1.1 Discrete Cosine Transform

The *JPEG* (Joint Photographic Experts Group) standard was designed to compress either full-color or grey-scale digital images. The main concept behind JPEG is to empirically adjust the compression algorithm to account for the characteristics of the human vision system. The empirical part consists in the development, through experimentation, of a matrix of parameters that synthesizes what our eyes actually see and what they discard. Let's start from the beginning.

The basic mathematical tool used in JPEG compression is the *discrete cosine transform* (DCT) which is defined by the expression

$$D(i, j) = \frac{2}{\sqrt{MN}} C(i)C(j) \sum_{m=0}^{m=M-1} \sum_{n=0}^{n=N-1} p(m, n) \cos\left[\frac{(2m + 1)i\pi}{2M}\right] \cos\left[\frac{2n + 1)j\pi}{2N}\right]$$

Figure 13.1   JPEG 8 × 8 Block Decomposition.

where the $C(i)$ and $C(j)$ coefficients are

$$
C(k) = \begin{cases} \dfrac{1}{\sqrt{2}} & \text{for } k = 0 \\ 1 & \text{for } k > 0 \end{cases}
$$

and $p(m,n)$ is the function to which the transform is applied. In our case, $p(m,n)$ represents pixel values, either intensity or color.

To recover the function $p(m,n)$ we apply the *Inverse Direct Cosine Transform* (IDCT) to $D(i,j)$.

$$
p(m, n) = \frac{2}{\sqrt{MN}} \sum_{i=0}^{i=M-1} \sum_{j=0}^{j=N-1} C(i)C(j)D(i, j) \cos\left[\frac{(2m + 1)i\pi}{2M}\right] \cos\left[\frac{(2n + 1)j\pi}{2N}\right]
$$

JPEG applies the DCT and all subsequent calculations on elementary image areas 8 pixels wide by 8 lines high (8 × 8) called *image blocks*.

One of the first operations performed by a JPEG encoder when presented with a new input image is to parcel it into 8 × 8 blocks (Figure 13.1).

For an 8 × 8 block the DCT and DCTI become

$$
D(i, j) = \frac{1}{4} C(i)C(j) \sum_{m=0}^{m=7} \sum_{n=0}^{n=7} p(m, n) \cos\left[\frac{(2m + 1)i\pi}{16}\right] \cos\left[\frac{(2n + 1)j\pi}{16}\right]
$$

$$
p(m, n) = \frac{1}{4} \sum_{i=0}^{i=7} \sum_{j=0}^{j=7} C(i)C(j)D(i, j) \cos\left[\frac{(2m + 1)i\pi}{16}\right] \cos\left[\frac{(2n + 1)j\pi}{16}\right]
$$

For clarity of presentation we limit our analysis to one 8 × 8 black and white intensity block. We use the following 8 × 8 luma block extracted from a stock black and white picture, as our test image block I:

$$
I = \begin{vmatrix}
170 & 153 & 153 & 153 & 160 & 160 & 153 & 134 \\
170 & 153 & 153 & 160 & 160 & 160 & 153 & 134 \\
170 & 110 & 153 & 160 & 160 & 153 & 153 & 134 \\
160 & 110 & 134 & 165 & 165 & 153 & 134 & 110 \\
160 & 134 & 134 & 165 & 160 & 134 & 134 & 110 \\
165 & 134 & 134 & 160 & 223 & 134 & 110 & 134 \\
165 & 134 & 160 & 196 & 223 & 223 & 110 & 134 \\
165 & 160 & 196 & 223 & 223 & 254 & 198 & 160
\end{vmatrix}
$$

We recall from matrix algebra that applying the DCT to the image array I is equivalent with multiplying the direct cosine transform matrix T by the matrix I; we then multiply the result with the transpose of T (the matrix obtained by switching rows to columns and columns to rows).

From the DCT definition statement we can readily identify the elements of the T matrix:

$$
T(i, j) = \sqrt{\frac{2}{M}} \cos\left[\frac{(2j + 1)i\pi}{2M}\right]
$$

where i and j are row and column numbers, from 0 to 7.

$$
T = \begin{vmatrix}
0.3536 & 0.3536 & 0.3536 & 0.3536 & 0.3536 & 0.3536 & 0.3536 & 0.3536 \\
0.4904 & 0.4157 & 0.2728 & 0.0975 & -0.0975 & -0.2778 & -0.4157 & -0.4904 \\
0.4619 & 0.1913 & -0.1913 & -0.4619 & -0.4619 & -0.1913 & 0.1913 & 0.4619 \\
0.4157 & -0.0975 & -0.4904 & -0.2778 & 0.2778 & 0.4904 & 0.0975 & -0.4157 \\
0.3536 & -0.3536 & -0.3536 & 0.3536 & 0.3536 & -0.3536 & -0.3536 & 0.3536 \\
0.2778 & -0.4904 & 0.0975 & 0.4157 & -0.4157 & -0.0975 & 0.4904 & -0.2778 \\
0.1913 & -0.4619 & 0.4619 & -0.1913 & -0.1913 & 0.4619 & -0.4619 & 0.1913 \\
0.0975 & -0.2778 & 0.4157 & -0.4904 & 0.4904 & -0.4157 & 0.2778 & -0.0975
\end{vmatrix}
$$

If we map the double cosine in the expression of $D(i,j)$ as m and n span the 0 to M-1 and 0 to N-1 intervals, we obtain a visual image of the $T(i,j)$ terms (Figure 13.2). Starting from the top left corner and going towards the bottom right we move from the lower spatial frequency components of the DCT to the higher ones (each box in Figure 13.2 corresponds to a unique set of i and j values).

When we apply the DCT matrix to an image we are actually calculating the amplitudes of the spatial sinusoids that, when superimposed, recreate that image. These are the harmonic components of the image. In many ways the DCT is similar to a two-dimensional Fourier series.

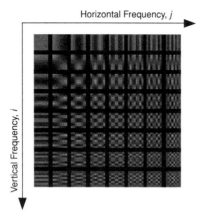

Horizontal Frequency, *j*

Vertical Frequency, *i*

Figure 13.2   DCT Coefficients.

Going back to the *I* matrix, we note that all its values are positive. Before we continue let us remove this apparent DC bias of the image by subtracting 128 from all $I(i,j)$.

$$I'(i,j) = I(i,j) - 128$$

The new image matrix *I'* is then

$$I' = \begin{vmatrix} 42 & 25 & 25 & 25 & 32 & 32 & 25 & 6 \\ 42 & 25 & 25 & 32 & 32 & 32 & 25 & 6 \\ 42 & -18 & 25 & 32 & 32 & 25 & 25 & 6 \\ 32 & -18 & 6 & 37 & 37 & 25 & 6 & -18 \\ 32 & 6 & 6 & 37 & 32 & 6 & 6 & -18 \\ 37 & 6 & 6 & 32 & 95 & 6 & -18 & 6 \\ 37 & 6 & 32 & 68 & 95 & 95 & -18 & 6 \\ 37 & 32 & 68 & 95 & 95 & 126 & 70 & 32 \end{vmatrix}$$

We apply the DCT to the source image by multiplying the *I'* matrix with *T* on the left

$$TI' = \begin{vmatrix} 106.43 & 5.71 & 6.93 & -0.86 & -5.30 & -0.58 & 2.87 & 1.18 \\ 22.63 & -4.45 & 40.10 & 13.67 & 9.19 & -23.58 & -6.66 & -4.17 \\ 68.24 & -18.84 & 42.39 & -26.51 & 6.01 & -6.66 & 3.49 & 5.65 \\ 126.59 & -49.45 & 28.13 & -25.59 & 10.61 & -1.79 & -7.83 & 3.18 \\ 159.12 & -74.26 & 26.79 & 9.46 & -20.51 & 9.33 & 11.10 & -17.28 \\ 122.70 & -65.28 & 77.03 & -47.53 & 10.96 & 14.53 & -20.05 & 6.92 \\ 42.79 & 7.63 & 38.34 & -43.99 & 32.88 & -29.40 & 15.88 & 1.54 \\ 9.19 & -12.78 & 34.18 & -10.81 & -7.78 & -7.22 & 14.16 & -2.54 \end{vmatrix}$$

and then by T^{τ} (transpose of T) on the right. If we round off the results we obtain:

$$I'' = TI'T^{\tau} = \begin{bmatrix} 233 & 21 & -103 & 78 & 51 & 18 & 25 & 8 \\ -75 & 19 & 71 & -21 & -18 & 26 & -18 & 12 \\ 104 & -22 & -14 & 5 & -36 & -11 & 16 & -18 \\ -47 & 31 & 10 & -2 & 27 & -38 & -19 & 11 \\ 13 & -7 & 3 & -3 & -29 & 25 & -12 & -10 \\ -16 & -1 & -19 & 16 & 16 & -8 & 25 & -4 \\ 5 & -10 & 11 & -9 & 10 & 2 & -9 & 24 \\ -2 & 1 & 3 & -3 & -9 & 12 & 9 & -9 \end{bmatrix}$$

Just as the coefficients of the Fourier series gave us the spectral components of the argument function, the I'' matrix gives us the horizontal and vertical spectral components of our image. As we can see, our 8×8 block is dominated by low-frequency components in both directions (the top left coefficients are the largest). Is this the case for most images? Yes and no. Naturally occurring scenes are generally a mixed bag of high and low frequencies. It is our vision that picks up on "image themes" rather than details. And *themes* compress very well. The *details* of a blooming lilac tree, for example, do not.

13.1.2 Quantization Matrices

And now for the empirical part. Through exhaustive research it was determined that loss of visual information in some frequency ranges is more acceptable than in others. In general, our eyes are more sensitive to low spatial frequencies than to high spatial frequencies. Therefore, if we want to conserve bandwidth, the effects of reducing the high-frequency content of an image will be less visible. As a result, a family of *quantization matrices* were developed as "sensitivity to frequency" charts. The bigger a certain matrix element is the less sensitive the eye is to that combination of horizontal and vertical spatial frequencies. The *quantization matrix* Q_{50} represents the best compromise between image quality and compression ratios.

$$Q_{50} = \begin{bmatrix} 16 & 11 & 10 & 16 & 24 & 40 & 51 & 61 \\ 12 & 12 & 14 & 19 & 26 & 58 & 60 & 55 \\ 14 & 13 & 16 & 24 & 40 & 57 & 69 & 56 \\ 14 & 17 & 22 & 29 & 51 & 87 & 80 & 62 \\ 18 & 22 & 37 & 56 & 68 & 109 & 103 & 77 \\ 24 & 35 & 55 & 64 & 81 & 104 & 113 & 92 \\ 49 & 64 & 78 & 87 & 103 & 121 & 120 & 101 \\ 72 & 92 & 95 & 98 & 112 & 100 & 103 & 99 \end{bmatrix}$$

For higher compression ratios but poorer image quality we multiply the Q_{50} matrix by a scalar larger than 1 and clip all the results larger than 255 to 255. For better quality levels we multiply Q_{50} by a scalar smaller than one. If we introduce a *quality factor* q between 1 and 99, with 1 for poorest and 99 for highest image quality, we can derive the quantization matrix for any desired quality level using the expression

$$[Q_q] = \begin{cases} \dfrac{50}{q} \, [Q_{50}] & \text{for } q < 50 \\ \dfrac{100-q}{50} \, [Q_{50}] & \text{for } q > 50 \end{cases}$$

For example, the quantization matrices for quality factors of 10 and 90 are easily derived to be:

$$Q_{10} = \begin{vmatrix} 80 & 55 & 50 & 80 & 120 & 200 & 255 & 255 \\ 60 & 60 & 70 & 95 & 130 & 255 & 255 & 255 \\ 70 & 65 & 80 & 120 & 200 & 255 & 255 & 255 \\ 70 & 85 & 110 & 145 & 255 & 255 & 255 & 255 \\ 90 & 110 & 185 & 255 & 255 & 255 & 255 & 255 \\ 120 & 175 & 255 & 255 & 255 & 255 & 255 & 255 \\ 245 & 255 & 255 & 255 & 255 & 255 & 255 & 255 \\ 360 & 255 & 255 & 255 & 255 & 255 & 255 & 255 \end{vmatrix}$$

$$Q_{90} = \begin{vmatrix} 3 & 2 & 2 & 3 & 5 & 8 & 10 & 12 \\ 2 & 2 & 3 & 4 & 5 & 12 & 12 & 11 \\ 3 & 3 & 3 & 5 & 8 & 11 & 14 & 11 \\ 3 & 3 & 4 & 6 & 10 & 17 & 16 & 12 \\ 4 & 4 & 7 & 11 & 14 & 22 & 21 & 15 \\ 5 & 7 & 11 & 13 & 16 & 21 & 23 & 18 \\ 10 & 13 & 16 & 17 & 21 & 24 & 24 & 20 \\ 14 & 18 & 19 & 20 & 22 & 20 & 21 & 20 \end{vmatrix}$$

Quantization matrices are used to artificially reduce the weights of the spatial frequency components of the DCT processed image. We divide each element of the I'' matrix by the corresponding element in the quantization matrix and round off the result.

$$I''_Q(i,j) = \text{ROUND} \left[\frac{I''(i,j)}{Q(i,j)} \right]$$

If we quantize the I'' using the Q_{50} quantization matrix we find:

$$I''Q_{50} = \begin{vmatrix} 15 & 2 & -10 & 5 & 2 & 0 & 0 & 0 \\ -6 & 2 & 5 & -1 & -1 & 0 & 0 & 0 \\ 7 & -2 & -1 & 0 & -1 & 0 & 0 & 0 \\ -3 & 2 & 0 & 0 & 1 & 0 & 0 & 0 \\ 1 & 0 & 0 & 0 & 0 & 0 & 0 & 0 \\ -1 & 0 & 0 & 0 & 0 & 0 & 0 & 0 \\ 0 & 0 & 0 & 0 & 0 & 0 & 0 & 0 \\ 0 & 0 & 0 & 0 & 0 & 0 & 0 & 0 \end{vmatrix}$$

We are left with only 19 frequency components from a total of 64. What we have done is to eliminate those components that were too small to overcome the eye's relative lack of sensitivity to their spatial frequency. As we can see, the remaining components are clustered towards the low-frequency range, the top left of the matrix.

If we drop the quality level to 10 we are left with only seven nonzero coefficients.

$$I''Q_{10} = \begin{vmatrix} 3 & 0 & -2 & 1 & 0 & 0 & 0 & 0 \\ -1 & 0 & 1 & 0 & 0 & 0 & 0 & 0 \\ 1 & 0 & 0 & 0 & 0 & 0 & 0 & 0 \\ -1 & 0 & 0 & 0 & 0 & 0 & 0 & 0 \\ 0 & 0 & 0 & 0 & 0 & 0 & 0 & 0 \\ 0 & 0 & 0 & 0 & 0 & 0 & 0 & 0 \\ 0 & 0 & 0 & 0 & 0 & 0 & 0 & 0 \\ 0 & 0 & 0 & 0 & 0 & 0 & 0 & 0 \end{vmatrix}$$

At a quality level of 90 we eliminate only 16 of the matrix elements, which indicates a relatively low compression ratio.

$$I''Q_{90} = \begin{vmatrix} 78 & 10 & -51 & 26 & 10 & 2 & 3 & 1 \\ -37 & 10 & 24 & -5 & -4 & 2 & -2 & 1 \\ 35 & -7 & -5 & 1 & -5 & -1 & 1 & -2 \\ -16 & 10 & 3 & 0 & 3 & -2 & -1 & 1 \\ 3 & -2 & 0 & 0 & -2 & 1 & -1 & -1 \\ -3 & 0 & -2 & 1 & 1 & 0 & 1 & 0 \\ 0 & -1 & 1 & -1 & 0 & 0 & 0 & 1 \\ 0 & 0 & 0 & 0 & 0 & 1 & 0 & 0 \end{vmatrix}$$

The larger the number of zeros the bigger the compression ratio we get from quantization. However, just by reducing the value of the remaining nonzero terms we increased compression efficiency. Smaller numbers require fewer bits to encode.

For a better evaluation of the algorithm let us recover our images from the compressed matrices I'' and compare them (numerically) with the original image. First we need to reverse the quantization step

$$I''_R(i,j) = I''_Q(i,j) \times Q(i,j)$$

and then, revisiting our matrix algebra, multiply the I''_R on the left by T^τ and on the right by T.

$$I'_R = T^\tau I''_R T$$

Each element of the resulting matrix is then rounded off to the closest integer value and the 128 bias we removed earlier is added back on.

$$I_R(i,j) = I'_R(i,j) + 128$$

Then I''_{Q50} yields:

$$I_{R50} =
\begin{vmatrix}
172 & 158 & 151 & 158 & 165 & 158 & 144 & 135 \\
164 & 160 & 154 & 155 & 164 & 169 & 156 & 138 \\
151 & 148 & 143 & 146 & 158 & 164 & 150 & 129 \\
156 & 134 & 130 & 153 & 165 & 149 & 129 & 122 \\
174 & 129 & 127 & 175 & 186 & 142 & 113 & 123 \\
171 & 126 & 129 & 187 & 201 & 150 & 114 & 121 \\
158 & 140 & 153 & 197 & 218 & 190 & 151 & 132 \\
159 & 170 & 189 & 216 & 243 & 242 & 202 & 157
\end{vmatrix}$$

If we form a differential matrix by subtracting from the original image I the corresponding elements from the reconstructed image I_{R50} we get:

$$\delta_{50} =
\begin{vmatrix}
-2 & -5 & 2 & -5 & -5 & 2 & 9 & -1 \\
6 & -7 & -1 & 5 & -4 & -9 & -3 & -4 \\
19 & -38 & 10 & 14 & 2 & -11 & 3 & 5 \\
4 & -24 & 4 & 12 & 0 & 4 & 5 & -12 \\
-14 & 5 & 7 & -10 & -26 & -8 & 21 & -13 \\
-6 & 8 & 5 & -27 & 22 & -16 & -4 & 13 \\
7 & -6 & 7 & -1 & 5 & 33 & -41 & 2 \\
6 & -10 & 7 & 7 & -20 & 12 & -4 & 3
\end{vmatrix}$$

The total difference between the two images is remarkably small when considering that 45 out of 64 spatial frequency components were discarded.

For the quality level $q = 10$ the results are:

$$
I_{R10} = \begin{vmatrix}
160 & 165 & 164 & 150 & 134 & 133 & 151 & 170 \\
146 & 152 & 154 & 143 & 131 & 134 & 154 & 174 \\
135 & 144 & 149 & 143 & 136 & 144 & 168 & 190 \\
141 & 151 & 159 & 157 & 153 & 164 & 191 & 214 \\
157 & 167 & 175 & 172 & 169 & 180 & 207 & 230 \\
163 & 171 & 177 & 171 & 164 & 171 & 195 & 217 \\
151 & 158 & 159 & 149 & 136 & 139 & 160 & 180 \\
137 & 142 & 140 & 126 & 111 & 110 & 128 & 146
\end{vmatrix}
$$

$$
\delta_{10} = \begin{vmatrix}
10 & -12 & -11 & 3 & 26 & 27 & 2 & -36 \\
24 & 1 & -1 & 17 & 29 & 26 & -1 & -40 \\
35 & -34 & 4 & 17 & 24 & 9 & -15 & -56 \\
19 & -41 & -25 & 8 & 12 & -11 & -57 & -104 \\
3 & -33 & -41 & -7 & -9 & -46 & -73 & -120 \\
2 & -37 & -43 & -11 & 59 & -37 & -85 & -83 \\
14 & -24 & 1 & 47 & 87 & 84 & -50 & -46 \\
28 & 18 & 56 & 97 & 112 & 144 & 70 & 14
\end{vmatrix}
$$

As expected, the differences between the original and the reconstructed image are much larger. Still, with almost 90 percent of the coefficients gone, the image is recognizable enough for security monitoring applications. At a 90 percent quality level the images are practically indistinguishable, as the difference matrix δ_{90} shows.

$$
I_{R90} = \begin{vmatrix}
168 & 155 & 152 & 154 & 159 & 160 & 157 & 132 \\
172 & 158 & 152 & 157 & 162 & 163 & 151 & 135 \\
168 & 109 & 159 & 157 & 159 & 151 & 155 & 136 \\
158 & 113 & 135 & 164 & 166 & 154 & 132 & 110 \\
164 & 130 & 128 & 166 & 169 & 131 & 136 & 110 \\
164 & 130 & 143 & 152 & 213 & 141 & 103 & 140 \\
172 & 126 & 163 & 200 & 230 & 218 & 115 & 131 \\
160 & 164 & 197 & 221 & 221 & 259 & 194 & 161
\end{vmatrix}
$$

$$
\delta_{90} = \begin{vmatrix}
2 & -2 & 1 & -1 & 1 & 0 & -4 & 2 \\
-2 & -5 & 1 & 3 & -2 & -3 & 2 & -1 \\
2 & 1 & -6 & 3 & 1 & 2 & -2 & -2 \\
2 & -3 & -1 & 1 & -1 & -1 & 2 & 0 \\
-4 & 4 & 6 & -1 & -9 & 3 & -2 & 0 \\
1 & 4 & -9 & 8 & 10 & -7 & 7 & -6 \\
-7 & 8 & -3 & -4 & -7 & 5 & -5 & 3 \\
5 & -4 & -1 & 2 & 2 & -5 & 4 & -1
\end{vmatrix}
$$

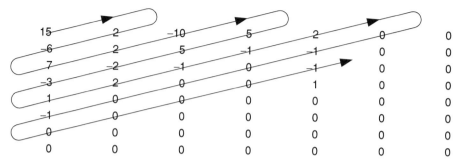

Figure 13.3   Zig-Zag Scanning.

One interesting fact about quantization matrices is that they are not symmetric with respect to the main diagonal. This means that our eyes are more sensitive to horizontal detail than to vertical detail, a fact no doubt due to the horizontal orientation of our binocular vision.

13.1.3 Zig-Zag Scanning

When we calculated the contents of our compressed image matrices I″ we noted that many terms are zeroed out and those that survived the process had a definite position bias towards the top left of the matrix. These results are inherent to the nature of quantization. The image elements closer to the top left are divided by smaller numbers then the elements located further out.

We will take advantage of this feature during our next step: the conversion of the compressed image matrix into a compressed data stream. In serializing the matrix we replace the normal line by line raster scan with a diagonal zig-zag scan starting with the top left corner and continuing to the opposite end of the matrix (Figure 13.3).

The first 31 terms of the scan are shown in Figure 13.4. All the terms following them are zero.

13.1.4 Huffman Coding

Up to this stage we have exploited the differential sensitivity of our vision system to the spectral content of an image. Further improvement in compression ratios has to come from standard data compression techniques. The one used by JPEG is *Huffman coding*. This algorithm is based on the observation that within any body of information some blocks of data are encountered more

Figure 13.4   Serial Readout of Zig-Zag Scan.

Table 13.1

Letter	Probability	Letter	Probability	Letter	Probability
Space	0.1859	H	0.0467	P	0.0152
E	0.1031	L	0.0321	G	0.0152
T	0.0796	D	0.0317	B	0.0127
A	0.0642	U	0.0228	V	0.0083
O	0.0632	C	0.0218	K	0.0049
I	0.0575	F	0.0208	X	0.0013
N	0.0574	M	0.0198	Q	0.0008
S	0.0514	W	0.0175	J	0.0008
R	0.0484	Y	0.0164	Z	0.0005

often than others. In the English language, for example, the letter E appears over 12 times more often than letter V. The probability of finding a particular letter of the English alphabet at a given position within an arbitrarily selected text is given in Table 13.1.

If we were to design a new written English language for a civilization in which writing space is a rare commodity, we would assign to "E" the graphic symbol with the smallest possible (but still visible) footprint—a dot, for example. Then the symbols would get progressively more complicated and space-consuming as the probability of them being used decreases. This is the idea behind Huffman coding. Since in the digital domain the unit of "space" is the bit, the Huffman coding process tries to minimize the total number of bits used in a data block by assigning shorter bit sequences to the more common data units (characters).

If we look at the Huffman encoded alphabet (Figure 13.5), we see that the least common letters X, Q, J, and Z are assigned 8-bit codes while the most common letters E, T, A, and O only 4. In contrast, the ASCII code assigns 8 bits to all letters. For a block of text 1000 letters long (about two written pages) ASCII encoding will use 8,000 bits, ignoring spaces. Huffman coding reduces the count to roughly 4,400. Huffman coding is "lossless," an important fact when dealing with an already quantized image. All the information compressed using Huffman coding can be recovered in its entirety.

With each block of information received by the Huffman decoder, the frequency for each character can be measured and a new probability table can be sent back to the encoder. Therefore the Huffman coding process can be dynamically changed to match the content of the files being compressed; Huffman coding is adaptive.

A more advanced type of probability based compression is *arithmetic coding*. In the alphabet example, Huffman coding assigned combinations of bits to single characters. Arithmetic coding assigns combination of bits to whole words (character sequences).

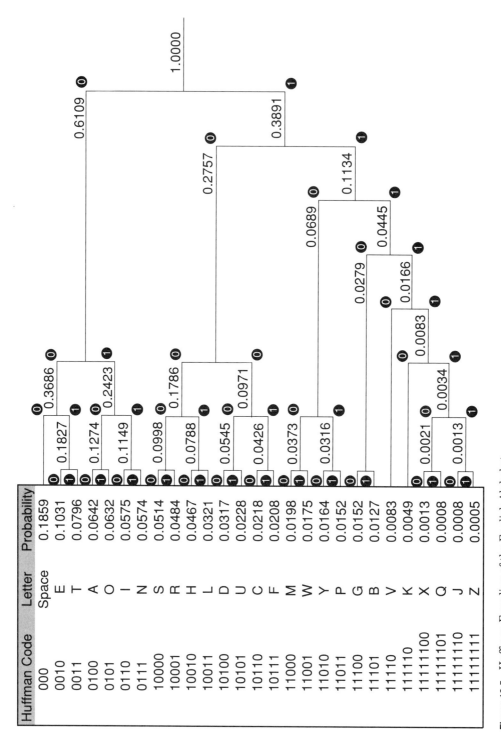

Huffman Code	Letter	Probability
000	Space	0.1859
0010	E	0.1031
0011	T	0.0796
0100	A	0.0642
0101	O	0.0632
0110	I	0.0575
0111	N	0.0574
10000	S	0.0514
10001	R	0.0484
10010	H	0.0467
10011	L	0.0321
10100	D	0.0317
10101	U	0.0228
10110	C	0.0218
10111	F	0.0208
11000	M	0.0198
11001	W	0.0175
11010	Y	0.0164
11011	P	0.0152
11100	G	0.0152
11101	B	0.0127
11110	V	0.0083
111110	K	0.0049
1111100	X	0.0013
1111101	Q	0.0008
1111110	J	0.0008
1111111	Z	0.0005

Figure 13.5 Huffman Encoding of the English Alphabet.

Figure 13.6   RLE/RLL Packet.

13.1.5 Run Length Encoding

Assume that we are processing a landscape image in the big sky country, or a portrait in front of a uniform backdrop. After using all the compression methods we covered so far, we still are left with long sequences of identical "blue sky" codes, or color codes for the studio backdrop. Since they are of low spatial frequency they will not be zeroed out by quantization, although based on their repetition, Huffman encoding would assign them high efficiency codes.

Run Length Encoding (RLE), also called *Run Length Limited* (RLL) encoding, compresses the redundant codes by replacing them with three segment packets. The first segment is a flag announcing that the packet payload is RLE/RLL–encoded data, the second segment is the code itself, and the third segment is a count of how many times the code is sequentially repeated (Figure 13.6).

If, for instance, we apply RLE/RLL to the sequence of 54 characters

00000000000ERRPPPPPSRTYYYYYYYYYYYYYYYIWWWWWWWWWWWWWERRBN $\rightarrow$

we obtain the 24-character sequence

$$@ERR5P\$SRT41Y\$I21W\$ERRBN \rightarrow$$

where $ is used as RLE/RLL flag and @ marks that all remaining characters are 0.

It is interesting to note that the RLE/RLL method is especially effective when compressing low spatial frequency data, such as quantized image data. As a matter of fact, the JPEG algorithm includes provisions for splitting the input data into frequency bands for highest possible compression efficiency.

For even higher compression ratios JPEG recommends down-scaling the image before compression (see Chapter 12).

Figure 13.7 shows the block diagrams for a JPEG encoder device (or software package). First the image is converted from its native color space to a 4:2:2 or 4:2:0 YCbCr space. As mentioned in Chapter 21, this is a data com-

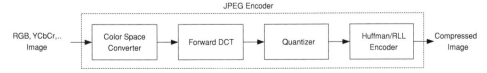

Figure 13.7   JPEG Encoder.

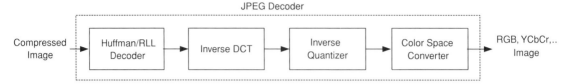

Figure 13.8 JPEG Decoder.

pression technique that makes use of our lower sensitivity to color details than to intensity details.

The luma and chroma matrices are then passed through a DCT block that separates them into spatial frequency components. Because of the lower eye sensitivity to color details, the range of the elements in the chroma matrices is limited to 99 instead of 255. The image is then quantized using separate quantization matrices for the luma and the chroma. As we saw before, the quantization matrices act as spatial frequency filters that remove or limit those components to which our vision system is less sensitive. Higher or lower compression ratios can then be obtained by multiplying the quantization matrix by the image quality factor q.

The lossless combination of a Huffman encoder and an RLA/RLL encoder convert the image matrix into a further compressed data sequence. The JPEG decoder in Figure 13.8 simply reverses the process. It first decodes the RLA/RLL stream and then passes it through an inverse Huffman table. Following that the image is subjected to an inverse DC transformation, an inverse quantization, and then it is finally converted to the native color space.

13.2 MPEG Data Compression and Decompression

13.2.1 I, P, and B Frames

The *MPEG* (Motion Picture Experts Group) standards pick up data compression where JPEG left off; MPEG addresses the compression of digital video and audio sequences, not just still images. To start with, MPEG breaks down the incoming video into frame sequences; each frame is separated into horizontal image slices, then each slice is divided into 16 × 16 luma and chroma macroblocks. These macroblocks are finally split into elementary 8 × 8 pixel blocks (Figure 13.9). The compression algorithms operate at the block and macroblock levels.

The major contribution brought on by the MPEG compression method is in the area of image prediction. Based on the past history of a sequence of frames, an MPEG encoder can predict many features of future frames. It is a lot less magical than it sounds, however. MPEG makes full use of the 150 milliseconds or so that distinguishes what we call real time from a visibly delayed image. It is within this imperceptible delay that all the video frames exist including past,

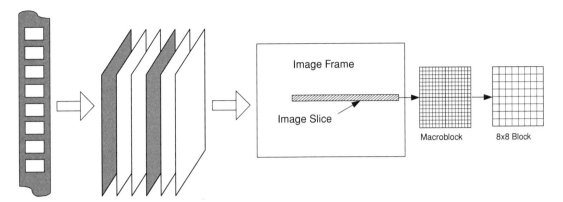

Figure 13.9 MPEG Video Decomposition.

present, and future images. For any one-way communication channel such as SDTV, HDTV, or streaming video, even the 150-msec restriction is lifted.

To better understanding the way MPEG compression operates let's look at the sequence of images in Figure 13.10, starting with the *I frame* (Intra frame). This picture is for all practical purposes a JPEG compressed image with an approximately two-bits-per-pixel data density. For a typical video stream there is at least one I frame in every 15-frame sequence, although the frequencies of different types of frames are user programmable (determined by MPEG encoder settings). Since they do not depend on any other frame for decompression, *I* frames are the entry and exit points of the MPEG stream.

P frames or *predicted frames* are obtained from I frames or other P frames through a process called *forward prediction*. It is based on the fact that, on average, images do not change much from frame to frame. And when they do, it is usually because some portions of the image have moved a small amount in an easily identifiable direction.

Forward prediction consists in comparing the current frame with the last I or P reference frame, calculating where in the past frame should different mac-

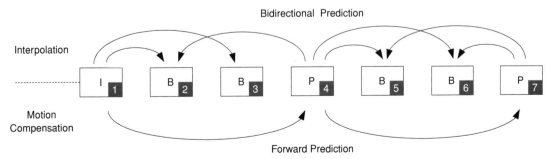

Figure 13.10

roblocks be taken from (displacement vectors), and packing the differences between source and the predicted macroblocks into a residual error frame.

The predicted picture can be reconstructed from the source frame by mixing the macroblocks according to the displacement vectors and correcting each macroblock using the information contained in the residual error frame.

13.2.2 Motion Estimation

The most difficult part in generating P frames is *motion estimation,* a key step in calculating the displacement vectors. The exact mechanism for finding matches in a past frame for macroblocks in the current frame and for the evaluation of macroblock movement is left to the discretion of the designer.

Figure 13.11 illustrates a possible motion-evaluation process. The macroblock at future position $P(x, y, t + \delta t)$ will take its pixel values from the macroblock currently at $(x - \delta x, y - \delta y)$. The translation vector $(\delta x, \delta y)$ is derived by multiplying the speed vector (v_x, v_y) by δt.

$$P(x, y, t + \delta t) = I(x - \delta x, y - \delta y, t) = I(x - v_x \delta t, y - v_y \delta t, t)$$

The speed vector (v_x, v_y) is obtained by comparing the current frame with the past reference frame. Comparing it with the past two frames can reveal linear acceleration vectors or rotational movements. If the motion reveals image details covered up in the reference frame, the encoder includes them in the residual error frame.

Evaluating the speed is not as simple as subtracting frame-to-frame pixel values. Direct subtraction would yield the same result for a fast-moving small pixel cluster as for a large pixel grouping slowly crawling around on the screen. A more reasonable result is given by the following formula:

$$(v_x, v_x) = \frac{\Sigma \; |\text{Frame-to-Frame Pixel Difference}|}{\Sigma \, |\text{Current Frame Pixel-to-Pixel Difference}\,|}$$

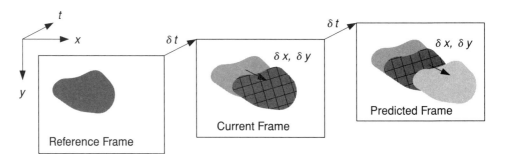

Figure 13.11 Motion Estimation.

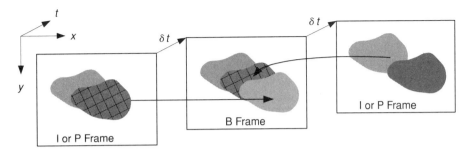

Figure 13.12   Bidirectional Prediction.

Its relative accuracy is due to the fact that the sum of the frame-to-frame pixel differences is divided by a normalization sum. Motion estimation is performed only on the luma macroblocks. The chroma macroblocks are just assumed to follow the luma. Matching macroblocks starts generally as a heuristic search, but once a match is found it switches to a neighborhood scan pattern.

P frame compression is "lossy" and, since P frames serve as sources for other P frames as well as B frames, their errors will propagate until the next I frame resets the reference image. As a rough average, every third frame is a P frame, and each P frame requires about half the data of an I frame.

B frames or *bidirectional frames* use both past and future I and P frames as references, and use a compression method called *bidirectional prediction*. The process of bidirectional prediction is similar to forward prediction. It generates displacement vectors from past and future macroblocks and produces residual error correction frames. However, when the motion estimator unveils a new area not seen in the past reference frame the MPEG encoder will pick it up from the future frame (Figure 13.12). This makes the B frames very efficient, requiring only a quarter of the data space needed by the I frames. Residual error frames for P and B frames are encoded and decoded in the same way as the I frames, but they use different quantization matrices.

13.2.3 Frame Reordering

How do B-frames access future reference frames? Since there are two B frames in between every pair of I or P frames, each B frame is, at any given time, two frames away at most from a future reference frame.

Frame Reordering

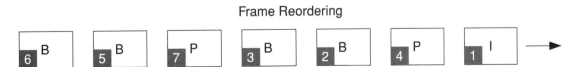

Figure 13.13   Frame Reordering.

At a rate of 30 frames per second, buffering three image frames will delay the stream about 100 milliseconds, which is within our tolerance limit for video delays. On the encoder side such a buffer is used to store the two B frames "under construction" until both *anchor frames* (the reference I or P frames) become available. Data is sent out as a reordered succession of frames with the I and P frames leading the set followed by the B frames (Figure 13.13). On the decoder side a reconstruction buffer stores the anchor frames until the B frames needed for frame decompression arrive.

13.2.4 MPEG Data Hierarchy

An MPEG video consists of a continuous stream of data sequences, each containing multiple images in a compressed and packetized format. The *video sequence* is the data structure at the top level of the MPEG hierarchy (Figure 13.14).

A new video sequence begins with a 32-bit unique *sequence start code*, followed by a number of sequence parameters including vertical and horizontal resolution, pixel shape (square ...), aspect ratio, bit rate, and decoder buffer size. A one-bit flag instructs the decoder to look for custom quantization matrices for the I and P frames or to use the JPEG matrix as default.

A *profiles and levels* block completes the video sequence header. A profile consists of a subset of the MPEG syntax. A *simple profile*, for example, does not use B frames and therefore frame buffering is not required. This makes the simple profile very useful for applications that do not tolerate delays. In normal *main profile* operation we do use B frames and buffers.

Levels are groups of parameters that define the type of video being processed. A *low level* is associated with 352×288 pixels, 30 frames per second, 4 Mbit/s data rate, and a 475KB buffer. With a resolution of 720×576 and a frame rate of 30 Hz (15 Mbit/s data rate, 1.8 MB buffer), the *main level* is the most common. The most advanced level, the *high level*, supports 1920×1152 pixels, 60 frames per second, 80 Mbit/s data rate, and a 10 MB buffer. MPEG-2 defines five different profiles and four different levels.

The payload of the video sequence consists in a series of *groups of pictures* (GOP), each containing at least one I frame. The header of a group of pictures is comprised of a start code, a SMPTE time stamp, and two GOP parameters—a "closed GOP" bit indicates if the decoder needs pictures from the previous GOP or not and a "broken link" bit indicates if the previous GOP is available or not. The GOP payload is a series of images or *pictures*.

A picture packet begins with its own start code followed by the picture type (I, P, B), and a set of picture parameters which includes picture display order, decoder buffer settings, and motion vector sizes. What follows is the set of *picture slices* that make up the frame. Each slice consists of a start code, the coordinates of its first macroblock (slice address), the quantization scale for that slice, and a number of macroblocks. At the lowest level in the MPEG hierarchy are the macroblock packets each carrying an address, type

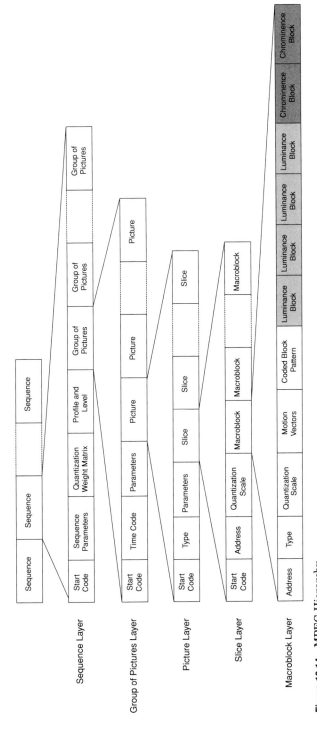

Figure 13.14 MPEG Hierarchy.

(based on type of motion vector used, quantization scale, and so on), quantization scale, motion vectors for each macroblock, a *coded block pattern* (CBP), four luminance blocks, and two chrominance blocks. The CBP is a bit pattern that indicates which block is coded. It allows the decoder to skip whole blocks if they are null, which is often the case when coding residual error frames.

13.3 MPEG Encoders (Coders) and Decoders

In a system with no predictive capabilities the video input is digitized then processed by a DCT block, quantized, and RLA/RLL/Huffman coded. What we obtain is a sequence of JPEG compressed frames. What differentiates MPEG from JPEG is its use of predictive techniques to achieve higher compression ratios. Figure 13.15 shows the block diagram of a low-level predictive MPEG encoder. The forward path of the encoder consists of the expected elements: the DCT, the quantization block, and the variable length encoder (RLA/RLL/Huffman). Unique to MPEG is its predictive double feedback loop.

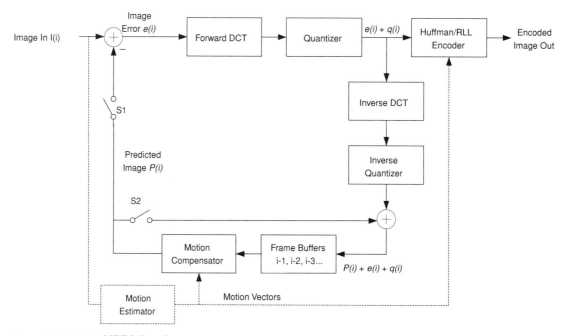

Figure 13.15   Basic MPEG Encoder.

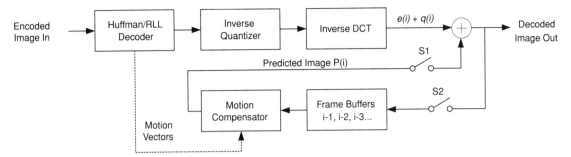

Figure 13.16 Basic MPEG Decoder

To generate an I frame the outer loop is opened (S1 open) and the device operates as a nonpredictive encoder. Following an I frame, the outer loop is closed (S1 closed), the output of the DCT–quantizer combination is inverse-transformed (through the inverse DCT and inverse quantizer), and a decoded version of the input image is generated. The decoded image is not identical to the input since it includes quantization errors $q(i)$. The decoded image is then motion compensated by the inner feedback loop and emerges as the predicted image $P(i)$. The motion compensation may or may not use information from an optional motion estimator block.

The predicted image $P(i)$ is then subtracted from the next incoming image $I(i)$ and the error frame $e(i)$ is created. From now on, until the next I frame, the outer loop remains closed and only the error frame $e(i)$ is transformed, quantized, variable length encoded, and sent out. Besides creating the predicted image $P(i)$, the smaller feedback loop also maintains the frame buffer that stores the I and P reference frames. These frames are reconstituted by adding the difference errors $e(i)$ and quantization errors $q(i)$ to $P(i)$. The B frames are not stored since they are not referenced by any other frames. For B frames the encoder keeps the inner loop opened (S2 open).

When the MPEG decoder in Figure 13.16 reconstructs an I frame the S1 switch is opened. But since the decoder will need the I frame as future reference for other P and B frames, the bottom switch is kept closed and the I frame is stored in the frame buffer. To reconstruct P and B frames the S1 switch is closed and the predicted frame $P(i)$ is added to the incoming error frame $e(i)$. If the reconstructed picture is a P frame switch S2 is closed and the frame is stored; if the picture is a B frame, S2 is kept open.

13.3.1 Scalable SNR Designs

MPEG-2 is a scalable video coding scheme; it allows decoders of various sophistications to decode the same MPEG-2–encoded stream to different levels of

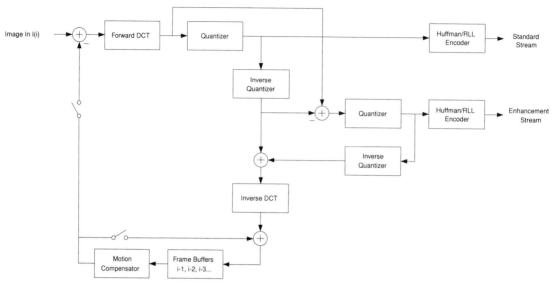

Figure 13.17   SNR Scalable MPEG Encoder.

image quality. Figures 13.17 and 13.18 show the block diagrams for an SNR (signal-to-noise ratio) scalable encoder and decoder pair.

The encoder is in many respects similar to the nonscalable encoder in Figure 13.15, with the addition of a third feedback loop that quantizes and variable length encodes the first quantization error. The output of the third loop forms an enhancement stream that is sent out along with the standard stream.

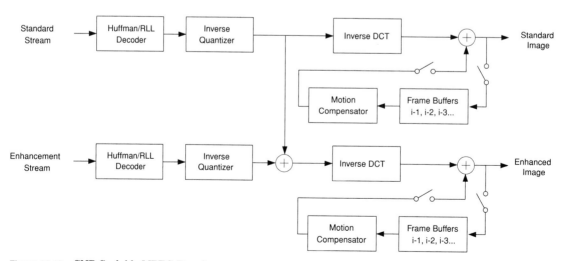

Figure 13.18   SNR Scalable MPEG Decoder.

The decoder reconstructs the standard signal the usual way. The elements of the image matrix provided by the standard inverse quantizer are further refined using the data arriving with the enhancement stream. The resulting matrix is then inverse transformed by the IDCT on the second branch of the decoder. The output of the second branch is a higher quality image than the standard image; it has the same resolution but a superior SNR performance.

13.3.2 Scalable Resolution Designs

MPEG-2 also accommodates spatial (resolution) scalability. A possible architecture for a spatially scalable encoder-decoder pair is presented in Figures 13.19 and 13.20. In this design the encoder operates a high-resolution loop in

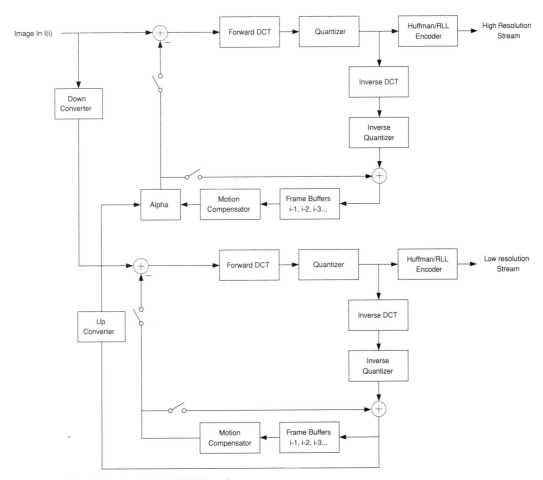

Figure 13.19   Resolution Scalable MPEG Encoder.

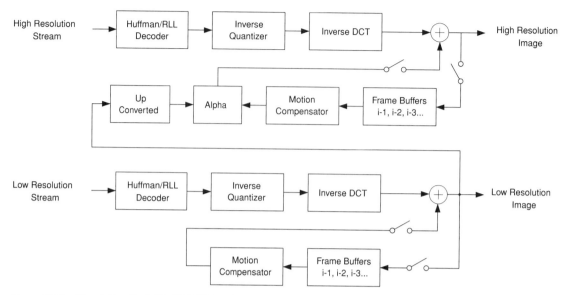

Figure 13.20   Resolution Scalable MPEG Decoder.

parallel to a low-resolution loop, with the lower resolution image being created by a down-converter (down-scaler). The high-resolution predicted image is a weighted average between the motion compensated, high-resolution predicted frame and the scaled-up version of the decoded low-resolution frame.

The decoder uses a nonscalable branch to recover the low, resolution image first. The high-resolution predicted image is then obtained as a weighted average of the up-converted low-resolution image and the output of the high-resolution motion compensator. This architecture allows for the same high-resolution image to be encoded for and decoded by both standard-definition and high-definition television.

13.4 MPEG Encoder Circuit

MPEG-2 video streams can be readily decoded by standard software decoders already built into your net browser. The decoding process itself is a lot less complex since it does not have to perform block matching or motion evaluation. By the time the stream reaches the decoder all the information regarding what goes where and when is encoded into the layers of the stream.

The encoder is a different matter, and until recently there was a notable absence of MPEG-2–encoding single chip ICs. As of late the situation changed and a number of such solutions have become available.

The general structure of a single chip MPEG-2 encoder is illustrated in Figure 13.21. On the diagram the video compression section consists of a single block. However, this block, usually a large ASIC or DSP core, occupies the majority of the silicon die. A smaller DSP section provides the audio compression function. The frame buffers are implemented in external SDRAM with only the memory manager being internal to the chip. The local controller is a CPU or DSP with some modest on-board resources and at least one communication interface to the host controller.

The MPEG-2 module circuit (Figure 13.22) is built around the SAA6752HS encoder from Philips. This is a highly integrated single-chip audio and video real-time MPEG-2 encoder. All video and audio encoding algorithms and software are run on an internal MIPS® processor core (MIPS is a registered trademark of MIPS Technologies).

The module expects the digital video input to be 8-bit, 27 MHz, YUV. However any other luma/chroma format such as YCbCr will also perform well. The video input interface also requires HSYNC, VSYNC, FIELD signals as well as 27 MHz pixel clock (CLKX2), all available from the Bt835 decoder module. The SAA6752HS provides the internal 4:2:2 to 4:2:0 color format conversion and means for inserting line 23 VBI (vertical blanking interval) codes including Closed Caption (CC), Wide Screen Signaling (WSS—aspect ratio, audio and data services), and copyright information. It is compliant with MPEG-2 *main profile* at *main level* (MP@ML) for 625 and 525 interlaced line systems. This translates into a maximum resolution of 720×576 at 30 frames per second and 15 Mbit/s data rate. The 1.8 Mbyte frame buffer requirement is addressed by an external MT48LCM16A2 (1M X32 × 4 SDRAM) memory

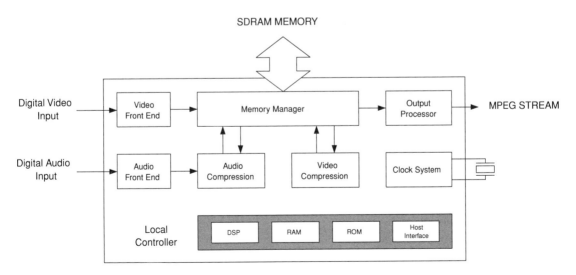

Figure 13.21   MPEG Encoder Block Diagram.

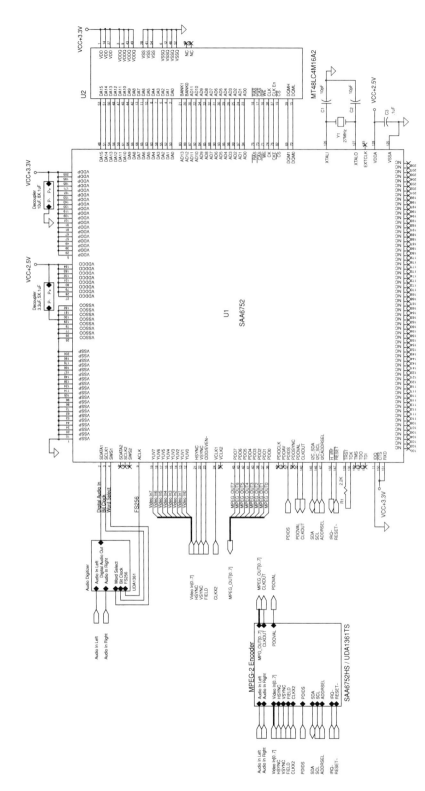

Figure 13.22   MPEG-2 Encoder Circuit.

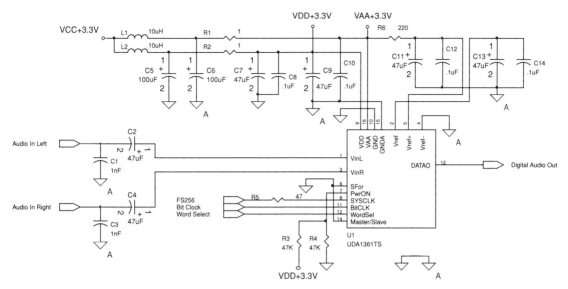

Figure 13.23   Audio ADC.

chip. Other resolutions supported by the encoder are DVD (D1 – 704 × 480 NTSC, 704 × 576 PAL), SVCD (2/3D1 – 480 × 480 NTSC, 480 × 576 PAL), SIF (1/2D1 – 352 × 480 NTSC, 352 × 576 PAL) and VCD (352 × 240 NTSC, 352 × 288 PAL). Supported modes include IPB frame, IP frame, and I frame only; also supported is variable video bit rate mode for constant picture quality or constant bit rate mode.

Communication with the local controller or host is done over a two wire I²C bus. The outputs of the encoder consist of an 8-bit parallel data stream MPEG_OUT[..], an output data clock CLKOUT, and a valid data flag PDOVAL. The host interface consists of an I²C link operational to 400 Kbit/s. The audio front processor is a 96KHz 24-bit stereo analog to digital converter (Figure 13.23) characterized by high linearity, good dynamic range, and low distortion (UDA1361TS). It operates as a slave to the SAA6752HS encoder which provides it with the *bit clock, system clock FS256* (256 times sampling frequency f_s), and *word select* (left/right audio select). It returns an I2S coded *digital audio in* stream.

PCB Layout of Video

Digital video electronics is, by its nature, mixed-signal. Most of the video sources in use today are analog and so are most of the displays. Since most video sources and displays in use today are analog, a significant part of the video processing chain is also analog. This will no doubt change in the future, as the means for transporting video signals in digital format become more economical. The main problem with analog circuits is their susceptibility to noise.

The offending noise source can be either analog or digital. Analog video outputs, for example, carry peak currents of 13 mA or more, and have bandwidths in excess of 4 MHz. Digital signals are also inherently noisy and power-hungry. It is not the frequency but the rate of change in output current, that produces the noise. And digital video consistently churns out high-current, high-frequency digital signals over wide internal data buses. The problem is made worse by the various mixed-signal ICs that needed to be intergrated into our mixed-signal printed circuit board.

Some seemingly digital ICs we believe to be noise-immune are actually analog and quite sensitive to it. Clocks are one such example. Even fundamental frequency crystal oscillators can easily be "pulled" off their center frequency by neighboring PCB traces. If for a microprocessor a .01 percent frequency "chirp" is irrelevant, it will wreck havoc with a video decoder IC. PLL clocks are even more noise-sensitive as their external loop filter makes a perfect antenna for stray EM emissions.

14.1 Ground Bounce

One of the fundamental problems in PCB design is that of ground bounce. Assume that our mixed-signal design can be separated into an analog section, a digital section, and an optional power section. In an effort to minimize interference we start by implementing separate power and ground distribution networks for each section. However, the grounds of all sections have to even-

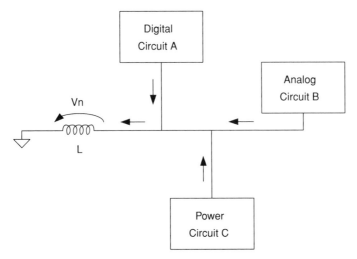

Figure 14.1 Basic Ground Bounce Effect.

tually be connected to a common zero potential. This can be done using the daisy chain or tree interconnection topologies shown in Figures 14.1 and 14.2. Both of them, however, are to be avoided. This is why: The ground currents of the analog section are of low amplitude and are likely to have a relatively moderate bandwidth, at most 4.2 MHz.

The power section, which is responsible for driving LED backlights and alike, will generate a large, mostly DC, ground current. Neither the analog or the power section ground currents are of great concern. The digital section ground current is. At every clock transition a number of digital lines will change their state and therefore generates current spikes. If we assume a ground lead inductance of L and a total resistive load equal to R_L, then the noise generated across the ground lead by the switching circuits is

$$V_n = \frac{L}{R_L} \frac{\Delta V}{T_{10-90}}$$

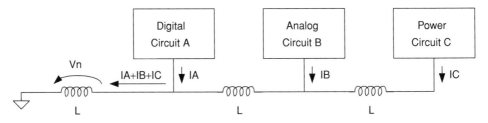

Figure 14.2 Ground Bounce in Daisy-Chained Circuits.

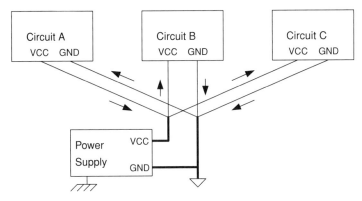

Figure 14.3 Star Power Connection.

For a capacitive load C_L the noise voltage becomes

$$V_n = LC_L \frac{1.52\Delta V}{T_{10-90}^2}$$

where T_{10-90} is the average rise time of the digital circuit outputs from 10% to 90% of final value, and ΔV is the voltage swing at the output.

This voltage is in addition to the voltage developed by the ground currents of the analog and power sections across the ground lead *resistance*. So even if the ground lead resistance is very low or zero we may still have noise. Since V_n is shifting the analog ground, V_n will appear as noise in any input signal referenced to the analog ground. For poorly laid out boards V_n can easily exceed 1 V!

Part of the solution is to use a star topology as shown in Figure 14.3. In this arrangement the ground currents of the three sections are kept separate almost all the way to the output of the power supply. In essence we have three ground leads and, therefore, three separate lead inductances. This way the digital ground "bounce" is seen only by the digital section.

Here are some additional measures you can take to avoid ground bounce. It is strongly recommended that digital video circuits use separate analog and digital ground planes, not split if possible. The connections from component pins to ground should be made close to each pin, using a separate via for each pin. Multiple ground pins should not be first connected by traces and then connected to the ground.

14.2 Board Partitioning

The best strategy for laying out a mixed signal board is to group components with similar noise and sensitivity characteristics and to minimize the interaction between the sensitive analog sections and the noise-generating digital sections. What follows is a general procedure for mixed signal board partitioning.

Step 1. Assign positions to all landmark connectors. These are connectors that cannot be moved, such as power connectors, video connectors, audio jacks, and so on. They can also be edge connectors for PCI computer cards, memory cards, and so forth.

Step 2. Group analog and mixed signal ICs according to their proximity to landmark connectors and frequency of operation. Graphics digitizers and video decoders should be placed close to the corresponding input connectors. Video encoders and DAC converters should also be placed close to their connectors, although they are less sensitive to noise than the decoders and digitizers.

Step 3. Group all the clocks and place them centrally between the analog and digital sections. This ensures that no digital clock line will meander through the analog section and it also minimizes the chances of clocks signals crossing each other. Treat oscillators and PLLs as analog circuits and buffer their outputs before distributing them on the board. For high-frequency clocks use clock distribution ICs (Chapter 4). Terminate high-frequency clock lines. Termination makes the line less sensitive to noise but unfortunately it also makes it a better noise emitter for the surrounding circuits. Keep such circuits away from clock terminators.

Step 4. Group the digital circuits according to their frequency of operation. Keep the highest-frequency parts close to the output filter of the power supply, if the power supply is on board. If not place them close to the input ground connector. Next place the digital power circuits and the low-frequency low power circuits as shown in Figure 14.4.

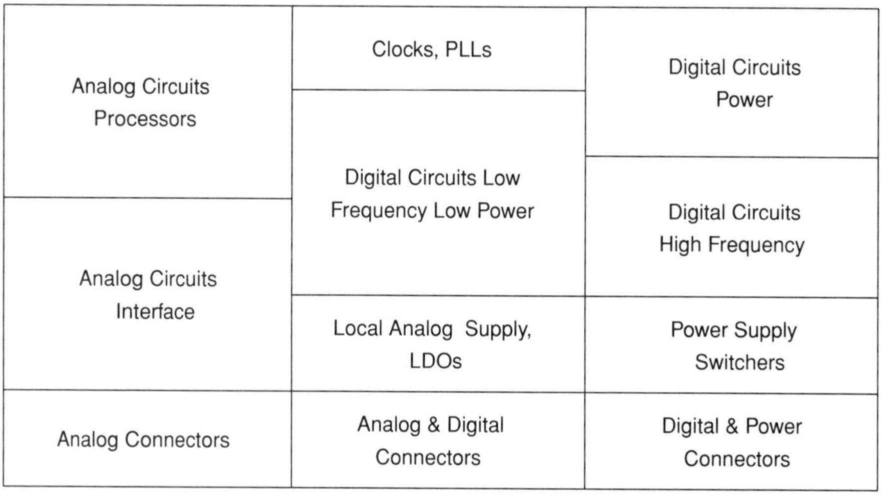

Analog Circuits Processors	Clocks, PLLs	Digital Circuits Power
	Digital Circuits Low Frequency Low Power	Digital Circuits High Frequency
Analog Circuits Interface		
	Local Analog Supply, LDOs	Power Supply Switchers
Analog Connectors	Analog & Digital Connectors	Digital & Power Connectors

Figure 14.4 Mixed-Voltage Board Partition.

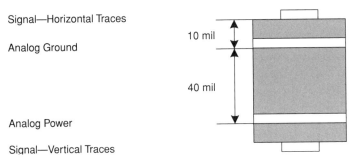

Signal—Horizontal Traces

Analog Ground

10 mil

40 mil

Analog Power

Signal—Vertical Traces

Figure 14.5 Four-Layer PCB Stack.

14.3 Ground and Power Planes, Plane Stacking

The next step after partitioning the board is to specify the ground and power plane arrangement. For a simple, purely analog design, such as a distribution amplifier or a video switch, a four-layer stack is sufficient (Figure 14.5). It can even accommodate moderate, low-frequency digital interfaces such as those required to change switch routing or gain settings. If the switch handles multiple random signals, copper pours should be used to separate the lines on the signal layers, and a generous amount of vias should connect the copper pour areas to the ground layer. The two copper planes making up the power and ground layers also form an ideal low-inductance power filter capacitor.

For a digital video circuit, when the frequency of operation reaches the transmission line range, a six-layer stack is more appropriate (Figure 14.6). This allows us to maintain a lower trace density for each layer and therefore minimize crosstalk.

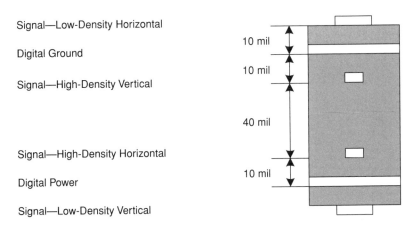

Signal—Low-Density Horizontal

Digital Ground

10 mil

10 mil

Signal—High-Density Vertical

40 mil

Signal—High-Density Horizontal

Digital Power

10 mil

Signal—Low-Density Vertical

Figure 14.6 Six-Layer PCB Stack.

Analog and Digital Signal—Low-Density

Split Analog and Digital Ground

Split Analog and Digital Power

Digital Signal—High-Density Vertical

Digital Signal—High-Density Horizontal

Digital Power

Digital Ground

Digital Signal—Low-Density

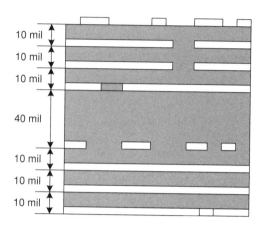

Figure 14.7 Eight-Layer PCB Stack

In the case of a mixed signal video design we may need to use eight layers: four for power and ground and four for routing (Figure 14.7). If the analog real estate is a relatively small fraction of the total circuit (10 to 20 percent), the best approach is to use a split analog/digital ground layer and a split analog/digital power layer in addition to a separate pair of digital power and ground planes. All the analog parts should overlay the analog ground plane, on one side of the board.

In a split-plane arrangement the analog power plane has to match the analog power ground, and the digital power plane has to match the digital ground (Figure 14.8). Under no circumstances should the analog power plane or an analog ground be allowed to overlap the digital ground or the digital power plane. (Figure 14.9). Since even split planes form ideal coupling capacitors, any overlap would inject digital noise into one or the other of the analog power rails. For the same reason we must also keep a minimum distance of at least 2 mm between the analog and digital ground planes (even larger for big size boards). The two digital ground planes need to be connected together by a suf-

Analog Power

Analog Ground

Digital Power

Digital Ground

Figure 14.8 Eight-Layer PCB Stack

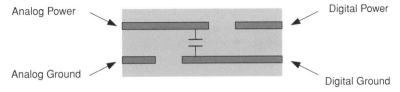

Analog Power   Digital Power

Analog Ground   Digital Ground

Figure 14.9 Incorrect Split Plane Alignment.

ficiently large number of vias to allow for the shortest possible return currents for all the top digital pins.

Sensitive small circuits that do not justify a full plane split can use small power islands instead (Figure 14.10). The connection between the islands and the main power plane can be made using an LDO (low drop-off regulator) or a ferrite bead (Figure 14.11). No clock or other high-frequency signals can be routed over the analog ground or power planes. If a high-frequency clock must reach a mixed-signal IC, one solution is to bring it in, close to the clock pin of the IC, using a via from a deeper routing layer. Mixed-signal ICs should be placed over the analog ground, close to the split, with the digital pins facing the split.

The power to the digital section can be derived from the analog power rail through an LDO (if the digital section has a lower core voltage than the analog supply) or a ferrite bead (Figure 14.12).

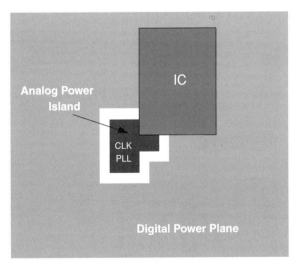

Figure 14.10 Power Island.

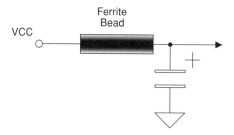

Figure 14.11   Island Power Supply Connection.

As can be seen in Figure 14.7, the power and ground planes are always paired. This lowers the impedance between them and reduces power supply noise. In addition, symmetry in the selection of power and ground planes adds to board stiffness and reduces thermal warping.

In most applications the electronic system ground is connected directly to the chassis ground through a ground screw or solder joint. Sometimes, however, this is not possible. Instead of letting the ground float and therefore be subject to noise pickup, a chassis ground layer can be placed underneath the system ground plane (Figure 14.13). At low frequencies the two are electrically isolated, while at high frequencies the two planes are shorted together by the interlayer capacitance.

ESD protection circuitry should also tie the two planes, especially if the chassis is connected to an outside ground through a long cable. The chassis ground has to be mechanically counterbalanced by another solid ground symmetrically located in the stack. Using a large capacitor instead of a chassis

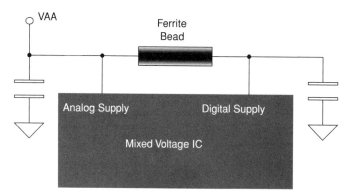

Figure 14.12   Mixed Voltage IC Power Connection.

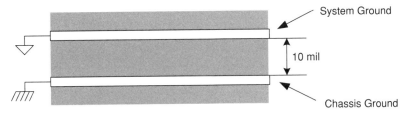

Figure 14.13 Floating Chassis Ground.

plane does not work. The series inductance of large capacitors creates severe ground bounce problems, even at moderate current levels.

All grounds must meet at a single low-impedance point. Otherwise multiple connections between different grounds, or between grounds and chassis, will create ground current loops. The best location for the ground connection point cannot be easily calculated. Not only do we need to know the topology of all the return current flows, but we must also know how to correlate that information to the noise sensitivity and location of the analog parts, heat dissipation, and so on. A good first guess is almost always a connection point in the ground pin area of the power connector. All connectors—but especially the power connectors—should have redundant ground pins as protection against wear and corrosion.

14.4 Return Currents and Loops

If we distribute power to a circuit by means of a power and ground network of traces (Figure 14.14), the return currents for all board interconnections will

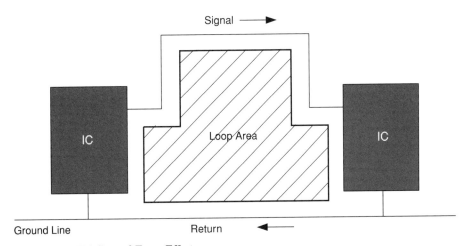

Figure 14.14 EM Ground Trace Effects.

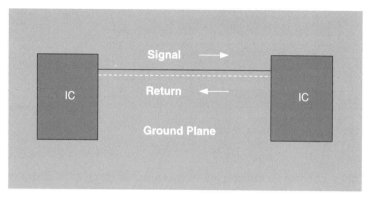

Figure 14.15   Return Currents in a Ground Plane.

be concentrated in a small number of ground lines. This creates a number of loops which, depending on the nature of the circuit, could radiate or pick up noise. Even more important, the return currents flowing through the ground trace impedances become multiple sources of noise.

Using a continuous ground plane solves the problem. The current through each PCB trace is accompanied by an equal return current traveling in the opposite direction through the ground plane, right underneath that trace (Figure 14.15). This minimizes the cross-section of the current loop and with it the EM noise. The low inductance of the ground plane also makes the ground bounce negligible. However, cutting slots in the ground plane brings the problem back, as can be seen in Figure 14.16. If the connecting trace crosses the slot on the top layer the return current must go around the slot in the ground plane. This creates both an EM loop and the potential for additional ground bounce. The problem becomes even more serious when multiple

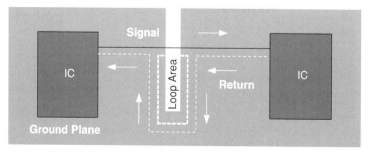

Figure 14.16   EM Effects of Ground Plane Slots.

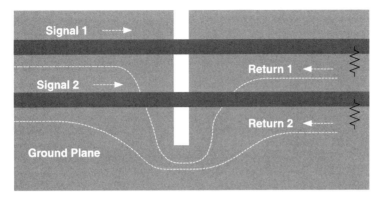

Figure 14.17   Trace Interference over Ground Plane Slots.

lines cross the same ground slot (Figure 14.17). In this case the return currents bunch together around the contour of the slot which results in signal crosstalk.

Although the most common reason to make a ground slot is to gain room for a few forgotten traces, sometimes the slot is inadvertently created when we build a new connector part (Figure 14.18). The presence of a continuous gap in the ground plane around the connector pins forces the pin return currents to flow together around the gap contour. This gives rise to crosstalk. The correct approach is to fill in the space around the pins with a ground copper pour so each pin return current has a separate path through the ground plane (Figure 14.19).

Properly designed ground slots can also be used to our advantage. If our circuit has two or more asynchronous clocks, ground "trenches" can be used to

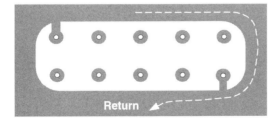

Figure 14.18   Inadvertent Connector Ground Slot.

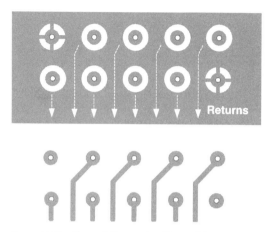

Figure 14.19   Correct Connector Ground Layout.

guide the return current of the digital clock away from the return current of the analog clock, for example.

Another approach is to isolate clocks and other potentially noisy, high-frequency signals by using guard traces connected to the ground plane by closely spaced vias (Figure 14.20).

14.5 Signal Traces

Most of the traces used in digital video circuits are routed either on top of or in between power planes. At high frequency the first should be treated as microstrip components [Figure 14.21(a)], the latter as stripline components [Figure 14.21(b)]. At moderate frequencies this arrangement minimizes crosstalk and shields the traces against noise. The PCB dielectric material will

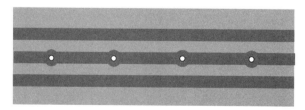

Figure 14.20   Ground Guard.

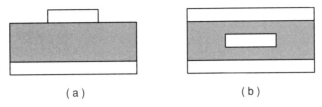

Figure 14.21   Microstrips (a) and Striplines (b).

also slow down fast signals (which may or may not be desirable) and soften signal edges.

Signal traces should be as straight as possible, with minimal layer jumps. For high-density circuits, a good approach is to design the low-frequency routes last, giving priority to the requirements of the fast signals. We must avoid using two consecutive routing layers, especially if separated by only 10 mils of dielectric. If necessary we can add ground planes as crosstalk shields. Additional power planes do not work well as shields since any AC return current going through power planes and decoupler caps to the ground planes will generate noise voltages across the decouplers.

We can minimize crosstalk in parallel runs by maintaining a clearance of at least four trace widths between traces (Figure 14.22). For differential lines the clearance between the line pair and other traces has to be at least twice the line to line spacing (Figure 14.23). The electrical length of the lines must be the same; this requires constant line spacing, identical terminations, equal number of vias in the same relative position (if vias are unavoidable).

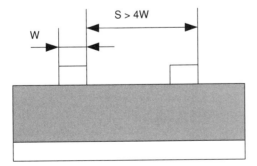

Figure 14.22   Parallel Lines Clearance.

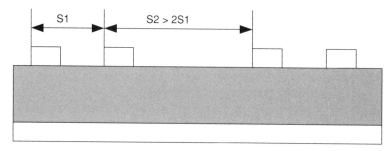

Figure 14.23 Differential Lines Clearance.

We should group the transmission line traces on the outer layers and route the rest of the traces internally. This eliminates the need for vias along the transmission lines; vias are impedance discontinuity points and contribute to line reflections. Placing components on the transmission line layers has to be done so the lines are allowed to run in clear straight paths.

For all high-frequency routes we need to keep the trace turns as gentle as possible (Figure 14.24). Because of their larger surface area sharp turns have a larger capacitance. This creates discontinuities and therefore reflections along the line. Right angle corners also make good noise antennas. If possible we should avoid the use of sockets and T connections to passives as any unnecessary pins and traces will only add inductance to the lines Figure 14.25.

When using programmable devices (FPGA/CPLD), we must configure all idle I/O lines as grounds or as inputs connected to the ground. All the unused opamp and comparator sections must be wired as virtual grounds (noninverting buffer with grounded input), and all unused logic gate inputs must be connected to power or ground.

Figure 14.24 PCB Corner Traces.

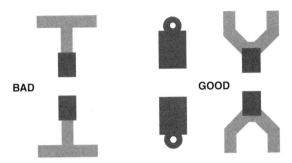

Figure 14.25   Traces to Passives.

14.6 Line Terminators

High-frequency lines need terminator impedances for two reasons; to avoid signal reflections in (electrically) long transmission lines and to dampen unwanted oscillations in parasitic line inductance/load capacitance resonators. The most common in transmission line applications is the *parallel terminator* (Figure 14.26). Without a terminator, the signal traveling along the transmission line will exhibit multiple reflections, and its rise time at the far end would increase proportional to the $\tau = Z_0 C$ time constant (C is the load capacitance). The use of a matched terminator ($R_t = Z_0$) minimizes reflection and also reduces the time constant to $\tau = (Z_0 \parallel R_t)C$.

To reduce the drive current we can replace the parallel terminator R_t with a pair of resistors with values $R_{t_1} = R_{t_2} = 2Z_0$ configured as a voltage divider (Figure 14.27). We can completely eliminate the DC component of the drive current by placing a capacitor in series with the load resistor (Figure 14.28). The value of the capacitor depends on the signal bandwidth but should be larger than 100 pF.

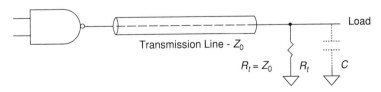

Figure 14.26   Parallel Terminator.

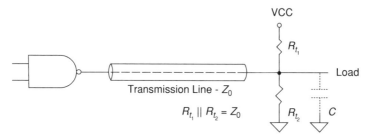

Figure 14.27 Voltage Divider Termination.

Terminators can also be used on the driver side, provided that the driver has a very low-output impedance (Figure 14.29). For line matching the impedance of the terminator plus the output impedance of the driver should equal to the characteristic impedance of the line. Even when they are not matched such series terminators are useful in dissipating line reflections and ringing, although they will increase signal rise time. Common values range from 22 to 100 Ω.

There are two ways to terminate differential lines. The first uses a single impedance with a value $R_t = 2Z_0$ connected across the two lines (Figure 14.30). The second method individually terminates the lines on a common DC blocking capacitor (Figure 14.31).

Next let us look at the case when a transmission line is split into secondary lines so it can reach multiple parallel loads (Figure 14.32). For a matched bifurcated system the characteristic impedance of the secondary lines must be twice Z_0. Each branch can then be individually matched by an $R_t = 2Z_0$ terminator. If these are traces on a PCB, for example, the widths of the secondary traces must be about half the width of the main trace. If the main trace

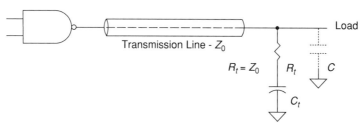

Figure 14.28 Termination with DC Block Capacitor.

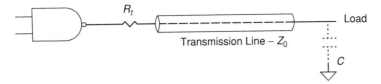

Figure 14.29 Driver Side Termination.

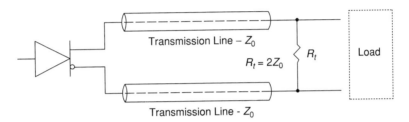

Figure 14.30 Differential Line Termination

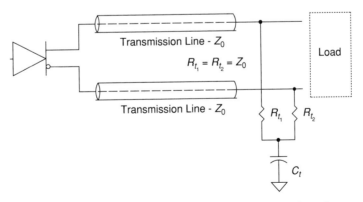

Figure 14.31 Differential Line Termination with DC Block Capacitor.

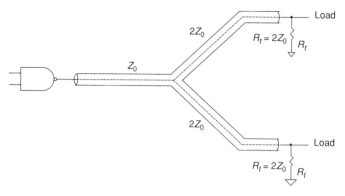

Figure 14.32 Termination for Bifurcated Line.

and each of the secondary traces have the same impedance, the line is impossible to match. A less demanding approach to multiple loads is to use a single terminator at the far end of the line, and high impedance "taps" along the way (Figure 14.33).

14.7 Power Supply Filters and Decouplers

Decoupling capacitors are used to minimize the noise injected into the power rails by digital integrated circuits or digital sections of mixed-signal ICs. When a digital circuit switches states its supply current suddenly changes, thus generating a transient voltage spike across the power-supply leads (Figure 14.34). The amplitude of this spike is proportional to the lead inductance and the supply current rate of change. Such current surges can inject significant noise in any neighboring analog circuit, while the accompanying voltage spike can even produce error pulses in downstream digital circuits.

A decoupling capacitor placed close to the supply pin of the digital IC eliminates this problem. When the supply current drawn by the IC suddenly changes the capacitor meets the increased current demand *locally*. This min-

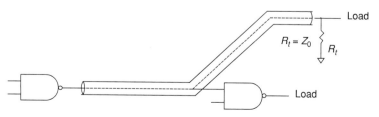

Figure 14.33 Line Termination with HiZ Taps.

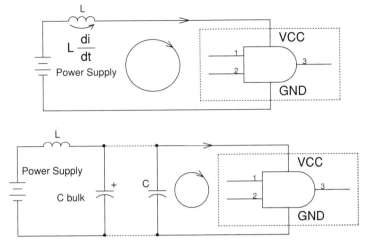

Figure 14.34 Decoupling Logic Power Pins.

imizes the rate of change of the current through the power-supply lead and therefore the associated current and voltage spikes. The decoupling capacitor is then recharged at a much lower rate from a large "bulk charge" capacitor located at the power-supply entry to the board. The "bulk charge" capacitor of choice is a large tantalum while decoupling caps are generally ceramic.

In contrast, analog circuits do not create power-supply noise; they need protection from it. A capacitor placed close to the power pins reduces the noise by completing, with the supply lead inductance, an LC lowpass filter (Figure 14.35).

The exact values of the decoupling capacitors depend on the characteristics of the IC. Too low of a value will make the cap ineffective in meeting the current demands of the IC, while a too-high value may bring the self-resonant frequency of the cap too close to the operational frequencies of the circuit. The "bulk charge" cap should be ten times larger then the sum of all decoupling capacitors in the area.

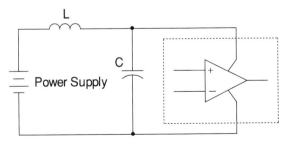

Figure 14.35 Filtering Analog Power Pins.

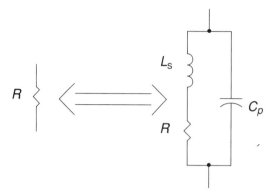

Figure 14.36   Equivalent Circuit of a Resistor

14.8 Component Selections

The equivalent impedance model for a resistor is shown in Figure 14.36. Besides the resistance R, the discontinuity in material at the end caps give rises to the parallel capacitance C_p. At high frequencies C_p lowers the overall impedance of the resistor. The series inductance L_s characterizes the way the resistive material is laid out. Any resistor made by coiling a conductor wire (wire-wound) or trace (film) will have a relatively high inductance. Bulk-type resistors (carbon) have a lower inductance and therefore are more suitable for high-frequency applications.

The equivalent impedance model for a capacitor is different (Figure 14.37). For film and electrolytic capacitors, which are both made by rolling multiple

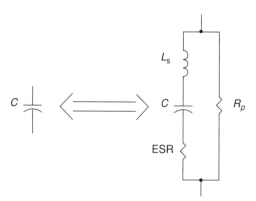

Figure 14.37   Equivalent Circuit of a Capacitor.

layers of conductors and dielectrics, the series inductance L_s is rather large. Ceramic caps have a much lower series inductance. For that reason they are a good choice for high-frequency signal operation.

As decouplers their low-capacitance values makes them ineffective by themselves. We also need large-capacitance, low-ESR (equivalent series resistance) devices such as tantalum electrolytics. The maximum operational frequency for a tantalum electrolytic is 10 MHz while for a ceramic cap it exceeds 1 GHz. The maximum capacitance for tantalum caps is about 1000 µF. For ceramics it tops off at about 5 µF. A common decoupler combination consists of a 10-µF tantalum in parallel to a .01-µF ceramic.

15

Digital Video Projects

15.1 Project 1—Video Encoder and Color Bar Generator

In any new endeavor the most rewarding route is the one that provides the quickest recognizable results. Our first project, building an encoder and color bar generator, satisfies this criterion. This board is based on the video encoder circuit covered in Chapter 9. The core of the circuit is the Bt864 video encoder from Conexant (Figure 15.1).

Power is supplied to the board through the JPWR1 connector (Figure 15.2). The input voltage should be provided by a well-regulated 1 A, + 5 V source. Care must be taken when selecting the power supply since voltages above 7 V, as well as sustained reverse voltages, will damage the board.

The 5 V is further regulated and dropped to the 3.3 V level of the Bt864 by U3, a low drop-out regulator from Linear Technologies (LT1585). The output of the regulator is split into a VCC + 3.3V branch which feeds the digital sections of U1, and a second branch which is first filtered by L4, C18, C19, and then is applied to the analog section of U1 (VAA+3.3V). Connectors JPWR2 and JPWR3 are provided so a number of boards can be daisy-chained off a single power supply. The presence or absence of the 5 V input voltage is signaled by the DL1 LED.

In minimal configuration (internal color bar mode) the only support circuits required by the Bt864 are an "active low" reset IC, and an oscillator capable of generating a frequency equal to twice the pixel rate. The power-up reset is generated automatically by U2 (TPS3103), which also reissues resets every time the 3.3 V power supply drops below its threshold voltage value of 2.941 V. In addition, the user can trigger manual resets by pressing the S1 switch. The pixel clock can be picked up locally (from the U4 oscillator) or can be supplied by external circuitry and routed to the board through pin 23 (ECLK) of the input connector JP1. Selection is made via the JP2 jumper.

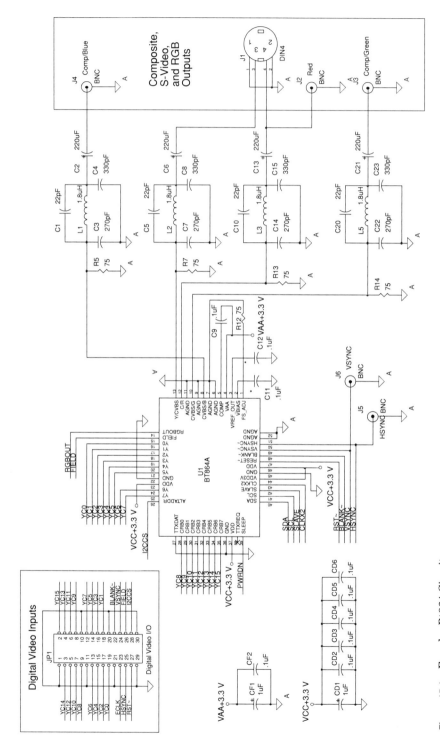

Figure 15.1 Encoder Bt864 Circuit.

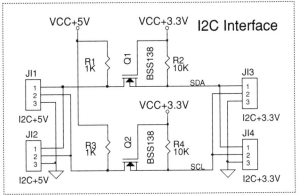

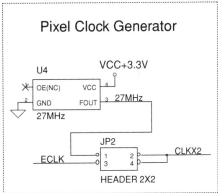

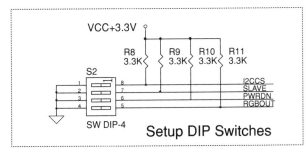

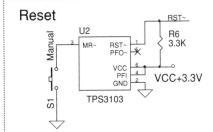

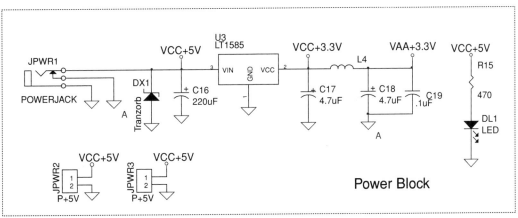

Figure 15.2 Encoder Ancillary Circuitry.

DIP switch S2 programs the hardware operating modes of the Bt864. In master mode operation the SLAVE signal must be set to zero. In this case the Bt864 generates all video timing internally, except for the pixel clock.

The resulting HSYNC and VSYNC outputs are available for external use through the JP1 connector. BLANK~ is always an input and its assertion blanks the video out regardless of the state of the Bt864. FIELD is always an

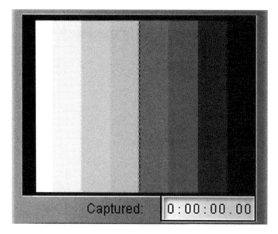

Figure 15.3   Color Bar pattern captured with Dazzle DVC80.

output and it identifies the current field as ODD or EVEN, both for master and slave operation.

The Bt864 is capable of generating composite, s-video, and standard resolution RGB outputs. For composite and s-video signals RGBOUT needs to be set to zero. Then the J3 and J4 connectors will provide the board's composite outputs and J1 the S-video. When RGBOUT is set to one, J2, J4, and J3 correspond to the red, green, and blue outputs while J5 and J6 provide the HSYNC and VSYNC reference signals.

Another position on the S2 switch takes the Bt864 into sleep mode, a function especially useful for portable electronics applications.

The supervisory control of the board is done through one of the I^{2}C interface connectors. If the board is controlled by a 3.3 V I^{2}C master we use the JI3 or JI4 ports, while if the master is 5 V we must use the JI1 or JI2 ports. Depending on the logic level of I2CCS (S2 DIP switch), the I^{2}C address of the Bt864 can be programmed to be either 0×88 (I2CCS high) or 0×8A (I2CCS low).

If a system has multiple Bt864s they can be controlled by holding all ICs to one address, 0×88 for example, except for the IC we want to program which is set to 0×8A. Then all the commands on the 0×8A channel will be accepted by

Figure 15.4   PC-Based I^{2}C/SPI Controller.

the addressed Bt864 and ignored by the rest. Afterwards we switch the next IC to 0×8A with the rest at 0×88, and so on.

Following power-up reset the Bt864 enters into an interlaced, black burst mode with no active video. To activate the video output we must set the EAC-TIVE register bit to one. In order to obtain color bars we also must set bit ECBAR to one. These bits are the D1 and D4 position of the internal register at subaddress 0×67 (see Chapter 7 for details on the operation of the I²C bus). A complete description of all the registers can be found in the Bt864A datasheet which can be downloaded from www.conexant.com and this book's website (www.digital-video-circuits.com).

Although most designers will use their own controllers to program the Bt864, for testing and troubleshooting purposes we can also use a PC based I²C interface such as the Aardvark from Total Phase, Inc (www. totalphase.com). This small unit is about 3.5″ × 1.5″ × 0.75″, connects to your PC via the USB port, requires no external power supply (it is powered by the USB port) and provides both I²C and SPI bus control functions.

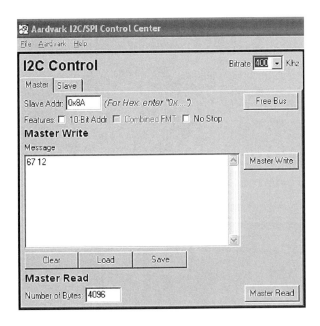

Total Phase, Inc. also offers an easy-to-use GUI, free of charge. The screen layout is straightforward. Once the user configures the I²C/SPI mode (the Aardvark also supports a General Purpose I/O mode), the Aardvark Control Center screen prompt the user for the slave address (Bt864 is always an I²C slave device), and the I²C message. This message should contain the register subaddress followed by the data to be transferred. Clicking on the Master Write icon initiates the actual transfer. Since the Aardvark is a 3.3 V device we must connect it to either the JI3 or JI4 ports.

And since we are discussing PC-based equipment, another useful tool is an image-capture appliance such as the USB Dazzle DVC80 from Pinnacle Systems. It allows us to store video clips captured during different stages of product development, each clip with the appropriate voice annotation. Such records are invaluable when chasing field and color inversions, image tears, and other video artifacts.

The actual circuit board layout for the encoder is shown in Figure 15.5 and is also available in Gerber and Acrobat format at www.digital-video-electronics. com. Since encoders are mixed-signal devices special precautions must be taken when laying out the board. In this case we use a four-layer stack, with two split power planes on the inner layers, and all signal routing on the outer

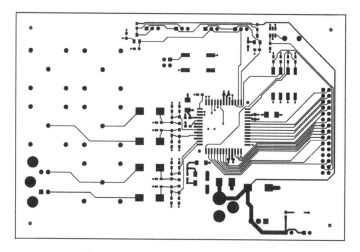

Figure 15.5(a)   Encoder Board PCB Layout Top Layer.

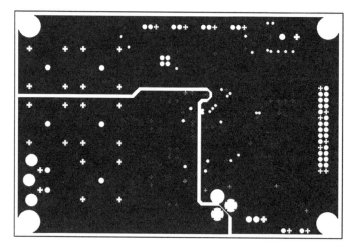

Figure 15.5(b)   Encoder Board PCB Layout Split Ground Layer.

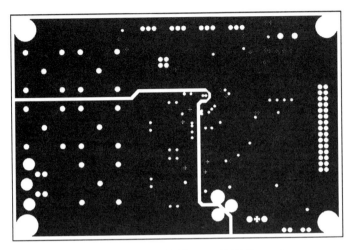

Figure 15.5(c)   Encoder Board PCB Layout Split Power Layer.

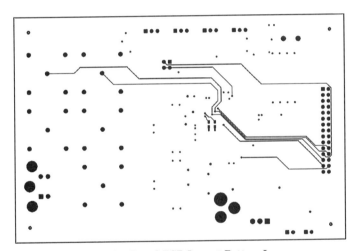

Figure 15.5(d)   Encoder Board PCB Layout Bottom Layer.

layers. All analog circuitry is kept over the analog power and ground planes. As discussed in Chapter 14 the analog and digital grounds meet at one point— at the input power connector site JPWR1. Since all traces are less than 5″ in length and the maximum clock frequency is only 27 MHz, we are not concerned with transmission line effects. But we are concerned with noise, and therefore all analog traces have been kept short and far from noisy digital lines.

The Bt864 is capable of much more than just generating color bars. For operation with external digital pixel sources the ECBAR has to be reset to zero, and 16 or 8 bits of pixel data must be provided at the right time on the YC[0..15] ports of the IC. To choose the 16-bit wide mode, the YC16 bit (D2 in register 0×5E) is set to 1. Then the 8-bit luma information is applied on the

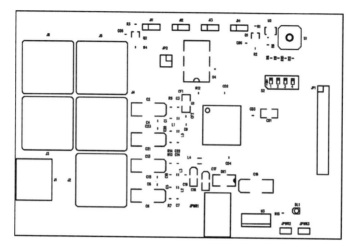

Figure 15.5(e)   Encoder Board PCB Layout Top Layer Silkscreen.

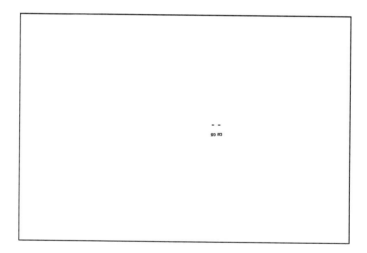

Figure 15.5(f)   Encoder Board PCB Layout Bottom Layer Silkscreen.

YC[0..7] lines and the chroma on the YC[8..15]. YC7 and YC15 are the MSB of their respective bytes. For 8-bit operation the YC16 bit is set to zero and the luma and chroma bytes are applied sequentially (...Y – Cb – Y – Cr...) to the YC[8..15] lines.

The advantage of 8-bit video buses is that fewer digital lines are needed to carry the data across generally noise-sensitive video boards. On the flip side, these signals have a higher frequency: 27 Mhz for 8-bit operation compared with 13.5 MHz for 16 bits.

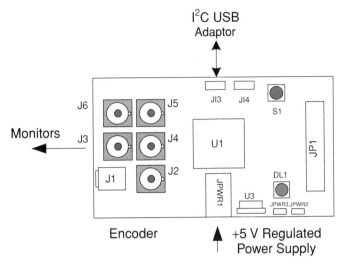

I^2C USB
Adaptor

Monitors

Encoder

+5 V Regulated
Power Supply

Figure 15.6   Encoder Board Wiring.

If the Bt864 operates in master mode, the external pixel source must supply the data synchronous to the HSYNC, VSYNC, CLKX2, and FIELD signals provided by the internal raster generator of the IC. In slave mode, the responsibility of generating the HSYNC and VSYNC signals rests with the external source. All digital video transfers for the Encoder board are handled through the Digital Video I/O connector JP1.

Bill of Materials

Reference	Value	PCB	Manufacturer	Part# / Details
C1, 5, 10, 20	22 pF	0603	Generic	Ceramic
C9, 11, 12, 19, CD2, 3, 4, 5, 6, CF2	.1 μF	0603	Generic	Ceramic
C2, 6, 13, 16, 21	220 μF	7343	Generic	Electro
C8, 15	330 pF	0603	Generic	Ceramic
C17, 18	4.7 μF	3216	Generic	Tantalum
C3, 7, 14, 22	270 pF	0603	Generic	Ceramic
C4, 23	330 pF	0603	Generic	Ceramic
CD1, CF1	1 μF	1216	Generic	Ceramic
DL1	LED		Lumex	SSL-LXA228GC-TR11
DX1			STMicro	SMBJ5.0A-TR
J1			CUI	MD-40SM
J2, 3, 4, 5, 6			AMP	413986-1
JI1, 2, 3, 4	Header		Generic	3/2mm
JP1	Header		Generic	15X2/2mm
JP2	Header		Generic	2X2/2mm
JPWR1			CUI	PJ-202A
JPWR2, 3	Header	Generic	2/2mm	
L1, 2, 3, 5	1.8 μH	2012	Panasonic	ELJRF1N8DF2

Bill of Materials *(continued)*

Reference	Value	PCB	Manufacturer	Part# / Details
L4			Allied Comp.	MLB20-700
Q1,2			Diodes Inc.	BSS138
R1, 3	1K	0603	Generic	
R6, 7, 8, 10, 11	3.3K	0603	Generic	
R5, 7, 12, 13, 14	75	0603	Generic	
R15	470	0603	Generic	
R2, 4	10K	0603	Generic	
S1			Panasonic	EVQ11U04M
S2		SW DIP-4	C&K	SD04H0SKD
U1 BT864A			Conexant	Bt864AKPF
U2			TI	TPS3103H20DVB
U3			Linear Tech	LT1585CM-3.3
U4			Epson	SG-615PCN-27.0000M

15.2 Project 2—General Purpose CPLD Processor Board

As we mentioned throughout this book, Complex Programmable Logic Devices (CPLDs) are well suited for the implementation of complex digital video circuits such as filters or raster generators. In this section we introduce a hardware platform that will allow us to experiment with Altera's popular MAX7000 and MAX3000 families of CPLDs. These high-performance devices have the added advantage of being easy to use and program. The software required to design and compile application circuits, as well as to program the chips themselves, is available free of charge from Altera (www.altera.com). In order to further simplify the use of the board, we designed in all the circuitry required to program the CPLD. All you need is a PC and a printer cable.

The CPLD board can be populated with one of the following, almost pin-compatible, ICs: EPM3256AQC208, EPM3512AQC208, or EPM7512BQC208. The first two are truly pin-compatible but differ in the number of basic CPLD "blocks" or "macrocells" they contain; EPM3512AQC208 has 512 while EPM3256AQC208 has only 256. To use the higher-performance 512 macrocells EPM7512BQC208 we must first install the zero-ohm jumpers R1, R2, R3, and R4 (Figure 15.7). A simple user interface allows for basic level interaction between the CPLD and the designer (Figure 15.8). This interface consists of an 8-position DIP switch (S2) and a nine-segment bar-graph LED (U5).

Like the encoder board, the CPLD board is powered by an external regulated 5 V power supply, with a minimum current rating of 1.5 A. The board also uses the same LT1585 IC to drop the 5 V input voltage to the 3.3 V required by the CPLD. Although nominally 3.3 V parts, the CPLDs we selected have all the I/O lines 5 V compatible which allows for the direct connection of 3.3 V as well as 5 V logic. The user interface, for example, operates from the 5 V side of the power block but is controlled through the 3.3 V I/O lines of the MAX devices. A 27 MHz pixel clock oscillator and a power-up reset IC (with manual reset input) round up the available local resources (Figure 15.9).

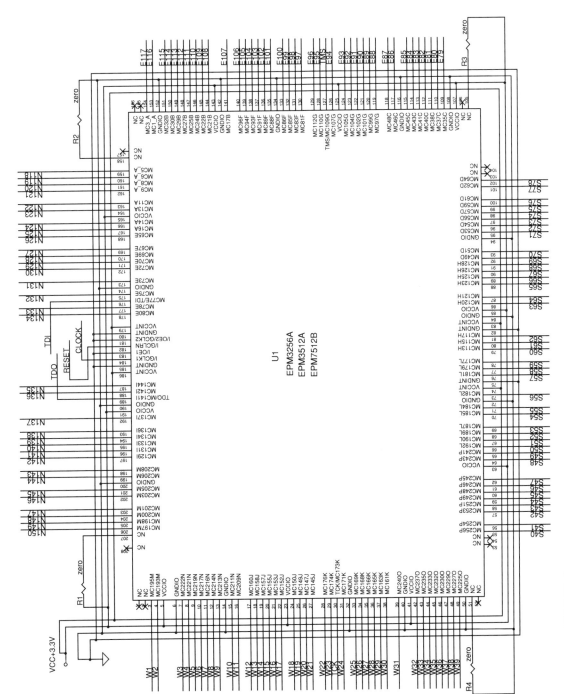

Figure 15.7 CPLD Block.

377

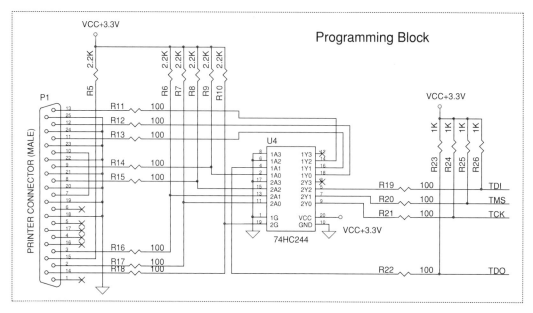

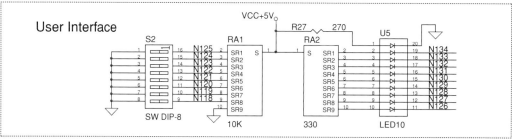

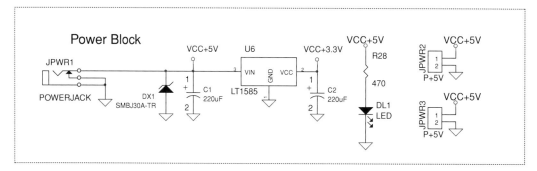

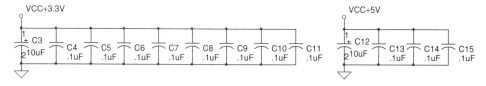

Figure 15.8   CPLD Ancillary Circuits 1.

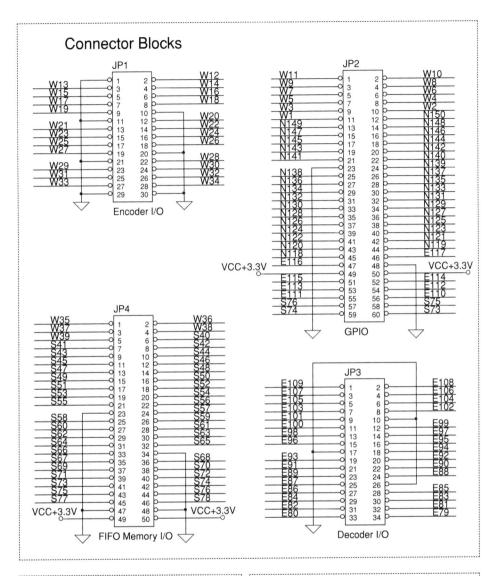

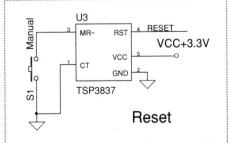

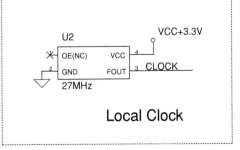

Figure 15.9   CPLD Ancillary Circuits 2.

Access to the CPLD board is provided through four 2-mm connectors: JP1, JP2, JP3, and JP4. They bring most of the CPLD I/O lines outside the board, together with 3.3 V power and ground lines. For most tasks related to the design and programming of Altera CPLDs the capabilities of the MAX+Plus II Baseline / Quartus II Web Edition software (free from Altera) are sufficient. However, more advanced features such as simulation and multichip partitioning are only available with the paid license version.

The operation of the programmer requires the installation of the ByteBlasterMV printer driver, also available on the Altera website.

The PCB board has a four-layer stack structure with the inner layers power and ground; the outer layers are used for routing (Figure 15.10). As with most purely digital boards operating at moderate frequencies, there is only minimal concern regarding noise or transmission line effects. The wiring of the CPLD board to power and printer cable is illustrated in Figure 15.11. PCB files are available at www.digital_video_electronics.com.

Next, let's get accustomed to the Max+PlusII software by developing a rather trivial CPLD function; we will program the CPLD to drive the eight LEDs with the information provided by the S2 DIP switch.

First we have to download the software and install our free license file (see instructions at the Altera website). Licenses have to be renewed every six months, a small price to pay for software of this caliber. When we open the program we are greeted by the MaxPlus II graphics interface.

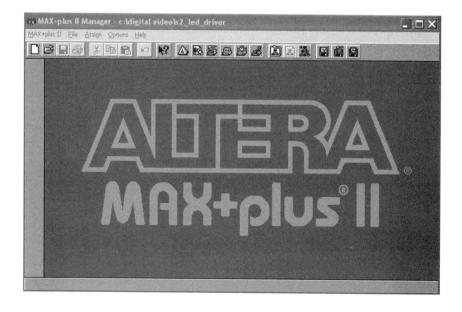

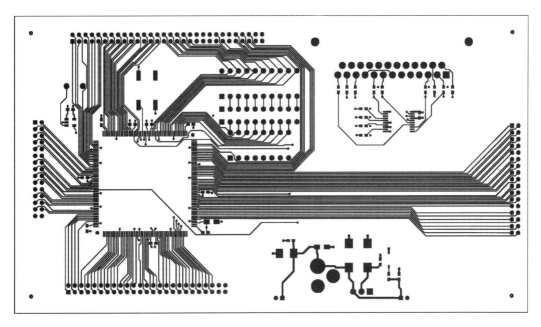

Figure 15.10(a)   CPLD Board PCB Layout Top Layer.

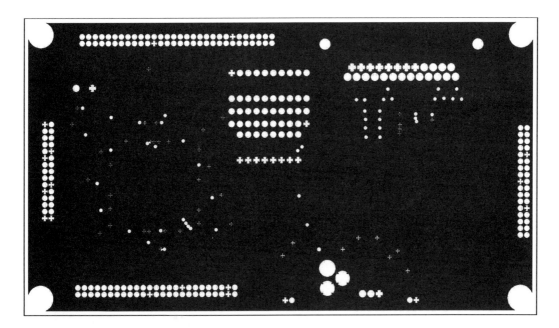

Figure 15.10(b)   CPLD Board PCB Layout Ground Layer.

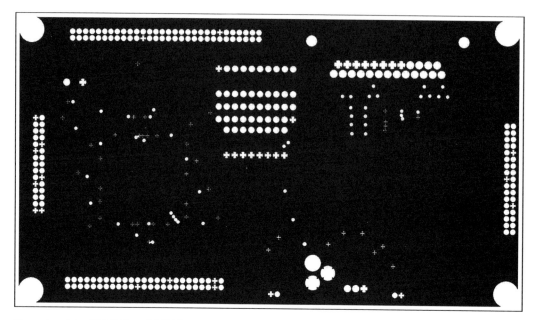

Figure 15.10(c)   CPLD Board PCB Layout Power Layer.

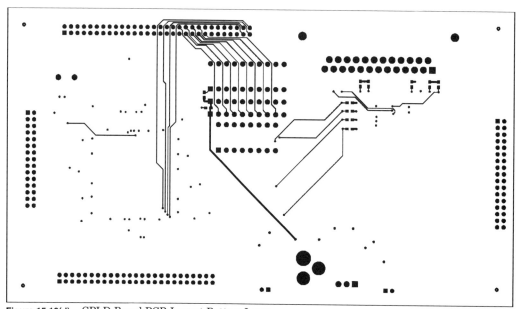

Figure 15.10(d)   CPLD Board PCB Layout Bottom Layer.

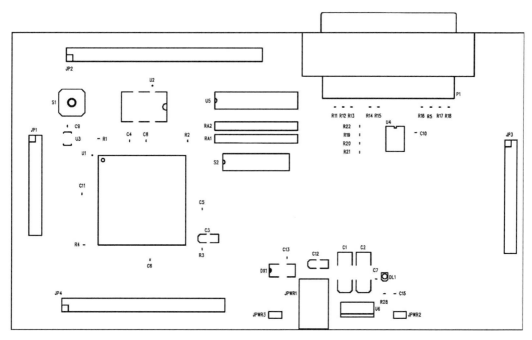

Figure 15.10(e)   CPLD Board PCB Layout Top Layer Silkscreen.

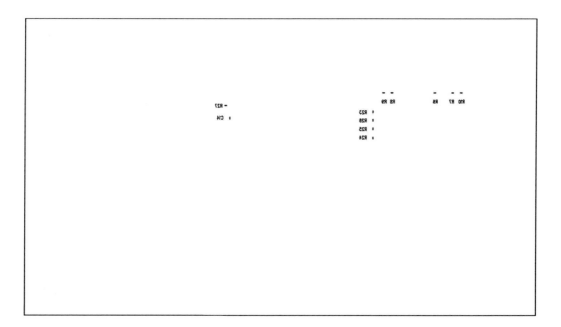

Figure 15.10(f)   CPLD Board PCB Layout Bottom Layer Silkscreen.

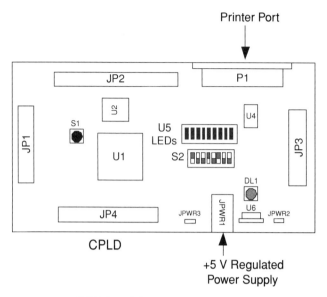

Figure 15.11   CPLD Board Wiring.

Then we start by creating a new file according to the normal Windows protocol; click on File, then New

We are next queried about the type of file we want to create. For this exercise we will select a Graphic Editor file type (extension .gdf), and name it s2_led_driver.gdf.

Once the design is named the Graphic Editor is ready to accept our schematics. The process is simple. First we bring into the schematic all the components we want to use, then we interconnect them. We start with the Graphics Editor window.

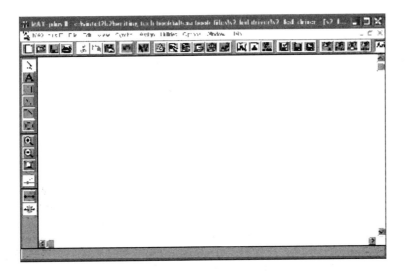

To bring logic signals into the schematics we have to declare them as inputs. From the Symbol Menu select Enter Symbol, then type "input" and press the OK button.

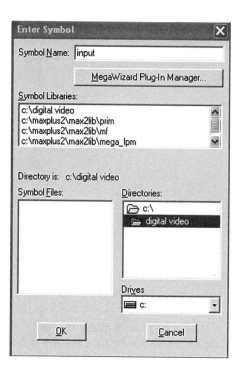

The INPUT library component will become visible on the screen. You can move it where you want it by simply using the mouse. Since we have a total of eight inputs (corresponding to the eight switches of S2) we need to use eight INPUT symbols.

The eight outputs that drive the eight LEDs of U5 must also be entered into the editor. To do so at the Enter Symbol prompt we type "output" and the OUTPUT port symbol will appear. Instead of repeatedly calling the OUTPUT port from the library we can also just copy and paste it seven more times. Finally, eight inverters (symbol NOT) are also introduced into the editor. They are needed to convert the high input levels provided by the S2 switch into the active low levels needed to turn on the LEDs. Once entered into the schematics the different components are interconnected using the Line icon.

Completion of this generic buffer requires identifying the input and output ports by giving them appropriate node names. Click twice on the PIN_NAME label of the first INPUT port and enter S2_1. This gives the port and the net to which it is attached the name S2_1. In a similar fashion, label the first OUTPUT port LED_1 and continue until all ports are marked (S2_2, LED_2 ... S2_8, LED_8). This buffer can be implemented in any Altera CPLD.

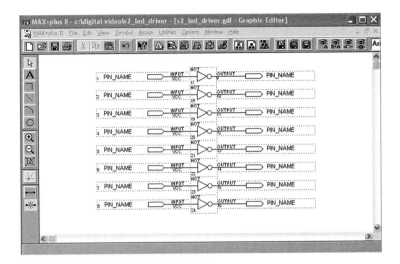

However, if we want to use it with a MAX EPM3512AQC208, for example, we need to assign this device to the design. To do it we pull down the Assign Menu and select Device. The family of interest is MAX3000A. Within this family we highlight EPM3512AQC208-7 and click on OK.

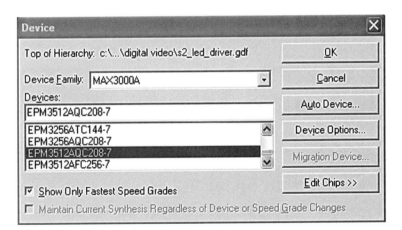

Now we need to match the CPLD schematics to the board hardware by assigning pin numbers to the input and output ports. This is accomplished by going back to the Assign Menu and this time selecting Pin/Location/Chip. After clicking on Pin we enter the node name (same as the INPUT or OUTPUT

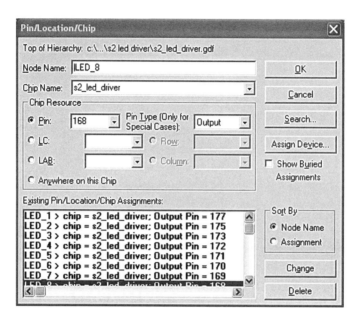

port name), the pin type (Input for an INPUT port and Output for an OUT-PUT port), and the pin number. We repeat the process until all the input and output ports have been assigned the desired pin numbers.

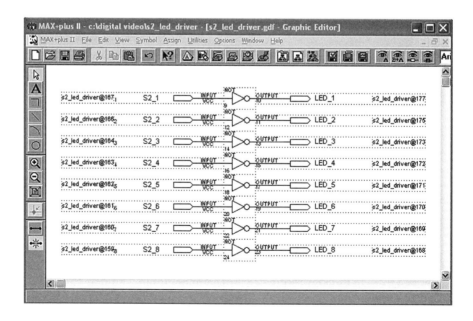

The last step in the design process is compiling the project. Successful compilation results in the generation of a number of files, including a Programmer Object File (extension .pof) which we need for programming the chip. The other files are the Compiler Netlist File (.cnf) which contains the connectivity information for the design, the Report file (.rpt) which includes partitioning, resource usage, inputs, outputs, and project timing information, and the Simulator Netlist File (.snf) necessary for the optional logic simulation package.

To compile the s2_led_driver project, first pull down the MAX+plus II Menu, then select Compiler and click on Start. If the Compiler detects any errors the software will guide you in determining what the problem is and will aid you in solving it.

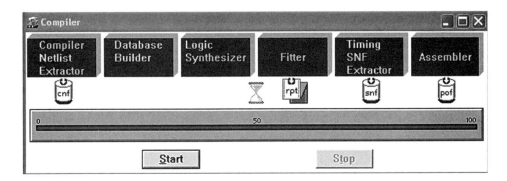

To program the EPM3512AQC208 we connect the CPLD port to the printer port of the PC using a 25-pin cable. From the MAX+plus II Menu we click on the Programmer entry. This brings up the programmer menu. If the name

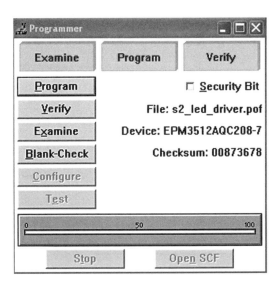

shown after "File:" is not s2_led_driver.pof then click twice on File then find and select s2_led_driver.pof. Proceed the same way if the Device entry is not correct.

If you do not want the device to be back-readable check the Security Bit box. However, this is generally done during production—not while a product is under development. Assuming everything went smoothly the output LEDs will follow the status of the input DIP switches, and you have completed (possibly) your first CPLD design.

The PCB files for this board can be found at www.digital-video-electronics.com.

Bill of Materials

Reference	Value	PCB	Manufacturer	Part# / Details
C1, 2	220 μF	7343	Generic	Electro
C3, C12	10 μF	3216	Generic	Tantalum
C4, 5, 6, 7, 8, 9, 10, 11, 13, 14, 15	.1 μF	0603	Generic	Ceramic
C12	10 μF	3216	Generic	Tantalum
DL1	LED		Lumex	SSL-LXA228GC-TR11
DX1	TZORB		STMicro	SMBJ30A-TR
JPWR1	POWERJACK		CUI	PJ-202A
JPWR2, JPWR3	Header		Generic	2/2mm
JP1	Header		Generic	15X2/2mm
JP2	Header		Generic	30X2/2mm
JP3	Header		Generic	17X2/2mm
JP4	Header		Generic	25X2/2mm
P1			Norcomp	177-025-110-071
R1, 2, 3, 4	zero	0603	Generic	
R5, 6, 7, 8, 9, 10	2.2K	0603	Generic	
R11, 12, 13, 14, 15, 16, 17, 18, 19, 20	100	0603	Generic	
R21, 22	100	0603	Generic	
R23, 24, 25, 26	1K	0603	Generic	
R27, 28	270	0603	Generic	
RA1	10K		CTS	770-10-1-103
RA2	330		CTS	770-10-1-331
S1	Pushbutton		Panasonic	EVQ11U04M
S2	SW DIP-8		C&K	SD03H0SKD
U1			Altera	EPM3256AQC208
				EPM3512AQC208
				EPM7512BQC208
U2 27MHz			Epson	SG615PTJ27.0000MC3
U3 TSP3837			TI	TPS3837J25DBV
U4 74HC244			TI	SN74HC244DBR
U5 LED10	20DIP300		Fairchild	MV57164
U6 LT1585			Linear Tech	LT1585CM-3.3

15.3 Project 3—Color Raster Generator

For a more challenging exercise we can connect the CPLD and Encoder boards together and design a simple CPLD-based color generator. Its function is to fill the screen with a specific color. What makes this design interesting is that we will program the CPLD to generate the YCbCr stream instead of the data being produced internally by the Bt864.

Since the pin-out of JP1 on the CPLD board matches exactly the pin-out of the JP1 connector on the encoder board, we can connect the two boards either by directly plugging the encoder board into the JP1 connector of the CPLD board, or by means of a short 30-pin cable (Figure 15.12).

The encoder board can be powered separately or by daisy chaining it to the CPLD board using the JPWR connectors. Although separate supplies are preferred, if we have to power the encoder from the CPLD board the connecting wire should be as thick as possible in order to avoid ground bounce. To maintain synchronicity between the boards, the CPLD will use the same 27 MHz clock as the encoder. To use the oscillator on the encoder board we jumper pins 1, 2 and 3, 4 on the JP2 header of the encoder board.

Control of the encoder is exercised through one of the 3.3 V I²C connectors, using the Aardvark adaptor or a similar device. For this application the Bt864 has to be set for normal operation (no color bars), master mode, and 8-bit bus input interface (default). Using the master mode simplifies our task by allowing us to use the reference signals (HSYNC, VSYNC) generated internally by the Bt864.

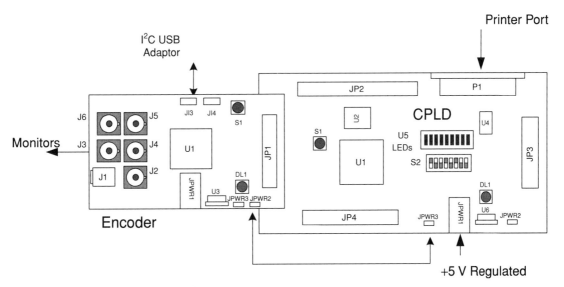

Figure 15.12   Color Generator Assembly.

The corresponding settings for the S2 DIP switch are SLAVE and RGBOUT low (master mode, composite/s-video outputs). Normal operation is achieved by clearing the ECBAR bit and setting to one the EACTIVE bit (D4 and D1 bits of the 0×67 subregister).

The CPLD design contains two major blocks; the YCbCr Sequence Generator and the Colors block. Following each HSYNC pulse received from the Bt864 the YCbCr Sequence Generator produces a continuous succession of binary outputs 1000, 0100, 0010, 0001, 1000, 0100...These signals drive the Colors block which in response to the 1000 and 0010 strings outputs a Y luma byte. For a 0100 input it generates a Cb chroma byte, and for a 0010 string outputs a Cr chroma byte. The result is a continuous stream of YCbCr bytes (YCO[7..0]) synchronized to the HSYNC signal provided by the Bt864. Since the YCbCr Sequence Generator is driven by a clock twice the pixel frequency (EXT_CLK), and the state machine of the sequence generator advances one state for each input clock pulse the Y, Cb, and Cr outputs advance at the required rate of twice the pixel clock.

The advantage of using the master mode for the Bt864 is that we do not need to use the VSYNC, FIELD, or BLANK~ information at all. The internal raster generator automatically overrides the inputs, when necessary, in order to produce a valid NTSC output. This also happens during the vertical and horizontal blanking periods when we still provide YCbCr information to the Bt864, but the data is conveniently ignored.

As can be seen from Fig. 5.13, when working with CPLDs we must account for all the active pins of the device. Since we connect the JP1 of the CPLD board to an external system we must assume that all the pins on JP1, and the corresponding pins on the MAX chip, are active. Therefore all the outputs are defined and connected to a logic source or logic level, and all inputs are defined, even if only some are actually used. Any undefined I/O may be assigned by the compiler a state which may contradict its external wiring.

Some further observations about this design. Every time we assign a name to an I/O port, the net to which the I/O is connected is given the same name. Directly connecting multiple I/O ports will result in an error, since the software assumes that we inadvertently shorted the nets together. To circumvent this problem we need to buffer the I/Os (using the SOFT buffer, for example) before tying them together. A different problem occurs in designs using multiple clocks. The CPLD software assumes that global clocks are present only at two designated pins: 181 and 184 for the chips we are using. We can still implement local clocks but we need to use a clock buffer, such as an AND gate, before distributing the signal. We must also note that the I2CCS output is set low, which corresponds to a Bt864 I^{2}C address of 0×8A.

The code for the YCbCr Sequence Generator is written using the MAX+Plus II AHDL language, an Altera version of the standard VHDL. Its operation is based on the use of a state machine with four states: s0, s1, s2, and s3. The state machine has four outputs (y1, y2, y3, y4), each driving a color component

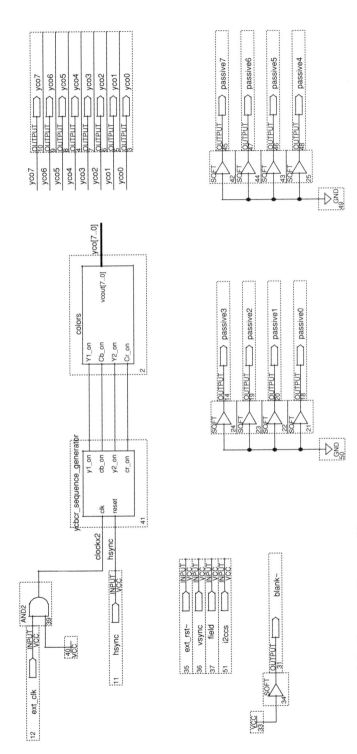

Figure 15.13 Color Generator Firmware Diagram.

enabling signal: y1_on = y1, cb_on = y2, y2_on = y3, and cr_on = y4. After each reset (!HSYNC) the machine cycles through states (s1→ s2 → s3 → s0), which in turn enables the desired sequence of color components Y (1000), Cb (0100), Y (0010), Cr (0001).

YCbCr_sequence_generator

```
SUBDESIGN YCbCr_Sequence_Generator
(
        clk, reset                              : INPUT;
        y1_on, cb_on, y2_on, cr_on              : OUTPUT;
)
VARIABLE
        sequence_generator:     MACHINE OF BITS (y1, y2, y3, y4)
                                WITH STATES(
                        s0      =       B"0001",
                        s1      =       B"1000",
                        s2      =       B"0100",
                        s3      =       B"0010");

BEGIN
        sequence_generator.clk                  =       clk;
        sequence_generator.reset                =       !reset;

        TABLE
        sequence_generator                      =>      sequence_generator;
                s0                              =>              s1;
                s1                              =>              s2;
                s2                              =>              s3;
                s3                              =>              s0;

        END TABLE;
        y1_on   =       y1;
        cb_on   =       y2;
        y2_on   =       y3;
        cr_on   =       y4;
END;
```

The Colors block reads the four bits provided by the YCbCr Sequence Generator and outputs the appropriate 8-bits digital output (yout[7..0]). The processing equations are simple four-term sums of products, where "#" is the OR operator and "&" is the AND operator.

To change the color from blue

```
Y1[]              =        Y_2;
Cr[]              =        Cr_2;
Cb[]              =        Cb_2;
Y2[]              =        Y2_2;
```

to red, for example, we chance our constants assignment to

```
Y1[]              =        Y_1;
Cr[]              =        Cr_1;
Cb[]              =        Cb_1;
Y2[]              =        Y2_1;
```

The CONSTANT header also provides the correct luma and chroma values for yellow, cyan, white, and black. Of course, you can also experiment with your own values.

```
colors

%RED%
CONSTANT  Y1_1        = 65;
CONSTANT  Cb_1        = 100;
CONSTANT  Y2_1        = 65;
CONSTANT  Cr_1        = 212;
%BLUE%
CONSTANT  Y1_2        = 35;
CONSTANT  Cb_2        = 212;
CONSTANT  Y2_2        = 35;
CONSTANT  Cr_2        = 114;
%YELLOW%
CONSTANT  Y1_3        = 162;
CONSTANT  Cb_3        = 44;
CONSTANT  Y2_3        = 162;
CONSTANT  Cr_3        = 142;
%CYAN%
CONSTANT  Y1_4        = 131;
CONSTANT  Cb_4        = 156;
CONSTANT  Y2_4        = 131;
CONSTANT  Cr_4        = 44;
%WHITE%
CONSTANT  Y1_5        = 180;
CONSTANT  Cb_5        = 128;
CONSTANT  Y2_5        = 180;
CONSTANT  Cr_5        = 128;
%BLACK%
```

```
CONSTANT Y1_6          = 16;
CONSTANT Cb_6          = 128;
CONSTANT Y2_6          = 16;
CONSTANT Cr_6          = 128;

%EXPERIMENTAL%
CONSTANT Y1_7          = 75;
CONSTANT Cb_7          = 212;
CONSTANT Y2_7          = 75;
CONSTANT Cr_7          = 114;

SUBDESIGN colors
(
        Y1_on, Cb_on, Y2_on, Cr_on                    : INPUT;
        vcout[7..0]                                   : OUTPUT;
)
VARIABLE
        Y1[7..0], Cb[7..0], Y2[7..0], Cr[7..0]        : NODE;
BEGIN

        Y1[]                =          Y1_2;
        Cr[]                =          Cr_2;
        Cb[]                =          Cb_2;
        Y2[]                =          Y2_2;

        vcout[0] =          Y1[0]&Y1_on # Cb[0]&Cb_on # Y2[0]&Y2_on # Cr[0]&Cr_on;
        vcout[1] =          Y1[1]&Y1_on # Cb[1]&Cb_on # Y2[1]&Y2_on # Cr[1]&Cr_on;
        vcout[2] =          Y1[2]&Y1_on # Cb[2]&Cb_on # Y2[2]&Y2_on # Cr[2]&Cr_on;
        vcout[3] =          Y1[3]&Y1_on # Cb[3]&Cb_on # Y2[3]&Y2_on # Cr[3]&Cr_on;
        vcout[4] =          Y1[4]&Y1_on # Cb[4]&Cb_on # Y2[4]&Y2_on # Cr[4]&Cr_on;
        vcout[5] =          Y1[5]&Y1_on # Cb[5]&Cb_on # Y2[5]&Y2_on # Cr[5]&Cr_on;
        vcout[6] =          Y1[6]&Y1_on # Cb[6]&Cb_on # Y2[6]&Y2_on # Cr[6]&Cr_on;
        vcout[7] =          Y1[7]&Y1_on # Cb[7]&Cb_on # Y2[7]&Y2_on # Cr[7]&Cr_on;

%
 Y1_on = 1 => vcout[] = Y1[]
 Cb_on = 1 => vcout[] = Cb[]
 Y2_on = 1 => vcout[] = Y2[]
 Cr_on = 1 => vcout[] = Cr[]
%

END;
```

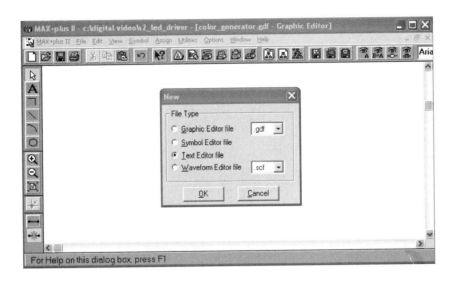

To enter an AHDL design we open a New file from the File pull-down menu and select Text Editor File. After the text is entered and checked for errors (File → Project → Save & Check), we can create a graphic library element that corresponds to the AHDL file by selecting Create Default Signal from the File menu. Further editing of the symbol can be done using the Symbol Editor from the Max+plus II menu.

After we type in the code for the YCbCr Sequence Generator and Colors, generate the appropriate graphics symbols, wire them according to Figure 5.13, and compile the overall design, we can program the CPLD. If we connect an NTSC monitor to the encoder board we will see a blue screen that was digitally generated by the CPLD board and converted to composite by the encoder board. We can then repeatedly change our design by changing colors, adding latches, or implementing a clock divider to cycle through different screens, all while we gain experience with the MAX+plus II and our new boards.

Note: We left out the device and pin assignment steps since they will depend on your IC choice for the CPLD board. However, these steps must be completed or the boards will not function properly.

15.4 Project 4—Framed Box Raster Generator

Control of the video raster itself requires a higher level of CPLD firmware complexity. As we saw in Chapter 11, the generation of a simple framed box display involves tracking the actual pixel and line counts, defining the inset and border areas, and comparing the counts to the boundaries of the inset and border regions.

The pixel_counter block is driven by a clockx2 (twice pixel rate) and an HSYNC reference signal, both originating on the encoder board (Figure 15.14).

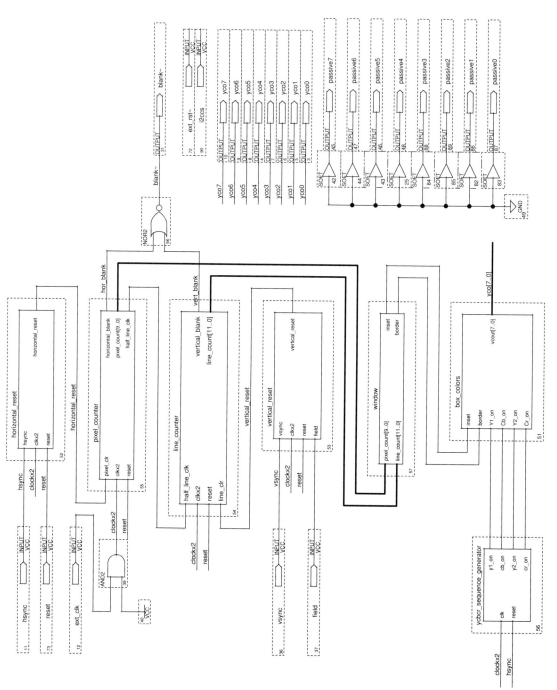

Figure 15.14 Framed Box Generator Firmware Diagram.

Figure 15.15 Framed Box Display.

A new pixel count starts at the beginning of each new HSYNC reference signal, when the horizontal_reset block resets the pixel counter. As the count progresses the counter output is compared with a number of landmarks. If the count is smaller than the front_blank value or larger than the end_of_line_blank value, the block generates an active low horizontal_blank signal. The half_line_clock needed by the line counter is produced by comparing the pixel count with the half_line_clk_q markers. As a result two half-line pulses are issued for each horizontal line, one at the beginning of the line and one in the middle.

The line_counter block uses the clockx2 input only for circuit synchronization. It is actually driven by the half_line_clk signal it receives from the pixel_counter and the vertical_reset pulse generated by the vertical_reset block at the beginning of each FIELD =1, VSYNC pulse. Both HSYNC and VSYNC are outputs of the Bt864 encoder (master mode). The vertical blanking signal is created by comparing the vertical line count (actually half-line count; see Chapter 11) with front, interfield, and end-of-frame vert_blank_q markers.

Pixel and line counts then make their way to the window block where they are compared against the coordinates of the corners of the inset and border rectangles. If the counts are within the inset rectangle the inset output is asserted, and if the counts are within the area of the border rectangle the border output is set high.

The inset and border signals act as color selection signals for the box_colors block, which is otherwise similar to the colors block encountered earlier. For inset the selected color is blue (Y1_2, Cb_2, Y2_2, Cr_2) while for the border the color is set to "experimental" (Y1_7, Cb_7, Y2_7, Cr_7). The numbers chosen for the border correspond to light blue, but they can be changed to any other color variation. As for the background the color shown in our example is white (Y1_5, Cb_5, Y2_5, Cr_5). The actual captured video output is shown in Figure 15.15.

pixel_counter

```
% NTSC pixel and horizontal grid parameters for 640 pixels per line %
CONSTANT front_blank                    =       151;
CONSTANT end_of_line_blank              =       791;
CONSTANT half_line_clk_q1               =       4;
CONSTANT half_line_clk_q2               =       471;
CONSTANT half_line_clk_q3               =       473;

SUBDESIGN pixel_counter
(
        pixel_clr, clkx2, reset         :       INPUT;
        horizontal_blank, pixel_count[9..0]  :  OUTPUT;
        half_line_clk                   :       OUTPUT;
)
VARIABLE
        count[9..0]                     :       DFF;
        clkx1, pixel_count[9..0]         :       DFF;
        half_line_clk                   :       DFF;
        horizontal_blank                :       DFF;

BEGIN
        pixel_count[].clk               =       clkx2;
        pixel_count[].clrn              =       !reset;
        clkx1.clk                       =       clkx2;
        clkx1.clrn                      =       !reset;
        count[].clk                     =       clkx2;
        count[].clrn                    =       !pixel_clr & !reset;

        half_line_clk.clk               =       clkx2;
        half_line_clk.clrn              =       !reset;
        horizontal_blank.clk            =       clkx2;
        horizontal_blank.clrn           =       !reset;

        half_line_clk                   =       !SOFT((count[].q < half_line_clk_q1)
                                                # ((half_line_clk_q2 < count[].q)
                                                & (count[].q < half_line_clk_q3)));
```

% The half_line_clock outputs a positive pulse twice during each video
line. First pulse is generated at the beginning of the line and
lasts for as long as count[].q < half_line_clk_q1. The second is in the
middle of the line, and lasts for as long as (half_line_clk_q2 < count[].q)&
    (count[].q < half_line_clk_q3) %

```
        horizontal_blank.d                              =           !SOFT((front_blank <= count[].q) &
                                                                    (count[].q <= (end_of_line_blank)));
```

% Horizontal blank is an active low signal which blanks the screen outside
the active line region - front_blank < count[].q < end_of_line_blank%

```
        clkx1.d                                         =           !clkx1.q;
```

% clkx1 is used to divide the input clock (clkX2) by 2. In 8 bit operation
one pixel corresponds to two input bytes (Y and Cb or Cr) and therefore two
input clock pulses. The division by 2 insures that we are counting full pixels
and not bytes%

```
        IF          clkx1       THEN
```

% If clkx1 is high proceed to pixel count evaluation%

```
                    IF          pixel_clr           THEN
                                count[].d           =       gnd;
```

% If pixel_clear is externally asserted reset the pixel count%
```
                    ELSE
                                count[].d           =       count[].q       +       1;
```

%...if not increment the pixel count%

```
                    END IF;
        ELSE
                    count[].d                       =       count[].q;
```

% If clkx1 is low wait for a clkX2 period (division by 2)%

```
        END IF;
        pixel_count[].d                             =       count[].q;
```

% Update the output pixel count%

```
END;
```


horizontal_reset

```
SUBDESIGN horizontal_reset
(
        hsync, clkx2, reset                             :           INPUT;
```

```
        horizontal_reset                              :        OUTPUT;
)
VARIABLE
        horizontal_res_state            : MACHINE OF BITS (horizontal_reset)
                                          WITH STATES (

                s0      =               B"0",
                s1      =               B"1",
                s2      =               B"0");

BEGIN
        horizontal_res_state.clk                =               clkx2;
        horizontal_res_state.reset              =               reset;
TABLE
        horizontal_res_state,   hsync    =>             horizontal_res_state;
                s0,             1        =>                     s0;
```

% If hsync is high (inactive) stay in state s0 with horizontal_reset output low%

```
                s0,             0        =>                     s1;
```

% When hsync drops to zero (active) jump to s1; horizontal_reset goes high.%

```
                s1,             x        =>                     s2;
```

% Go to the next state on the following input clock; this completes the output horizontal_reset pulse.%

```
                s2,             0        =>                     s2;
```

% Stay at s2 as long as hsync is low; horizontal_reset stays low.%

```
                s2,             1        =>                     s0;
```

% When hsync becomes 1 again go to the s0 (start-up) state; horizontal_reset stays low.%

```
END TABLE;

END;
```

line_counter

```
% NTSC line and vertical grid parameters for 480 lines frame format %
CONSTANT vert_blank_q1                   =        40;
CONSTANT vert_blank_q2                   =        516;
```

```
CONSTANT vert_blank_q3                      =           564;
CONSTANT vert_blank_q4                      =           1040;


SUBDESIGN line_counter
(
        half_line_clk, clkx2, reset         :           INPUT;
        line_clr                            :           INPUT;
        vertical_blank, line_count[11..0]   :           OUTPUT;
)
VARIABLE
        count[11..0]                        :           DFF;
        new_half_flag                       :           DFF;
        vertical_blank, line_count[11..0]   :           DFF;
BEGIN
        line_count[].clk                    =           clkx2;
        line_count[].clrn                   =           !reset
        count[].clk                         =           clkx2;
        count[].clrn                        =           !reset
        new_half_flag.clk                   =           clkx2;
        new_half_flag.clrn                  =           !reset;
        vertical_blank.clk                  =           clkx2;
        vertical_blank.clrn                 =           !reset;


        vertical_blank                      =           SOFT((count[].q < vert_blank_q1)
                                                        # ((count[].q > vert_blank_q2)
                                                        & (count[].q < vert_blank_q3))
                                                        # (count[].q > vert_blank_q4));
```

% There are two vertical blank regions in each video frame. The first
spans the bottom and the top of the frame and lasts from vert_blank_q4 to
vert_blank_q1, including the counter reset event. The second marks the
interframe segment and extends between vert_blank_q2 and and vert_blank_q3%

```
        IF      line_clr    THEN
                count[].d                   =           GND;
```

% If line_cnt_clr is asserted clear the line counter%

```
        ELSE
                IF      half_line_clk    THEN
```

% If the half_line_clk input becomes high, the (half) line counter
must be incremented but only once regardless of how long half_line_clk
stays high. This single-shot counter increment is done using an auxiliary
variable new_half_flag.%

```
                              IF        !new_half_flag       THEN
```

% If new_half_flag is low (new half line) then...%

```
                                  new_half_flag       =       VCC;
                                  count[].d           =       count[].q     +     1;
```

%...if this is NOT the last expected half line increment the count and set the
new_half_flag high%

```
                          ELSE
```

% If the new_half_flag is high (the counter has already been incremented during
this "new_half_line high" interval maintain the count and keep the new_half_flag
high%

```
                                  count[].d       =       count[].q;
                                  new_half_flag   =       VCC;
                          END IF;
              ELSE
                  new_half_flag       =       GND;
                  count[].d           =       count[].q;
```

% If the half_line_clk is low keep the new_half_flag low and maintain the count%

```
              END IF;
      END IF;
              line_count[].d          =       count[].q;
```

% Update the output line counter%

```
END;
```

vertical_reset

```
SUBDESIGN vertical_reset
(
      vsync, clkx2, reset, field           : INPUT;
      vertical_reset                       : OUTPUT;
)

VARIABLE
      vertical_res_state : MACHINE OF BITS (vertical_reset)
                              WITH STATES (
```

```
                    s0      =       B"0",
                    s1      =       B"1",
                    s2      =       B"0",
                    s3      =       B"0");
BEGIN
        vertical_res_state.clk              =       clkx2;
        vertical_res_state.reset            =       reset;

        TABLE
        vertical_res_state,     vsync,    field    =>    vertical_res_state;

            s0,                 1,        x        =>            s0;
```

% Regardless of the value of field, the state machine stays in s0
as long as vsync is high%

```
            s0,                 0,        0        =>            s2;
```

% When vsync becomes low the path is determined by the value of
the field input. If field = 0 we transfer directly to s2 keeping
the vertical_reset output low%

```
            s0,                 0,        1        =>            s1;
```

% For field = 1 we go to the state s1 which sets the vertical_reset
output high....%

```
            s1,                 x,        x        =>            s2;
```

%.... then go to state s2 thus completing vertical_reset output pulse.%

```
            s2,                 0,        x        =>            s2;
```

% We stay at s2 until the end of the input vsync pulse%

```
            s2,                 1,        x        =>            s3;
```

% Then we return to s0.%

```
            s3,                 x,        x        =>            s0;

        END TABLE;

END;
```

window

```
CONSTANT box_left_side                              = 300;
CONSTANT box_width                                  = 360;
CONSTANT box_right_side                             = box_left_side + box_width;
CONSTANT box_top_side                               = 160;
CONSTANT box_height                                 = 240;
CONSTANT box_bottom_side                            = box_top_side + box_height;
CONSTANT border_width                               = 16;
CONSTANT inter_field_offset                         = 524;

SUBDESIGN window
(
    pixel_count[9..0], line_count[11..0]    : INPUT;
    inset, border                           : OUTPUT;
)
BEGIN
    inset  =  SOFT(((((pixel_count[] > box_left_side)&(pixel_count[] <= box_right_side))&
                    ((line_count[] > box_top_side)&(line_count[] <= box_bottom_side)))#
                    (((pixel_count[] > box_left_side)&(pixel_count[] <= box_right_side))&
                    ((line_count[] > box_top_side + inter_field_offset)&
                    (line_count[] <= box_bottom_side + inter_field_offset)))));

% The output "inset" is high within the inset box (window) boundaries described by
box_left_side, box_right_side, box_top_side, and box_bottom_side; zero outside%

    border = SOFT(((((pixel_count[] > box_left_side - border_width)&
                    (pixel_count[] <= box_right_side + border_width))&
                    ((line_count[] > box_top_side - border_width)&
                    (line_count[] <= box_bottom_side + border_width)))#
                    (((pixel_count[] > box_left_side - border_width)&
                    (pixel_count[] <= box_right_side + border_width))&
                    ((line_count[] > box_top_side + inter_field_offset - border_width)&
                    (line_count[] <= box_bottom_side + inter_field_offset +
                        border_width)))));

% The output "border" is high within a box larger than the inset box window by an amount
equal to "border_width" in all four directions; zero outside%

END;
```

```
box_colors


%RED%
CONSTANT Y1_1                          =        65;
CONSTANT Cb_1                          =        100;
CONSTANT Y2_1                          =        65;
CONSTANT Cr_1                          =        212;
%BLUE%
CONSTANT Y1_2                          =        35;
CONSTANT Cb_2                          =        212;
CONSTANT Y2_2                          =        35;
CONSTANT Cr_2                          =        114;
%YELLOW%
CONSTANT Y1_3                          =        162;
CONSTANT Cb_3                          =        44;
CONSTANT Y2_3                          =        162;
CONSTANT Cr_3                          =        142;
%CYAN%
CONSTANT Y1_4                          =        131;
CONSTANT Cb_4                          =        156;
CONSTANT Y2_4                          =        131;
CONSTANT Cr_4                          =        44;
%WHITE%
CONSTANT Y1_5                          =        180;
CONSTANT Cb_5                          =        128;
CONSTANT Y2_5                          =        180;
CONSTANT Cr_5                          =        128;
%BLACK%
CONSTANT Y1_6                          =        16;
CONSTANT Cb_6                          =        128;
CONSTANT Y2_6                          =        16;
CONSTANT Cr_6                          =        128;
%EXPERIMENTAL%
CONSTANT Y1_7                          =        125;
CONSTANT Cb_7                          =        212;
CONSTANT Y2_7                          =        125;
CONSTANT Cr_7                          =        114;


SUBDESIGN box_colors
(
        inset, border, Y1_on           :        INPUT;
        Cb_on, Y2_on, Cr_on            :        INPUT;
        vcout[7..0]                    :        OUTPUT;
)
```

```
VARIABLE
      Y1[7..0], Cb[7..0], Y2[7..0], Cr[7..0] : NODE;

BEGIN

IF inset THEN
      Y1[]              =         Y1_2;
      Cr[]              =         Cr_2;
      Cb[]              =         Cb_2;
      Y2[]              =         Y2_2;

% If the pixel and line counters are within the range of the "inset" window
display "black"%

ELSIF border THEN
      Y1[]              =         Y1_7;
      Cr[]              =         Cr_7;
      Cb[]              =         Cb_7;
      Y2[]              =         Y2_7;

% If the pixel and line counters are not within the range of the "inset" window,
but are in the range of the "border box" display the experimental color - light blue%

ELSE
      Y1[]              =         Y1_5;
      Cr[]              =         Cr_5;
      Cb[]              =         Cb_5;
      Y2[]              =         Y2_5;

% If outside both windows display "white"%

END IF;
      vcout[0] =        Y1[0]&Y1_on # Cb[0]&Cb_on # Y2[0]&Y2_on # Cr[0]&Cr_on;
      vcout[1] =        Y1[1]&Y1_on # Cb[1]&Cb_on # Y2[1]&Y2_on # Cr[1]&Cr_on;
      vcout[2] =        Y1[2]&Y1_on # Cb[2]&Cb_on # Y2[2]&Y2_on # Cr[2]&Cr_on;
      vcout[3] =        Y1[3]&Y1_on # Cb[3]&Cb_on # Y2[3]&Y2_on # Cr[3]&Cr_on ;
      vcout[4] =        Y1[4]&Y1_on # Cb[4]&Cb_on # Y2[4]&Y2_on # Cr[4]&Cr_on;
      vcout[5] =        Y1[5]&Y1_on # Cb[5]&Cb_on # Y2[5]&Y2_on # Cr[5]&Cr_on;
      vcout[6] =        Y1[6]&Y1_on # Cb[6]&Cb_on # Y2[6]&Y2_on # Cr[6]&Cr_on;
      vcout[7] =        Y1[7]&Y1_on # Cb[7]&Cb_on # Y2[7]&Y2_on # Cr[7]&Cr_on;

%
 Y1_on = 1 => vcout[] = Y1[]
 Cb_on = 1 => vcout[] = Cb[]
 Y2_on = 1 => vcout[] = Y2[]
 Cr_on = 1 => vcout[] = Cr[]

%
END;
```

15.5 Project 5—Video Decoder-Digitizer

Video digitizers have evolved dramatically over the last decade. Not only do they incorporate all the blocks required to convert analog composite video into parallel digital formats, but they do it inexpensively enough to be used in consumer webcams and security quad boxes.

For this project we developed a stand-alone video decoder-digitizer board which uses our familiar Bt835 IC from Conexant. Control of the board is achieved by means of an I²C port operational both at 3.3 V and 5 V levels (see the encoder board). The I²C address of the Bt835 can be set using the I2CCS line from the S2 DIP switch. For I2CCS high the address is 0×8A and for I2CCS low the address is 0×88. The other positions of the DIP switch can tri-state the outputs of the Bt835 (OE~) or bring the IC in a low power sleep state (PWRDN).

Although the board has two separate inputs (J1 for composite signals and J2 for s-video), only one can be used at a given time. By default the Bt835 expects a composite signal at MUX0. To process s-video the SVID bit must be set high (D6 of the CONTROL_0 register, subaddress 0×15). Another bit that needs to be set is LEN (D4 of register CONTROL_3, subaddress 0×18) which determines the width of the output data bus. If LEN is one (default) the output bus is 16 bits wide, while if LEN is set to zero the bus width is 8 bits. And you will not see an output at all unless the output enable bit NOUTEN is set to zero (D7 of register CONTROL_3, subaddress 0×18).

Following the inputs, the video signals are first passed through coupling capacitors (C3, C4) and then through protection networks (D1, R3, D2, R5) before making their way to the MUX0 and CIN pins of the Bt835 (Figure 15.16). The system clock can be driven either by an on-board crystal oscillator or by an external source. A sophisticated PLL block allows the Bt835 to synthesize all the clock frequencies for all its formats (including PAL) off this single input source. Furthermore, it will detect the format of the incoming signal and adjust its processing parameters accordingly. Connector JP1 provides external access to most of the relevant outputs of the Bt835, as well as to its input digital port.

Power to the board can be applied either through its primary connector JPWR1 or through the JPWR2 or JPWR3 headers (Figure 15.17). The 5 V required by the analog front end of the decoder is then obtained by further filtering the input supply, while the digital 3.3 V is generated by an LT1585 LDO regulator. The LED diode DL1 indicates if power is present.

The schematics of the decoder circuit may be be relatively simple, but the PCB layout is another matter. The most care must be taken in the separation of the analog and digital power regions as well as in protecting the input paths from digital noise, especially between the input video connectors and the input pins of the IC (Figure 15.18).

At this stage, a properly functioning decoder board will output a steady stream of digital video data accompanied by appropriate reference signals. To

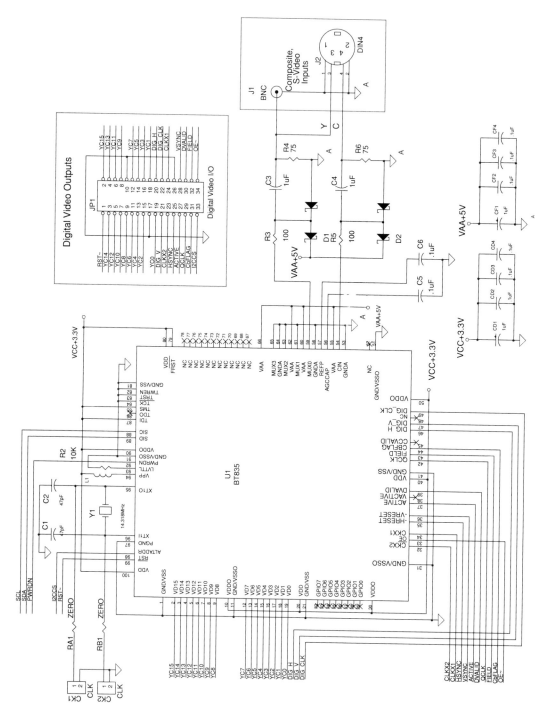

Figure 15.16 Decoder Bt835 Circuits.

410

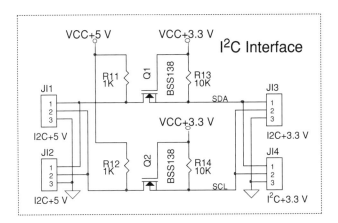

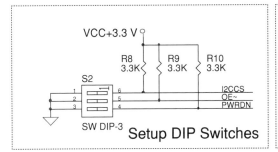

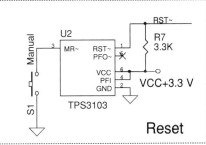

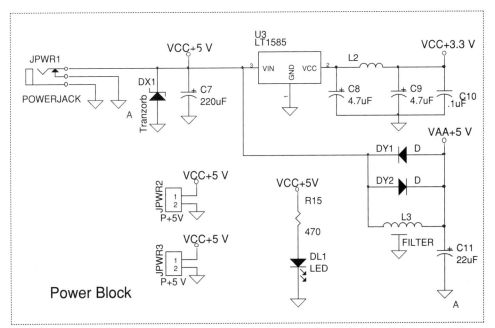

Figure 15.17 Decoder Ancillary Circuits.

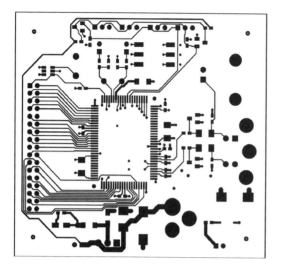 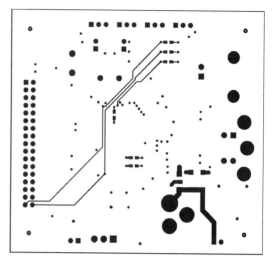

Figure 15.18(a)   Decoder Board PCB Top and Bottom Layers (1 and 4).

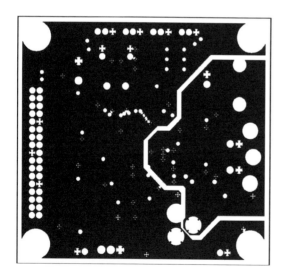 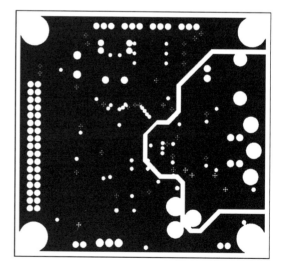

Figure 15.18(b)   Decoder Board PCB Split Ground and Power Layers (2 and 3).

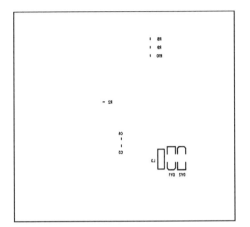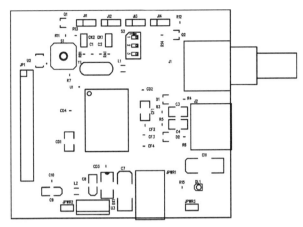

Figure 15.18(c)   Decoder Board PCB Bottom and Top Silkscreens.

observe this stream we have to hook up to the board's JP1 connector a logic analyzer with a reasonably wide bandwidth. Such logic analyzers are available from Agilent Technologies (HP), Tektronix, and many others. However, most command prices starting at about $2,000. A less expensive alternative which lacks some of the features of the big names but is affordable and surprisingly flexible is the $330 500 MHz, 16-channel, ANT-16 USB logic analyzer (Figure 15.19) from USB Instruments (www.usb-instruments.com).

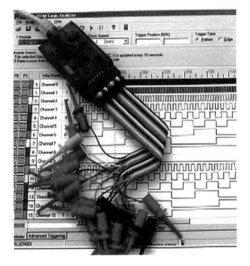

Figure 15.19   NT-16 Logic Analyzer.

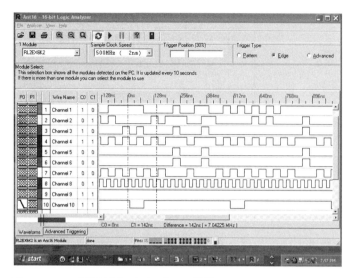

Figure 15.20   ANT-16 Logic Analyzer GUI.

In Figure 20 the ANT-16 was used to capture the 8 data outputs (Channel 1 through 8), the CLOCKX2 (Channel 9), FIELD (Channel 10), and VALID (Channel 11) signals generated by the Bt835 (8-bit mode). The general wiring of the decoder board is shown in Figure 15.21. The PCB files for the decoder board are available at www.digital-video-electronics.com

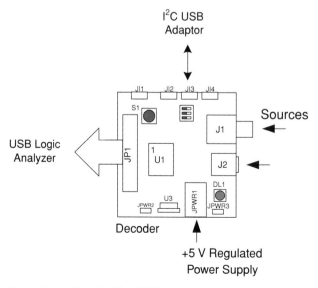

Figure 15.21   Decoder Board Wiring.

Bill of Materials

Reference	Value	PCB	Manufacturer	Part# / Details
C1, 2	47 pF	0603	Generic	Ceramic
C5, 6, 10, CD2, 3, 4, CF2, 3, 4	.1 µF	0603	Generic	Ceramic
C11	22 µF	7343	Generic	Tantalum
C3, 4, CD1, CF1	1 µF	1206	Generic	Ceramic
C7	220 µF	7343	Generic	Electro
C8, 9	4.7 µF	3216	Generic	Tantalum
CK1, 2	Header		Generic	2/2mm
D1			Diodes Inc.	BAT54ST
DL1	LED		Lumex	SSL-LXA228GC-TR11
DX1			STMicro	SMBJ5.0A-TR
DY1, 2			Diodes Inc.	S1A
J1			AMP	226990-1
J2			CUI	MD-40SM
JI1, 2, 3, 4	Header		Generic	3/2mm
JP1	Header		Generic	2X17/2mm
JPWR1			CUI	PJ-202A
JPWR2, 3	Header		Generic	2/2mm
L1, 2			Allied Corp.	MLB20-700
L3			Panasonic	EXCCET103U
Q1,2			Diodes Inc.	BSS138
R10	3.3K	0603	Generic	
R11, 12	1K	0603	Generic	
R2, 13, 14	10K	0603	Generic	
R15	470	0603	Generic	
R3, 5	100	0603	Generic	
R4, 6	75	0603	Generic	
R7, 8, 9	3.3K	0603	Generic	
RA1, RB1	zero	0603	Generic	
S1			Panasonic	EVQ11U04M
S2	SW DIP-3		C&K	SD03H0SKD
U1			Conexant	Bt835KRF
U2			TI	TPS3103H20DVB
U3			Linear Tech	LT1585CM-3.3
Y1	14.318 MHz		CTS	MP143

15.6 Video Buffer Board with Memory FIFOs

As noted earlier, the different clock speeds and timing requirements of the encoder and decoder boards precludes us from getting an output image simply by connecting the boards together. The output of the decoder has to be stored in a video buffer, synchronous with the decoder timing. The saved video information is then retrieved from the buffer in accordance with the timing information provided by the encoder. The simplest video buffers are those using FIFO memory ICs such as the MS81V10160 from OKI Semiconductor.

The capacity of each IC is 664,320 × 16 bits, more than sufficient to store two complete NTSC or PAL frames. However, since we cannot independently access the two frames (they are stored consecutively in the FIFO), any two-bank buffer architecture will need two separate memory chips. The board is designed to be used with the CPLD board from which it draws power through the JP1 connector.

As with most digital boards the only layout-related task is minimizing the radiated noise that may be picked up by other analog or mixed boards in the system. The circuit diagram, the CPLD-FIFO board assembly and the PCB layout are shown in Figures 15.22, 15.23, and 15.24, respectively.

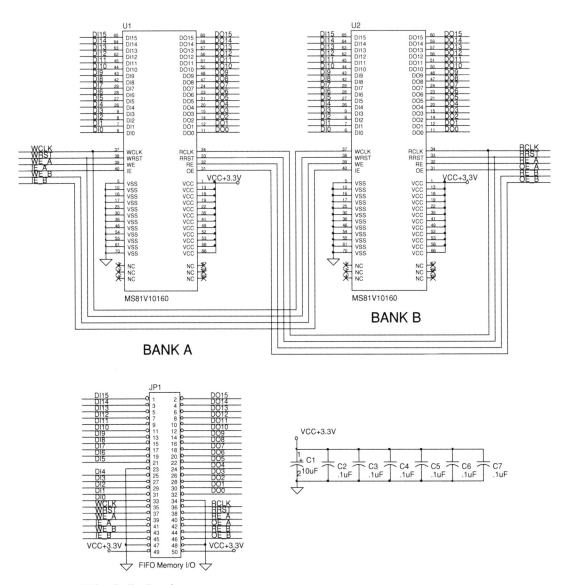

Figure 15.22   Video Buffer Board.

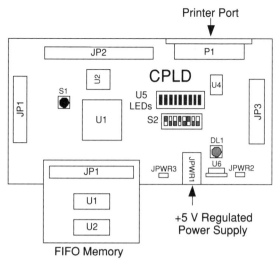

Figure 15.23 CPLD with FIFO Board Assembly.

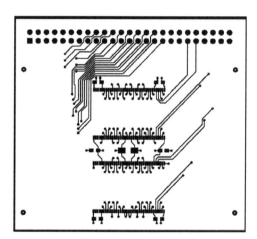
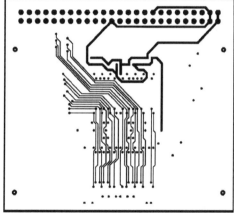

Figure 15.24(a) FIFO Board PCB Layers 1 and 2.

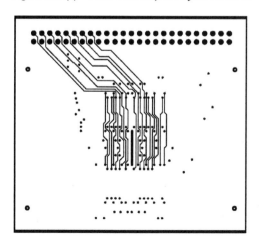
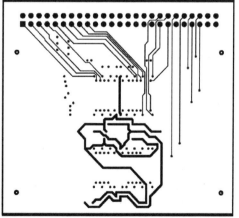

Figure 15.24(b) FIFO Board PCB Layers 3 and 4.

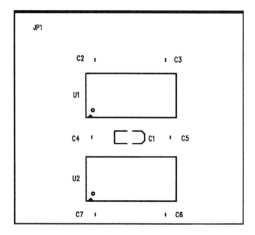

Figure 15.24(c)   FIFO Board PCB Top SIlkscreen.

Bill of Materials

Reference	Value	PCB	Manufacturer	Part# / Details
C1,	10 µF	3216	Generic	Tantalum
C2, 3, 4, 5, 6, 7	.1 µF	0603	Generic	Ceramic
JP1	Header		Generic	25X2/2mm
U1, 2			OKI	MS81V10160-12TA

15.7 Project 6—Digital Video Processor

Once the FIFO buffer is connected to the CPLD board we have all the elements necessary to complete the design of a simple digital video processor (Figure 15.25). Its function is to digitize an incoming video signal, store it in a local buffer, then retrieve it and display it in a format compatible with that of the input source. Processing options include freeze frame, control of brightness, contrast, saturation, and hue, color inversion, and others.

The memory architecture of choice is dual-bank, since it eliminates both field inversions and image tears. To implement it we will use each memory IC on the FIFO board as a stand-alone frame size memory bank. In this arrangement consecutive frames will always be read from different banks.

The video output is generated by the encoder board which is supplied with digital video data from the read side of the FIFO buffer banks, under the control of the CPLD. The same CPLD mediates the transfer of input video data from the decoder board into the write side of the FIFOs. The read-write conflict avoidance function is performed by the memory_write block using the bank_ab~ signal provided by the bank_ab~ block (Figure 15.26). The bank_ab~

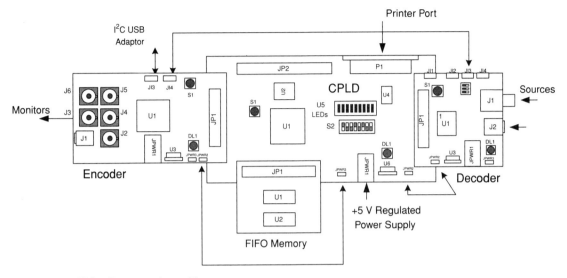

Figure 15.25 Video Processor Assembly.

signal toggles with each new frame passed on to the encoder board. When BANK A is read bank_ab~ is high, when BANK B is read bank_ab~ is low.

During the vertical interval preceding a new input frame the memory_write block locates the current position of the read pointer. If the encoder reads the odd field of BANK A or the even field of BANK B then the block will write the next frame in BANK B. Otherwise the next frame will be written in BANK A. The write_enable lines of the two FIFO memory chips (we_a, we_b) are then set accordingly. The writing itself consists in generating a write clock pulse (write_clock) when valid video data is present on the output bus of the decoder. The Boolean equation that qualifies write_clock is:

$$\text{write\_clock} = \text{vert\_res}{\sim}\ \&\ !\text{clock}\ \&\ \text{valid}\ \&\ \text{active}\ \#\ \text{write\_res\_clk};$$

As expected, write clock is generated outside vertical intervals during the active periods of horizontal lines, when the VALID Bt835 line is high. An additional write_res_clk pulse is added to reset the FIFO write pointer when the write_reset output of the memory_reset block is high.

Another memory_write function is freeze. If at the top of a new input frame the freeze input is high, that frame is skipped, thereby freezing the previous image in the memory buffer. Such a function is important if we want to transfer an image from the buffer to a PC using a slow communication channel such as a serial RS-232 port.

Reading the video data out of the FIFOs is quite straightforward. The clock signal that drives the encoder (clockx2) also drives the read clock for the FIFOs (rclk). The read enable (full_screen) and the read reset (vertical_reset)

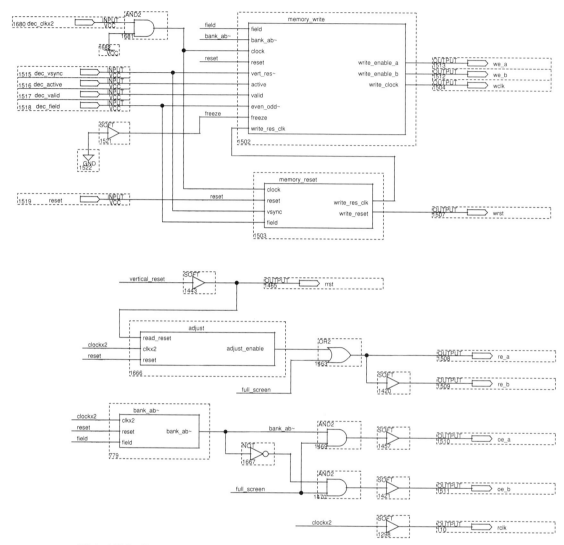

Figure 15.26 Digital Video Processor Firmware Diagram.

signals are produced by a full_screen_raster generator (Figures 15.27 and 15.28) similar to one developed for Project 4. The differences between the framed box raster generator of Project 4 and the full_screen_raster block consist in the elimination of the box_colors and the YCbCr_sequence_generator blocks, and the replacement of the window block with a new full_screen_ window block (Figure 15.29). The function of full_screen_window is to define the 720 × 480 "full screen" expands of the video output when the full_screen output signal is set high.

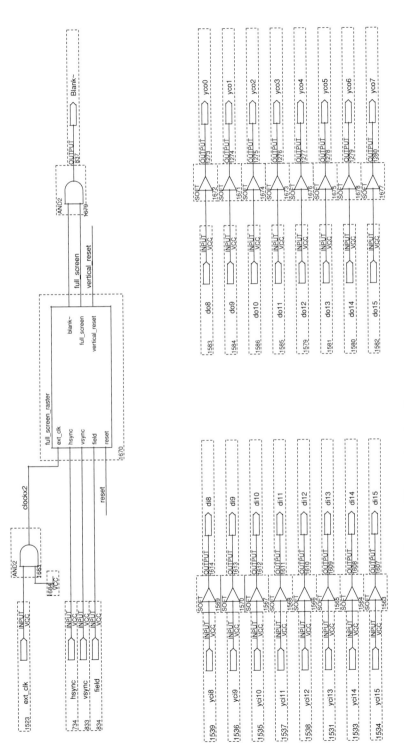

Figure 15.27 Digital Video Processor Firmware Diagram *(continued)*.

421

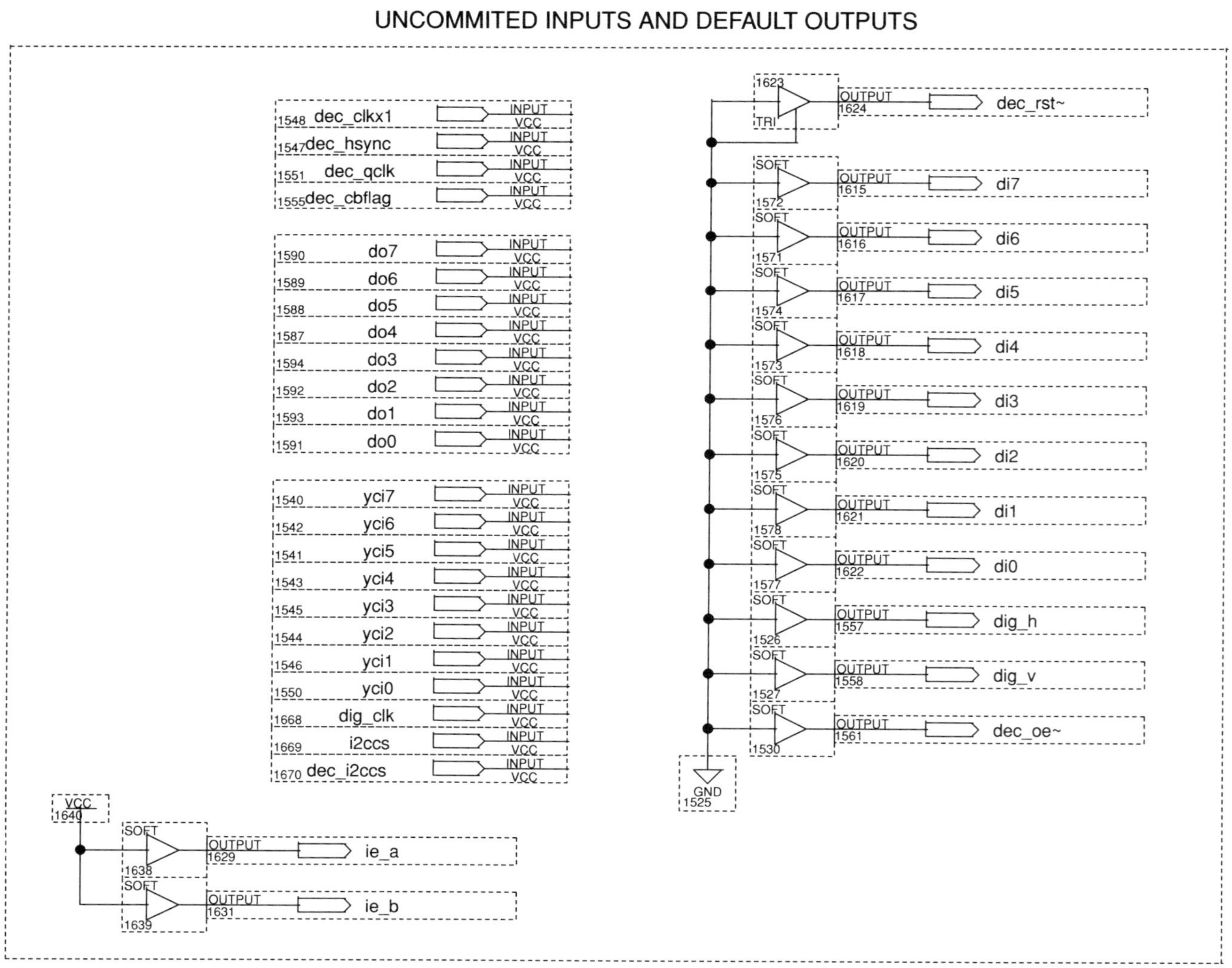

UNCOMMITED INPUTS AND DEFAULT OUTPUTS

Figure 15.28 Digital Video Processor Firmware Diagram (*continued*).

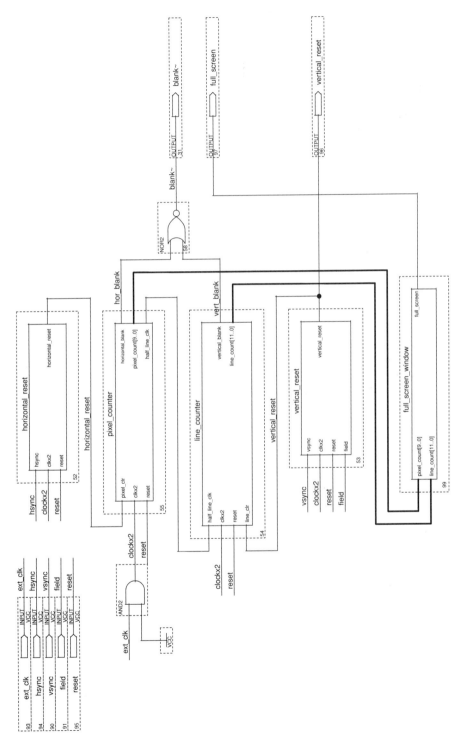

Figure 15.29 Full Screen Raster Firmware Diagram.

Figure 15.30 Live Output of the Digital Video Processor using Dazzle 80DVC.

The adjust block has been added to allow the user to experiment with the YCbCr structure of the digital video output. By delaying the output from the FIFOs by one clock pulse the sequence YCbYCr is changed to CbYCrY, thereby swapping the luma and chroma values. The output image then consists of various shades of green and purple, colors with equal Cr and Cb numbers. Delaying an extra clock pulse leads to color inversion where Cb and Cr are swapped. All the red color components are then replaced by blue and vice versa, resulting in a ghoulish, mainly blue live image.

There are two ways to control the video attributes of the output image. The first and significantly easier way is to use the internal functions of the decoder. The related user accessible registers of the Bt835 are BRIGHT (0×10), CONTRAST (0×11), SAT_U (0×12), SAT_V (0×13), and HUE (0×14). The U and V in SAT_U and SAT_V refer to the color components defined in section 8.3.2.

By changing these parameters you can modify the brightness, contrast, saturation, and hue of the output image. This can also be accomplished—the hard way—by using the CPLD resources to implement the equations in section 8.3.3.

Since you now have access to the digital form of the input image you can realize any number of other video effects like image strobing, conversion to black and white, partial freeze frames, and so on. In Figure 15.30 we have a captured image of the live output of the digital video processor.

```
full_screen_window
```

```
CONSTANT box_left_side        =    153;
CONSTANT box_width            =    720;
CONSTANT box_right_side       =    box_left_side + box_width;
CONSTANT box_top_side         =    38;
CONSTANT box_height           =    480;
CONSTANT box_bottom_side      =    box_top_side + box_height;
```

```
CONSTANT border_width              =    2;

CONSTANT inter_field_offset        =    524;

SUBDESIGN full_screen_window
(
    pixel_count[9..0], line_count[11..0]    :    INPUT;
    full_screen                             :    OUTPUT;
)
BEGIN
    full_screen = SOFT(((((pixel_count[]> box_left_side)&(pixel_count[] <= box_right_side))&
                ((line_count[] > box_top_side)&(line_count[] <= box_bottom_side)))#
                (((pixel_count[] > box_left_side)&(pixel_count[] <= box_right_side))&
                ((line_count[] > box_top_side + inter_field_offset)&
                (line_count[] <= box_bottom_side + inter_field_offset)))));

% The output is high within the active window boundaries described by box_left_side,
box_right_side, box_bottom_side, and box_top_side; zero outside%

END;
```

memory_write

```
SUBDESIGN memory_write
(
    field, bank_ab~                         :    INPUT;
    clock, reset, vert_res~, active, valid  :    INPUT;
    even_odd~, freeze, write_res_clk        :    INPUT;
    write_enable_a, write_enable_b          :    OUTPUT;
    write_clock                             :    OUTPUT;
)
VARIABLE
    freeze_flag                             :    DFF;
    new_frame_ab~                           :    DFF;
BEGIN
    freeze_flag.clk            =    clock;
    freeze_flag.clrn           =    !reset;
    new_frame_ab~.clk          =    clock;
    new_frame_ab~.clrn         =    !reset;
    write_clock                =    vert_res~ & !clock & valid & active # write_res_clk;
%
    field = 1  => The output field is ODD;
    field = 0  => The output field is EVEN;
```

```
      bank_ab~ = 1=> The output frame is BANK A;
      bank_ab~= 0=> The output frame is BANK B;

At the time when (!vert_res~_qi & even_odd~_qi) = 1

    If field = 1 (odd) and bank_ab~ = 1 (bank A) then store
    new frame at B (new_frame_ab~ = 0);
    If field = 0 (even) and bank_ab~ = 1 (bank A) then store
    new frame at A (new_frame_ab~ = 1);
    If field = 1 (odd) and bank_ab~ = 0 (bank B) then store
    new frame at A (new_frame_ab~ = 1);
    If field = 0 (even) and bank_ab~ = 0 (bank B) then store
    new frame at B (new_frame_ab~ = 0);
%
    IF    (!vert_res~ & even_odd~) THEN
%
All bank selection decisions are performed during the (decoder) even field
    vertical intervals
%
        IF    freeze THEN
                freeze_flag.d   =   VCC;
        ELSE
                freeze_flag.d   =   GND;
        END IF;
%
If the freeze control signal is asserted the freeze_flag is set high.
%
        IF ((field & bank_ab~) # (!field & !bank_ab~)) THEN
            new_frame_ab~.d = gnd;
        ELSE
            new_frame_ab~.d = vcc;
        END IF;

    ELSE
        freeze_flag.d   =   freeze_flag.q;
        new_frame_ab~.d =   new_frame_ab~.q;
    END IF;

    IF   (freeze_flag.q # ((!vert_res~) & even_odd~)) THEN
        write_enable_a   =   gnd;
        write_enable_b   =   gnd;

%
If the freeze_flag was previously set high all new write operations are disabled.
```

```
%
    ELSIF  (!new_frame_ab~.q)  THEN
            write_enable_a    =    vcc;
            write_enable_b    =    gnd;
    ELSE
            write_enable_b    =    vcc;
            write_enable_a    =    gnd;
    END IF;
END;
```

memory_reset

```
SUBDESIGN memory_reset
(
    clock, reset                :       INPUT;
    vsync, field                :       INPUT;
    write_res_clk, write_reset  :       OUTPUT;
)
VARIABLE
    write_state    : MACHINE OF BITS (write_res_clk, write_reset)
    WITH STATES (
        s0       =    B"00",
        s1       =    B"01",
        s2       =    B"11",
        s3       =    B"01");
BEGIN
    write_state.clk              =    clock;
    write_state.reset            =    reset;

    TABLE
      write_state, vsync,    field  =>   write_state;
          s0,        1,        x    =>   s0;
          s0,        0,        0    =>   s0;
          s0,        0,        1    =>   s1;
          s1,        x,        x    =>   s2;
          s2,        x,        x    =>   s3;
          s3,        0,        x    =>   s3;
          s3,        1,        x    =>   s0;

    END TABLE;
END;
```

```
adjust

CONSTANT pixel_delay               =   0;

SUBDESIGN adjust
(
    read_reset, clkx2, reset      :   INPUT;
    adjust_enable                 :   OUTPUT;
)
VARIABLE

    delay[2..0]              :   NODE;

    adjust_state   : MACHINE OF BITS (adjust_enable)
    WITH STATES (
        s0        =   B"0",
        s1        =   B"1",
        s2        =   B"1",
        s3        =   B"1",
        s4        =   B"1",
        s5        =   B"1",
        s6        =   B"0");

BEGIN
    adjust_state.clk         =   clkx2;
    adjust_state.reset       =   reset;

    delay[]                  =   pixel_delay;

TABLE
adjust_state,    read_reset,        delay[]   =>   adjust_state;
    s0,              0,              x         =>   s0;
    s0,              1,              x         =>   s1;
    s1,              x,              0         =>   s6;
    s1,              x,              1         =>   s5;
    s1,              x,              2         =>   s4;
    s1,              x,              3         =>   s3;
    s1,              x,              4         =>   s2;
    s2,              x,              x         =>   s3;
    s3,              x,              x         =>   s4;
    s4,              x,              x         =>   s5;
    s5,              x,              x         =>   s6;
    s6,              1,              x         =>   s6;
    s6,              0,              x         =>   s0;
```

```
END TABLE;

END;



bank_ab~

SUBDESIGN bank_ab~
(
    clkx2, reset, field        :    INPUT;
    bank_ab~                    :    OUTPUT;
)
VARIABLE
    bank_flag, bank_ab~        :    DFF;
BEGIN
    bank_ab~.clk               =    clkx2;
    bank_ab~.clrn              =    !reset;
    bank_flag.clk              =    clkx2;
    bank_flag.clrn             =    !reset;

    IF (!field) THEN
        IF !bank_flag.q THEN
            bank_flag.d        =    vcc;
            bank_ab~.d         =    !bank_ab~.q;
        ELSE
            bank_flag.d        =    vcc;
            bank_ab~.d         =    bank_ab~.q;
        END IF;
    ELSE
        bank_flag.d        =    gnd;
        bank_ab~.d         =    bank_ab~.q;
    END IF;
END;
```

15.8 Project 7—Simple Image Scaler

One of the most common video effects is image down-scaling. And as we saw
in Chapter 12 the Bt835 IC already has a built-in scaler engine of reasonable
quality. All we need to do is program the Bt835 with the necessary scaling
parameters for the desired window size and to create a matching window
template on the screen.

Just as an example, we will scale the video input image to fit into the
framed window we generated in Project 4. The size of the window is 360×240

and corresponds to the CIF video-telephony standard. The HSCALE and VSCALE parameters that have to be loaded into the 16-bit registers HSCALE (0x0B, 0x0A) and VSCALE (0x0D, 0x0C) are calculated using the Bt835 specific formulas given by Conexant.

$$\text{HSCALE} = 4096 \, [(910/\, H_{desired}) - 1]$$

For full screen the total output resolution (including active, blanking, and sync) is 858×525, while the resolution of the active pixels alone is 720×480. If we want to scale down horizontally by two (360 pixels), the Bt835 requires us to scale the whole line by two. Then the target horizontal length is 50 percent of 858 or $H_{desired} = 429$. The resulting HSCALE can be rounded off to 4592 or 11F0, a value which is stored in the HSCALE registers.

In the vertical direction you start with a desired scaling ratio $S_{desired}$, then calculate and store in VSCALE the value of

$$\text{VSCALE} = [0x10000 - 512 \, (S_{desired} - 1)] \, \& \, 0x1FFF$$

Since we want to scale by a factor of 2, then $S_{desired} = 2$ and we find

$$\text{VSCALE} = [0x10000 - 0x0600] \, \& \, 0x1FFF$$

which finally yields VSCALE = 0x1A00.

The output image can be seen in Figure 15.31.

On the CPLD side the design uses many of the elements introduced earlier in Projects 4 and 6. The write-to-memory procedure is identical to that used by the video processor (Figure 15.32). The read mechanism is somewhat different, although it uses the box_generator block developed for Project 4.

Figure 15.31 Live Output of the Scaler using Dazzle 80DVC.

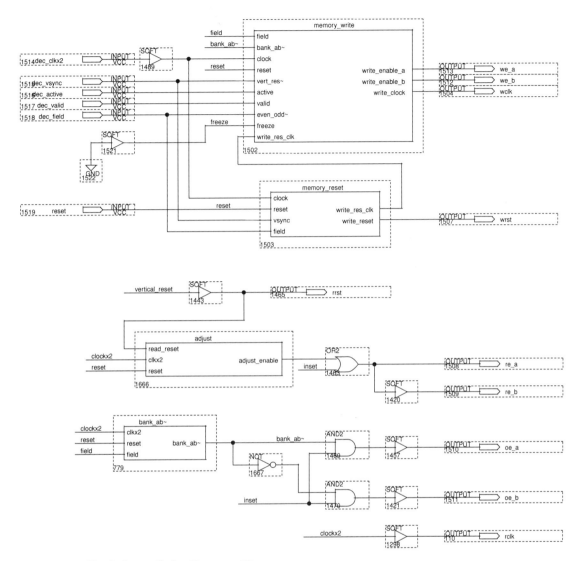

Figure 15.32   Simple Image Scaler Firmware Diagram.

The difference consists in the introduction of two tri-state 8-bit buffers to switch between the live video retrieved from the FIFOs (within the inset) and the border and background data produced by the box_generator. (Figures 15.33. 15.34 and 15.35)

This design can be modified to accommodate other scaling factors or incorporate image cropping. Two decoder boards, two CPLD boards, two FIFO boards, and one encoder can be used to develop PiPs, SbSs (side-by-side), or PoPs (Picture-on-Picture or stacker) prototypes.

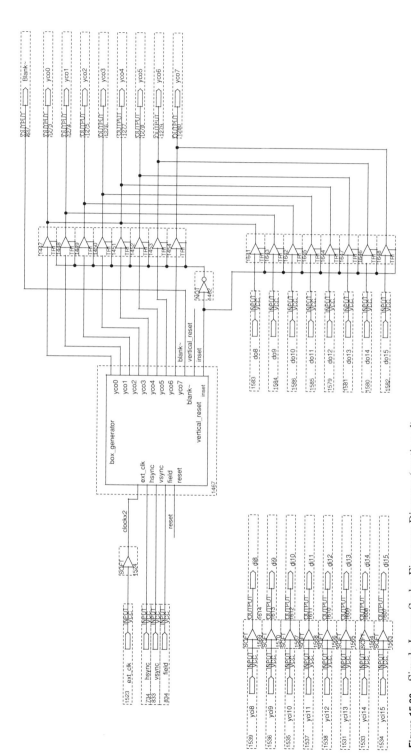

Figure 15.33   Simple Image Scaler Firmware Diagram (*continued*).

432

UNCOMMITED INPUTS AND DEFAULT OUTPUTS

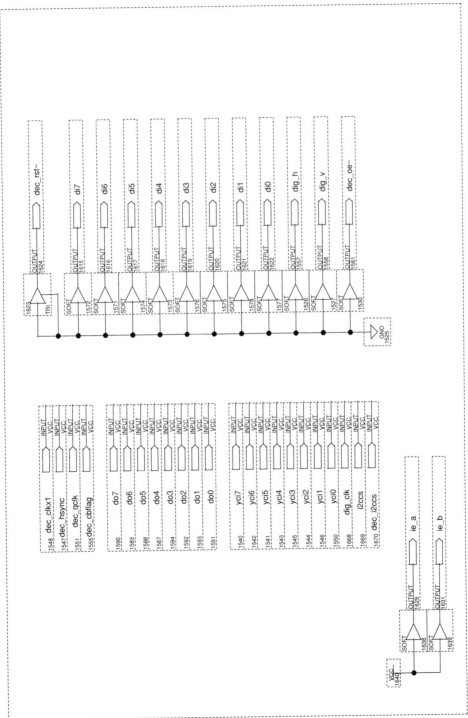

Figure 15.34 Simple Image Scaler Firmware Diagram (*continued*).

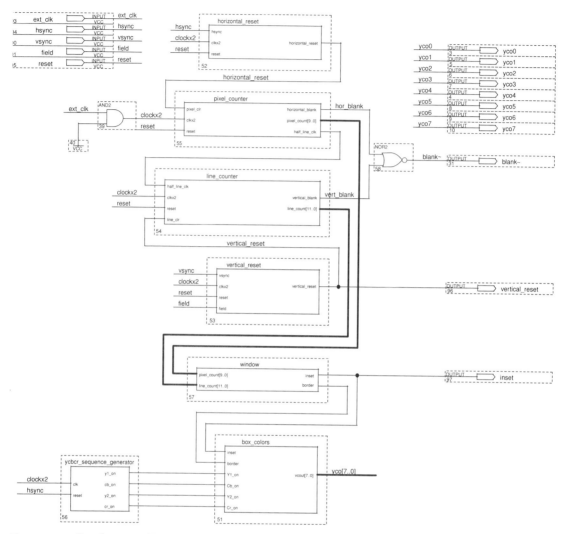

Figure 15.35 Box Generator Firmware Diagram.

15.9 Project 8—Controller Board

Up to this point all the parameters necessary for the proper operation of our systems were transferred to the decoder and encoder chips by means of a PC I²C adaptor. However, this path is not suitable for the transfer of images to and from the FIFO memory. What is needed is a processor that has access to the video buffers and has a high-speed link to the PC.

A very simple, general purpose processor can be built around an expanded mode 80C51 microcontroller. In this case the core processor consists of an 80C51 IC, 64 KB of program EPROM, and 32 KB of RAM (Figure 15.36).

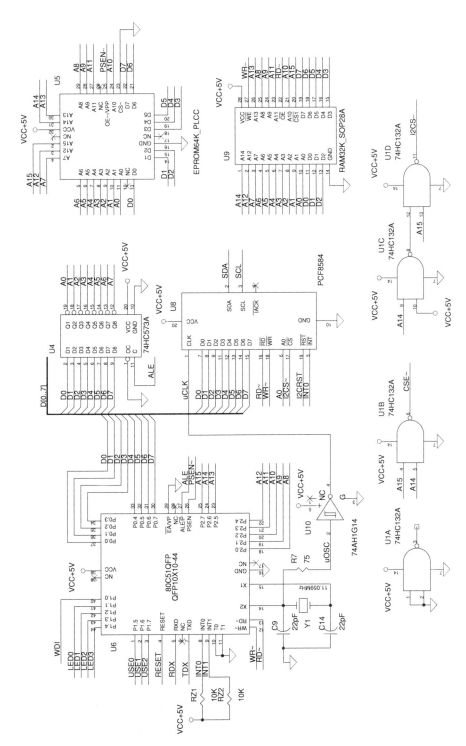

Figure 15.36 Controller Core Circuit.

An elementary user interface consisting of a seven-segment common cathode LED display and three momentary switches provides for local data entry and display services (Figure 15.37). Both are handled by I/O Port 1 of the micro. An alternate way to enter data is by means of an IR remote control. To support it we use an integrated IR receiver-demodulator from Sharp Semiconductors. Its open collector output is connected in parallel to the S1 switch and can be bit-banged through the P1.5 input port.

The 24-pin connector JP1 provides a complete parallel interface to the CPLD board. On the PC side communications are handled by an RS-232 port which, for a 12 MHz microcontroller, can operate at up to 115.2 Kbaud. A separate 5 V I^2C port is used to communicate with the I^2C peripherals in the system. The I^2C protocol is implemented by a separate Philips controller (U8), with minimal impact on microprocessor overhead. For system and user parameters the board is also equipped with an I^2C EEPROM (U11).

In normal operation, microprocessor firmware is programmed into the EPROM using a standard type programmer, which then is plugged into the board's memory socket. For development purposes, a more elegant solution consists in using an "in-system programmable" or ISP microcontroller with on-board FLASH memory. Such 80C51-type micros are available from companies like Atmel, and can be programmed directly through the built-in UART without the use of any additional hardware. Even the programming voltage is internally generated.

8051 C compilers are supplied by a number of vendors, Keil Software, Inc. being one of the leading companies in the field. A complete evaluation package of their C × 51 software tools can be downloaded from www.keil.com or www.digital-video-electronics.com. This software is great for getting a feel for the product, although it is limited to applications 2K or smaller. C FLASH drivers for the Keil compiler, supporting the AT89C51RD2 ISP processor, are available from Atmel.

The controller board is designed to work with the other project boards (Figure 15.38), together forming a complete development platform for standard-resolution digital systems. For proper operation, connector JP1 on the controller board is plugged into pins 1 to 24 of JP2 on the CPLD board, the encoder is plugged into the CPLD JP1, the decoder in JP3, and the FIFO in JP4.

Although we chose the 80C51 for our controller, the CPLD interface is general enough to allow for the use of almost any other microprocessor development board, as long as the board allows the user access to the data, address, and control buses. The mapping of the CPLD JP2 connector to the hardware connections on the micro development board can be done by a user-assembled "scrambling" cable which should be short and shielded with conductive (and grounded) heat shrink tubing. Even microprocessor modules, such as the powerful Ethernet/TCP-IP Rabbit from Rabbit Semiconductor or the easy-to-use Basic Stamp from Parallax, can be used as controllers.

The PCB files for this board are also available at www.digital-video-electronics.com

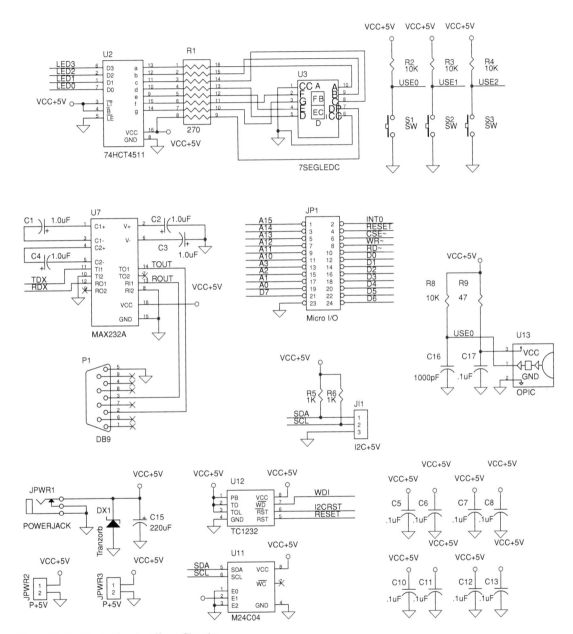

Figure 15.37   Controller Ancillary Circuits.

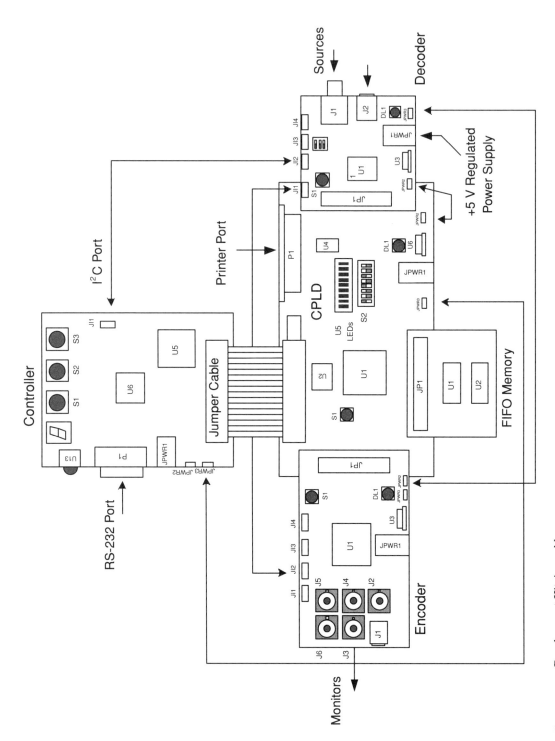

Figure 15.38 Development Kit Assembly.

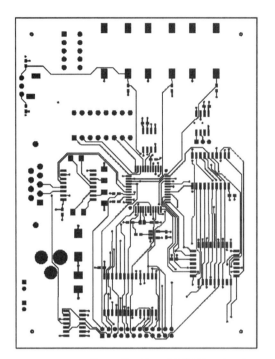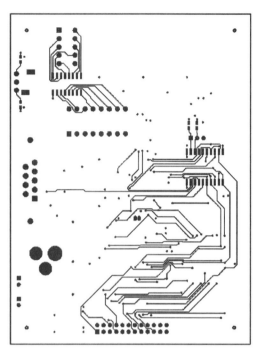

Figure 15.39(a)   Controller Board PCB Routing Layers (1 and 4).

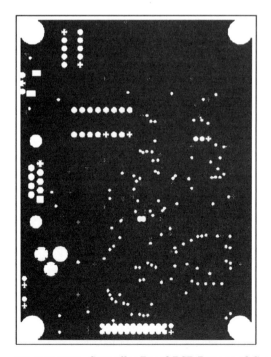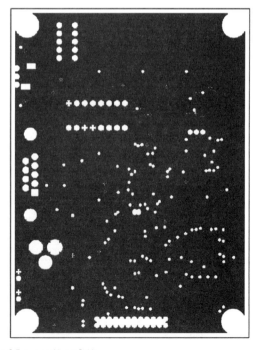

Figure 15.39(b)   Controller Board PCB Power and Ground Layers (2 and 3).

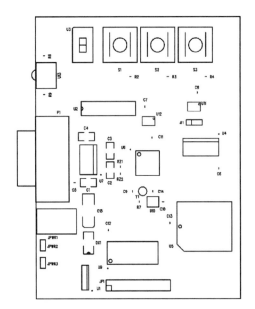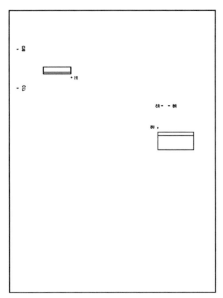

Figure 15.39(c) Controller Board PCB Silkscreen Layers.

Bill of Materials

Reference	Value	PCB	Manufacturer	Part# / Details
C1, 2, 3, 4	1 µF	1206	Generic	Ceramic
C5, 6, 7, 8, 10, 11, 12, 13, 17	.1 µF	0603	Generic	Ceramic
C9, 14	22 pF	0603	Generic	Ceramic
C15	220 µF	7343	Generic	Electro
C16	1000 pF	0603	Generic	Ceramic
DX1			STMicro	SMBJ5.0A-TR
JI1	Header		Generic	3/2mm
JP1	Header		Generic	15X2/2mm
JPWR1			CUI	PJ-202A
JPWR2, 3	Header		Generic	2/2mm
P1			Norcomp	177-009-210-071
R1	270	0603	Generic	Ceramic
R2, 3, 4, 8, RZ1, RZ2	10K	0603	Generic	Ceramic
R5, 6	1K	0603	Generic	Ceramic
R7	75	0603	Generic	Ceramic
R9	47	0603	Generic	Ceramic
S1, 2, 3			Shurter	1241.1612.11
U1			TI	CD74HC132M
U10			TI	SN74AHC1G14DBV
U11			ST	M24C04MN6
U12			TI	TC1232COA
U13			Sharp	GP1UD28YK
U2			TI	CD74HC4511E
U3			LiteOn	LSHD-A103
U4			TI	SN74HC573DW
U5			Assman	A-CCS32-Z-SM
U6			Atmel	AT89C51ED2-RLTIM
U7			Maxim	MAX232CWE
U8			Philips	PCF8584TD
U9			NEC	UPD43256BGU-70
Y1			Epson	CA-301 11.059M-C

15.10 Project 9—Image Capture Appliance

In this section we will use all the boards developed so far to produce a sample RS-232 image capture appliance. The CPLD firmware for this project starts where Project 6 left off: a complete decoder capture, FIFO store, and encoder retrieval design (Figures 15.26 through 15.29). What we need to add are facilities for the controller to freeze the image during the interframe vertical interval (freeze input to memory_write), bit-bang the control pins of the FIFO ICs (rclk, rrst, re_a, re_b, oe_a, oe_b), and read the outputs of the FIFOs (do15..do8). These functions are performed by a 74374 latch and a 74244 buffer block, both driven by a simple address decoder (Figure 15.40).

Regardless of the controller we use, besides initializing the Bt864 and Bt835, the software has to set the freeze bit (freeze), enable one of the FIFO banks (for BANK A set re_a and oe_a high, re_b and oe_b low), reset the FIFO address pointers (pulse rrst and rclk), increment the pointer (pulse rclk), read the output (load data and store in RAM), stop if the last FIFO image byte is detected, and clear the freeze bit.

To service the PC link a minimal controller firmware package has to initialize the local UART, decode the host "read" instruction, post a "busy condition" during image capture, clear "busy condition" when capture is complete, and transmit the image block when a host "upload" command is decoded.

decoder

```
SUBDESIGN decoder
(
    cs~, rd~, wr~, a3, a2, a1, a0    : INPUT;
    a15, a14, a13, a12, a11          : INPUT;
    write1, write2, read             : OUTPUT;
)
BEGIN
    read = !(a15 & a14 & !a13 & !a12 & !a11 & !a3 & !a2 & !a1 & !a0 & !cs~ & !rd~);
         %1100 0xxx xxxx 0000%
    write1 = (a15 & a14 & !a13 & !a12 & !a11 & !a3 & !a2 & !a1 & !a0 & !cs~ & !wr~);
           %1100 0xxx xxxx 0000%
    write2 = (a15 & a14 & !a13 & !a12 & !a11 & !a3 & !a2 & !a1 & a0 & !cs~ & !wr~);
           %1100 0xxx xxxx 0001%
END;
```

The upload function includes packetizing the data according to the file transmission protocol we choose (128 bytes plus checksum for XMODEM, for example). Many commercial and freeware C libraries have standard routines that handle such tasks. The easiest way for a Windows PC host to receive the image file is by using the Hyperterminal utility (from Start, Run type hyper-trm). After you create a connection to your COM port (File Menu) and enter

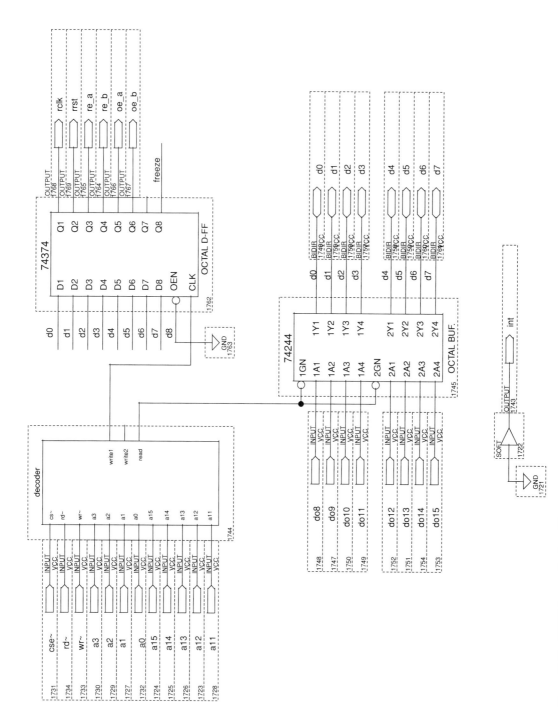

Figure 15.40 CPLD Image Capture Firmware.

the same settings as those used by the serial port on the controller board, communication with the board is similar to the familiar chat room format. If you type the read command that the controller is programmed to recognize, followed by the upload command the image file is read into the PC. Then you go to the Receive File entry on the Transfer Menu. On the Receive File pop-up select the place to store the file and the transfer protocol which, again, has to match the one used by the controller.

In order to see the image we have to convert it to RGB using the equations in section 2.3.2, further convert it to 4:4:4, and then append a standard bitmap header. Detailed information on the structure of bitmap and other file formats can be found in the book *Encyclopedia of Graphics File Formats*, 2nd edition, by James D. Murray and William van Ryper (O'Reilly & Associates). Once in a proper .BMP format the file can be read using Windows Paint or any other graphics package.

15.11 Project 10—Graphics Display Appliance

A different way of using the five-board kit, complementary to image capturing, is to load the FIFO buffers with a PC-generated or a PC-stored image. From the FIFOs the image is then read by the encoder board and displayed on the output monitor.

Unless the image has been captured using a "raw" YCbCr appliance such as the one described in the previous section, the image is likely to be structured in one of the common commercial formats, such as .gif or .jpeg. Before downloading it to the controller board the image has to be converted to a bitmap format using software such as Windows Picture Viewer, Photoshop, and so on. Next the header of the bitmap file has to be read, interpreted, and then stripped from the file. At this stage the graphic file consists exclusively of RGB data. This has to be converted to YCbCr using the equations in section 2.3.2, truncated to 8-bits, and finally sequenced in accordance with the 4:2:2 format used by the Bt864. Now the file is compatible with the format used by our boards. To download it to the controller board we select the Hyperterminal Send File entry of the Transfer Menu, and then specify the file and the appropriate file transfer protocol.

The controller firmware has to decode the "download" command, receive the file transmitted by the host and store it in the local RAM. Then it posts a "busy" flag, (returns a "busy" status byte for any host command) and proceeds to bit-bang the input side of the FIFOs—that is, it resets the FIFOs, selects the active bank, retrieves consecutive bytes from the RAM, places them on the data 74374 latch, pulses wclk, and repeats the cycle until the end of file is reached (Figure 15.41).

We must also make sure that the size of the graphics image we download is precisely 720×480, otherwise the output image will be scrambled.

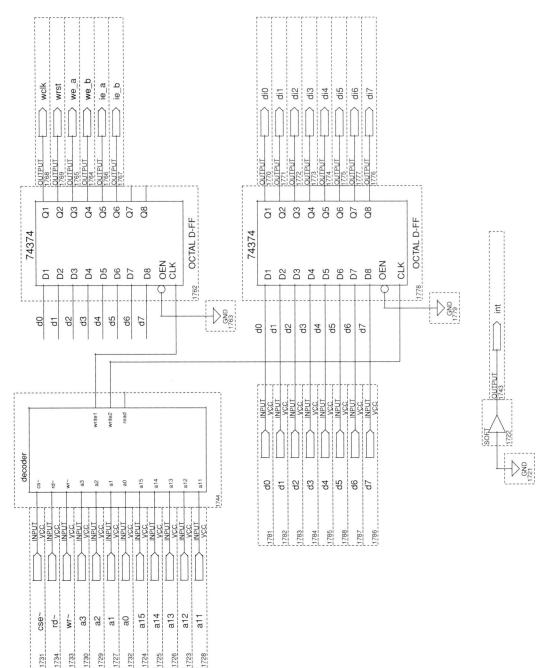

Figure 15.41 CPLD Graphics Download Firmware.

15.12 Project 11—Scan Converter

In many applications such as video conferencing, distance learning, and telemedicine the use of computer-generated graphics requires the images to be converted to composite video first. Since there are few high-resolution CODECs available at this time (outside HDTV), if you want to include an Exel spreadsheet or PowerPoint presentation in your teleconference you must convert it first to composite. Then you can use a standard MPEG CODEC for the telecom link.

The conversion to composite or s-video is done by scan converters. Although the reduced resolution of the composite signal degrades to some extent the quality of the image, the much smaller bandwidth required for its transmission makes its use indispensable in video telecommunications.

Figures 15.42 through 15.46 depicts a scan converter circuit using the FS403 IC from Focus Enhancements, the market leader in consumer scan converters. The FS403 is a highly integrated IC that accepts and digitizes computer-generated images from 640×480 to 1600×1200. Although the IC can operate in stand-alone mode, a number of options can be programmed by the local or host controller using its I^2C communications port. The scale-down operation is supported by an external SDRAM fully controlled by the FS403. The input signal resolution and rate are automatically detected, so no local firmware is needed.

The PCB layout is shown in Figure 15.47 and can be downloaded from www.digital-video-electronics.com.

Figure 15.49 shows the assembled scan converter board.

15.13 Project 12—Wire and Infrared (IR) Remote Control

Although not a video product in itself, the IR remote controller is almost universally associated with video products. This particular design also adds an RS-232 port, an I^2C port, and six general purpose I/O lines on a board small enough to fit inside a standard remote control enclosure. The serial port connects to the outside through a three-pin connector, while the I^2C and the GPIO can be accessed through a 14-pin header (Figure 15.50). The infrared LEDs are modulated on a 40 KHz carrier frequency generated by a standard design NAND crystal oscillator. Its output is keyed by the IR data signal of the micro. In order to minimize battery drain, when no key is pressed the main voltage regulator U9 is shut down. When an input key is pressed, U11 enables U9 which in turn powers up the microprocessor system.

Figure 15.51 shows the four layers of the IR remote PC board (also found at www.digital-video-electronics.com). The assembled unit is shown in Figure 15.52.

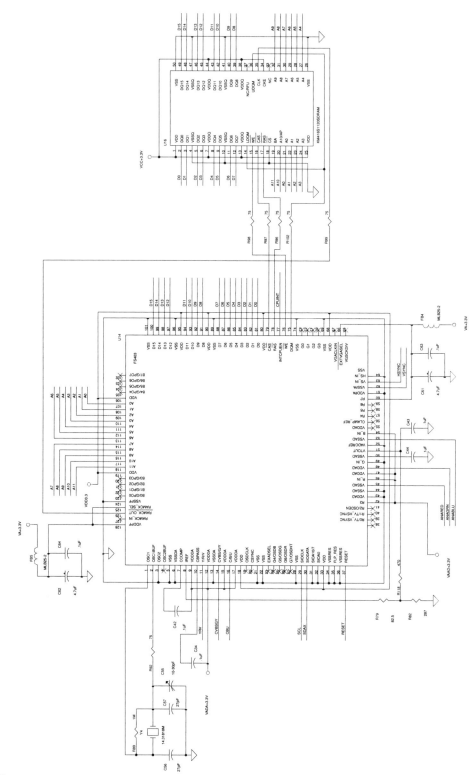

Figure 15.42 Scan Converter Circuit.

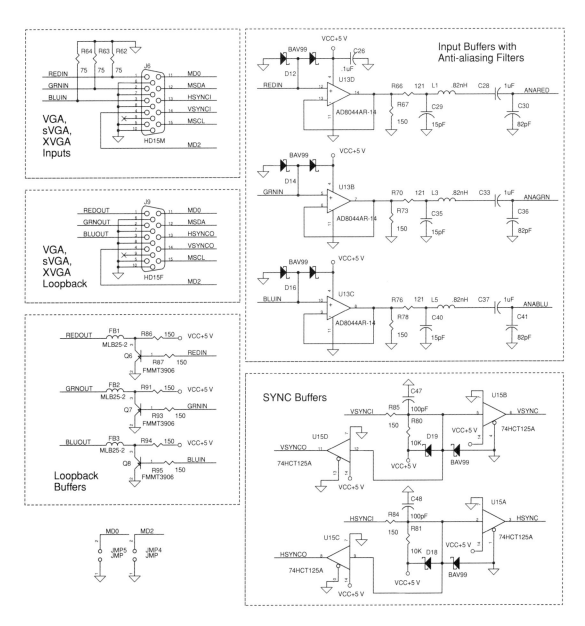

Figure 15.43   Scan Converter Ancillary Circuits 1.

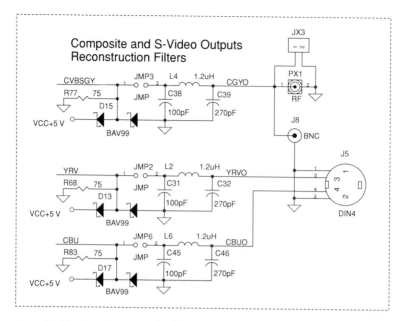

Figure 15.44   Scan Converter Ancillary Circuits 2.

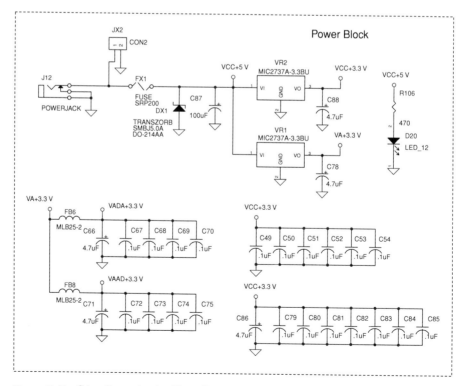

Figure 15.45   Scan Converter Ancillary Circuits 3.

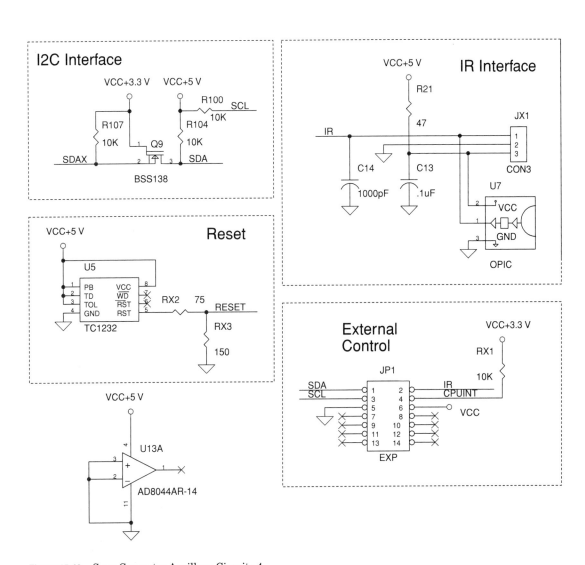

Figure 15.46   Scan Converter Ancillary Circuits 4.

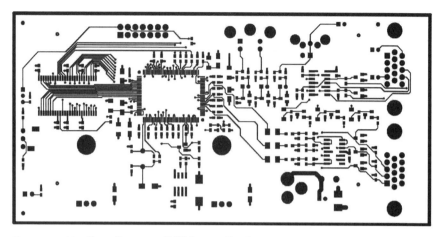

Figure 15.47(a)   Scan Converter PCB Layout Top Layer.

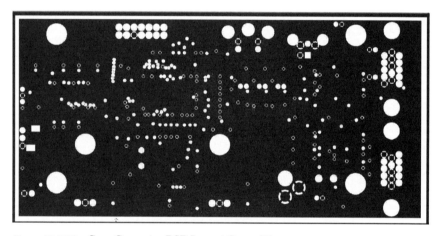

Figure 15.47(b)   Scan Converter PCB Layout Ground Layer.

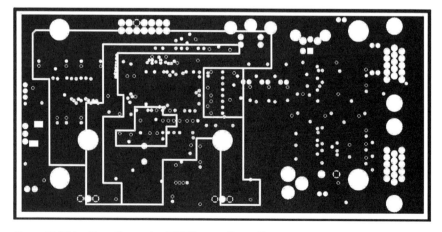

Figure 15.47(c)   Scan Converter PCB Layout Power Layer.

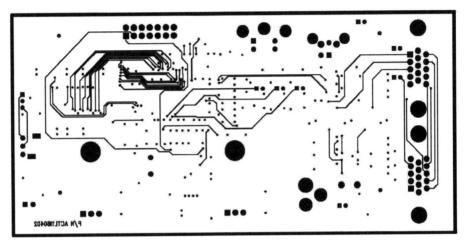

Figure 15.47(d)   Scan Converter PCB Layout Bottom Layer.

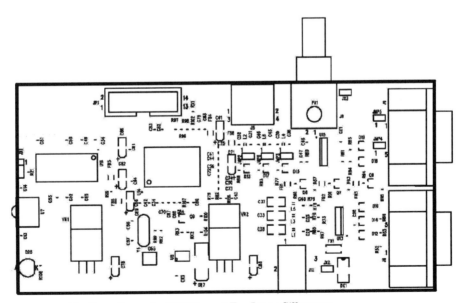

Figure 15.47(e)   Scan Converter PCB Layout Top Layer Silkscreen

Figure 15.48   Scan Converter Output captured with Dazzle 80DVC.

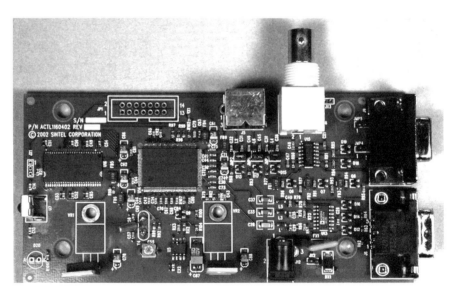

Figure 15.49   Scan Converter Board.

Bill of Materials

Reference	Value	PCB	Manufacturer	Part# / Details
C13, 26, 34, 42, 43, 44, 49, 50, 51, 52, 53, 54, 63, 64, 67, 68, 69, 70, 72, 73, 74, 75, 79, 80, 81, 82, 83, 84, 85	.1 µF	0603	Generic	Ceramic
C14	1000 pF	0603	Generic	Ceramic
C28, 33, 37	1 µF	1206	Generic	Ceramic
C29, 35, 40	15 pF	0603	Generic	Ceramic
C30, 36, 41	82 pF	0603	Generic	Ceramic
C31, 38, 45, 47, 48	100 pF	0603	Generic	Ceramic
C32, 39, 46	270 pF	0603	Generic	Ceramic
C55	10-30 pF		Sprague	GKG30066-07
C56, 57	27 pF	0603	Generic	Ceramic
C61, 62, 66, 71, 78, 86, C88	4.7 µF	3216	Generic	Tantalum
C87	100 µF	7343	Generic	Electro
D12, 13, 14, 15, 16, 17, 18, 19, 20			Generic	BAV99
LED			Lumex	SSL-LX3044DH-TR
FB1, 2, 3, 4, 5, 6, 8		3216	Allied Comp.	MLB20-700
JMP 2, 3, 4, 5, 6	Header		Generic	2/2mm
JP1	Header		Generic	7X2/.1"
JX1	Header		Generic	3/2mm
J5			CUI	MD-40SM
J6	HD15M		AMP	749767-1
J8			AMP	226990-1
J9	HD15F		AMP	787066-1
J12			CUI	PJ-202A
L1, 3, 5	.82 µH	2012	Panasonic	ELJ-FDR82KF PAN
L2, 4, 6	1.2 µH	2012	Panasonic	ELJ-FD1R2KF PAN
Q6, 7, 8			Zetex	FMMT3906ATA
Q9			Diodes Inc.	BSS138
R80, 81, 100, 104, 107, RX1	10K	0603	Generic	
R62, 63, 64, 68, 77, 83, 92, 96, 97, 98, 99,102, RX2	75	0603	Generic	
R67, 73, 78, 84, 85, 86, 87, 91, 93, 94, 95 RX3	150	0603	Generic	
R21	47	0603	Generic	
R66, 70, 76	121,1%	0603	Generic	
R79	82.5 1%	0603	Generic	
R82	287 1%	0603	Generic	
R89	1M	0603	Generic	
R118, 106	470	0603	Generic	
U5			T1	TC1232COA
U7			Sharp	GP1UD28YK
U13			Analog Devices	AD8044AR-14
U14			Focus Enhance	FS403
U15			TI	74HCT125DR
U16			Samsung	KM416S1120
VR2,VR1			Micrel	MIC2937A-3.3BU
Y4		14.318MHz	CTS	MP143

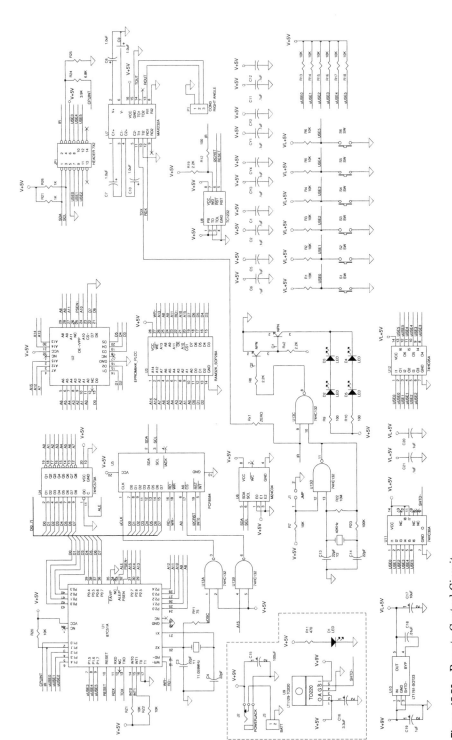

Figure 15.50 Remote Control Circuit.

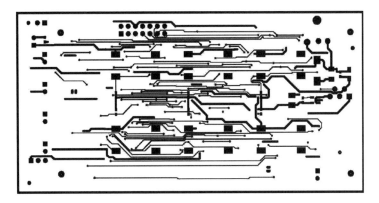

Figure 15.51(a)   Remote Control Unit PCB Layout Top Layer.

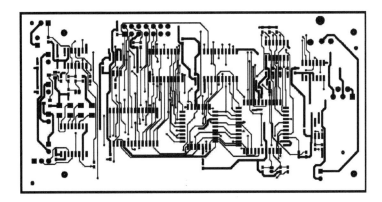

Figure 15.51(b)   Remote Control Unit PCB Layout Bottom Layer.

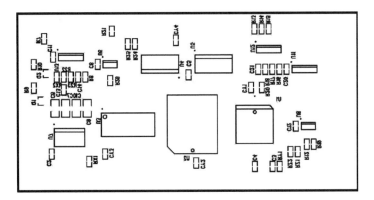

Figure 15.51(c)   Remote Control Unit Top Silkscreen.

Bill of Materials

Reference	Value	PCB	Manufacturer	Part# / Details
C1, 2, 5, 6, 11, 12, 20, 21, CY1, 2, 3, 4	.1 µF	0603	Generic	Ceramic
C3, 4, 13, 14	22 pF	0603	Generic	Ceramic
C7, 8, 9, 10, 19	1 µF	1206	Generic	Ceramic
C15	100 µF	7343	Generic	Tantalum
C16	3.3 µF	3216	Generic	Ceramic
C17	10 µF	3528	Generic	Tantalum
C18	.01 µF	0603	Generic	Ceramic
D1, 2, 3, 4			Photonic Detect	PDI-E803
D5			Lumex	SSL-LX3044DH-TR
P1	Header		Generic	7X2/.1″
J1	Header		Generic	2/2mm
J2			CUI	PJ003A
J3	Header		Generic	2/2mm
P1	Header		Generic	3/2mm 90 degrees
Q1			Zetex	FMMT2222ATA
RY1	75	0603	Generic	
R1, 2, 3, 4, 5, 6, 7, 13, 14, 15, 16, 17, 18, 20	10K	0603	Generic	
R8, 19	2.2K	0603	Generic	
R9, 10, 12	100	0603	Generic	
R11	470	0603	Generic	
R22	10M	0603	Generic	
R23	100K	0603	Generic	
R24	3.9K	0603	Generic	
R25	6.8K	0603	Generic	
S1, 2, 3, 4, 5, 6			Schurter	1241.1612.11
U1			Intel	N80C51FA1SF88
U2			Assman	A-CCS32-Z-SM
U3			NEC	UPD43256BGU-70
U4			TI	SN74HC573DW
U5			Philips	PCF8584TD
U6			ST	M24C04MN6
U7			Maxim	MAX232CWE
U8			TI	TC1232COA
U9			Linear Tech	LT1129CT-5
U10			Linear Tech	LT1761ES5-5
U11			TI	CD74HC30M
U12			TI	CD74HC05M
U13			TI	CD74HC132M
Y1			Epson	CA-301 11.059M-C
Y2			Citizen	CFV206-40.000KAZF

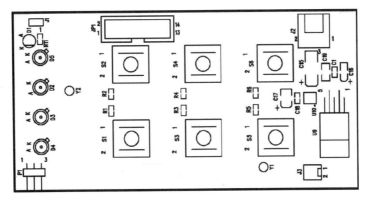

Figure 15.51(d)   Remote Control Unit Bottom Silkscreen.

Figure 15.52   Remote Control Board.

5.14 Notes

The manufacturing of well designed PC boards can be handled by many producers. However, we always ask them to verify the design and always make the corrections they ask for (if not detrimental to board operation).

All the boards presented in this chapter were produced by Advanced Circuits Inc. (www.4pcb.com). Their layout debug utility allows the designer to check the boards for manufacturing problems before they are built (www.freeDFM.com)

Assembling these boards is beyond the capabilities of most people (definitely beyond mine!). Such assembly jobs should be handled only by people experienced in the assembly and rework of surface mount boards and components. Fortunately, there are many small electronic assembly houses able and willing to do prototype jobs at a reasonable cost. They can be found either on the web or in your local industrial directory. If they do not accept such jobs, ask if they would allow their employees to help you. Most will agree.

Because the focus of this book is on video hardware, we have left the software design and coding to the reader. The software requirements of the projects are minimal and generally not related to the video characteristics of the system (with the notable exception of image file format conversion). Besides, coding details will depend on the processor and software tools the reader is familiar with.

Finally, we have used the building blocks detailed in this book for a number of projects including the four-channel scaler in Figures 15.53 and 15.54. It includes four decoder blocks, two CPLD blocks, one encoder block, and an 80C51-based local controller, all almost identical to the corresponding projects in this chapter.

Figure 15.53   Four Channel Scaler Board.

Figure 15.54   Four Channel Scaler Box ($4.5'' \times 6'' \times 1.25''$).

Index